1988

DISCRETE MATHEMATICS
A Unified Approach

DISCRETE MATHEMATICS
A Unified Approach

Stephen A. Wiitala
Associate Professor of Mathematics
Norwich University

McGraw-Hill Book Company
New York   St. Louis   San Francisco   Auckland   Bogotá   Hamburg
Johannesburg   London   Madrid   Mexico   Milan   Montreal   New Delhi
Panama   Paris   São Paulo   Singapore   Sydney   Tokyo   Toronto

DISCRETE MATHEMATICS: A Unified Approach

1234567890 HALHAL 89432109876

ISBN 0-07-070169-5

This book was set in Times Roman.
The editor was Kaye Pace;
the designer was Charles A. Carson;
the cover illustrator was John Hite;
the production supervisor was Diane Renda;
the drawings were done by Volt Information Sciences, Inc.
Arcata Graphics/Halliday was printer and binder.

The front and back covers of the book are meant to represent the Königsberg bridge problem.

Library of Congress Cataloging-in-Publication Data

Wiitala, Stephen A. (date).
   Discrete mathematics.

   1. Mathematics—1961–   .   2. Electronic data
processing—Mathematics. I. Title.
QA39.2.W53   1987   510   86-21031
ISBN   0-07-070169-5

ABOUT THE AUTHOR

Stephen A. Wiitala was born in Vancouver, Washington, in 1946. He attended Western Washington University, receiving a B.A. degree in 1968. Following a brief period as a high school mathematics teacher, he entered graduate school in 1970, receiving an M.A. in mathematics from Western in 1971. He continued his graduate work at Dartmouth College, receiving a Ph.D. in mathematics in 1975. His dissertation dealt with geometric algebras defined over finite fields; and in studying these topics, he first became involved with combinatorics and the interaction between computer science and mathematics.

His first college teaching position was at Nebraska Wesleyan University, where he taught mathematics and computer science. In 1980 he moved to Norwich University, in Northfield, Vermont, where he became responsible for the development of a mathematics-based computer science program and for a discrete mathematics course designed for sophomore mathematics and computer science majors. This book is the result of the time spent in developing that course.

Professor Wiitala currently serves as the program director for the computer science mathematics program at Norwich University and has also served as the director of academic computing.

CONTENTS

PREFACE

Discrete mathematics has become a popular introductory course for students contemplating the study of either computer science or mathematics.

As a topic in mathematics, discrete mathematics presents a convenient introduction to the techniques of proof and the process of logical reasoning but allows the student to focus attention on concepts which are relatively concrete, as opposed to some of the ideas in the calculus. As a topic for computer science, it provides much of the mathematical foundation for the later study of theoretical topics as well as a foundation for many of the more concrete ideas. Thus, the topics in discrete mathematics present the opportunity for many illustrative examples.

Students of mathematics should see some of the applications of mathematics in other disciplines, and computer science students need to see the mathematical foundations of their discipline from the beginnings of their study.

This book stresses two themes. The first is the application of the topics to computer science. In particular, whenever feasible, algorithms are included to accomplish the processes that are being described. The second is an introduction to the methods of theoretical mathematics. The mathematical topics covered stress the development of concepts and the process of proof.

There is a reason for my choice of this particular title, *Discrete Mathematics: A Unified Approach.* Although a strong relationship exists between most of the topics in discrete mathematics, in attempting to make their texts as flexible as possible, too many authors present the ideas as a compendium of separate ideas which have no relationship to each other. The selection of topics and order of presentation in this book is designed to allow for a more unified presentation of the material. It is possible to use the material in this book in a different order, but I feel that the present order provides the student with the best sense of unity.

Ample material is given here for a two-semester or three-quarter course in discrete mathematics. At Norwich we have covered Chapters 1, 2, 4, 5, and parts of 6 in the first semester, and the remainder of Chapter 6 and Chapters 7, 8, and 9 in the second semester. Chapter 3 is covered in another course but could be used in either semester. Chapter 4 can be omitted if the students are already familiar with matrix arithmetic.

Suggested course organizations:

One semester (*stressing Graph Theory*): Chapters 1, 2, 4 (if needed), and 5, and Sections 6.1 to 6.5.

One semester (*stressing Theory of Computation*): Chapter 1; Sections 2.1 to 2.5, 5.1 to 5.4, 6.1, 6.2, 6.6, 6.9, and 7.1 to 7.4; and Chapters 8 and 9.

Two semesters: First semester, Chapters 1 to 5; second semester, Chapters 6 to 9.

Brief Description of Chapters

Chapter 1: An introduction to the ideas of mathematical logic. This chapter lays the foundation for the techniques of proof used in the later chapters. It provides an introduction to predicate logic and mathematical induction. The presentation of mathematical induction is based on predicate logic.

Chapter 2: An introduction to mathematical set theory and combinatorial analysis. The combinatorial analysis is presented at a relatively low level and is used in the discussion of algorithm analysis later in the text.

Chapter 3: A brief introduction to boolean algebra. On the basis of similarities already observed between logic and set theory, the concept of a boolean algebra as a mathematical abstraction is introduced. The chapter begins with a discussion of logic circuits. This chapter is independent of those that follow, and can be used at any time after Chapter 2, or it may be omitted altogether.

Chapter 4: A brief introduction to matrix arithmetic. The barest essentials of matrix addition, scalar multiplication, and matrix multiplication are described for the student who has not encountered these topics before. They are needed in order to make use of the matrix algorithms in graph theory in the following chapter. Students who have already been introduced to matrices can skip this chapter.

Chapter 5: The theory of undirected graphs. This chapter and Chapter 6 form the heart of my discrete mathematics course. Chapter 5 presents the terminology and algorithms needed for the applications of graph theory to computer science. Several of the classic ideas, such as connectivity, chromatic number, and planarity are discussed.

Chapter 6: The theory of directed graphs. The ideas of undirected graphs are extended to that of graphs in which the edges have a specified direction. Discussion of applications of graphs to computer science and other areas is included, as well as the important algorithms of Dijkstra and Warshall. Directed (rooted) trees are presented in this chapter as a special case of the directed graph.

Chapter 7: An introduction to finite automatons and formal languages as described by them. The state diagrams of finite automatons are used to illustrate another example of the use of directed graphs. The famous syntax diagrams of Pascal are discussed as an example of the finite automaton.

Chapter 8: An introduction to Turing machines. Pushdown automata are briefly discussed as an extension of the finite automaton, and these ideas are extended to the Turing machine. A brief discussion of Church's thesis is included.

Chapter 9: The culmination of the book is a discussion of the pumping lemmas and Turing's halting problem. These theorems enable us to put most of the ideas in the book together to illustrate their use in producing some significant results.

The last three chapters are written at a level which is accessible to freshmen and sophomores, and as a consequence some of the mathematical details have been left out in some cases.

Acknowledgments

I would like to thank the many people who made the process of writing this book a lot easier than I thought it might be, particularly the staff at McGraw-Hill who provided an incredible amount of assistance.

I must thank those generations of Norwich University students who used my manuscript as a textbook and provided useful feedback to me in the process; and Prof. Gerard LaVarnway, who used an early draft of Chapters 5 and 6 in his section of discrete mathematics. I also thank Prof. Bruce Edwards of the University of Florida (and a graduate school colleague) for encouraging me to start this project when it was merely the glimmer of an idea.

I appreciate the help and direction given me by those who read my manuscript and provided extremely useful suggestions for revision: Lynne Doty, Marist College; Joe Hemmeter, University of Delaware; Burton Kleinman, Queensborough Community College; Robert McGuigan, Westfield State College; Kay Smith, St. Olaf College; J. W. Smith, University of Georgia; and Jerry Waxman, Queens College.

Finally, I thank my family for putting up with me through the process of writing and producing this book. They had to endure a number of late nights with the printer running on the computer, and then cope with a grouch the next morning on more than one occasion.

Stephen A. Wiitala

1

Mathematical Logic and Proofs

1.1 WHAT IS A PROOF?

The Basic Problem

In mathematics and computer science, as well as in many places in everyday life, we face the problem of determining whether something is true. Often the decision is easy. If we were to say that $2 + 2 = 5$, most people, no doubt, would immediately say that the statement was not true. (Actually, the more likely response would be, "What kind of dummy would think that?" or something even more insulting.) If we were to say $2 + 2 = 4$, then undoubtedly the response would be, "Of course" or "Everybody knows that." However, many statements are not so clear. A statement such as "The sum of the first n odd integers is equal to n^2," in addition to meeting with a good deal of consternation, might be greeted with a response of "Is that really true?" or "Why?" This natural response lies behind one of the most important concepts of mathematics, that of proof. In this chapter we explore the ideas of mathematical logic which lie behind the concept of proof.

If a statement is obviously true or false, we usually don't worry too much about proof or disproof (although some of the most difficult problems in mathematics are those whose truth may seem obvious). But for those statements whose truth is unknown, it is useful to be able to determine exactly what the situation is. Many people find that the process of discovering a proof is more interesting than the statement being proved. The well-known pythagorean theorem of geometry has been known to be true for thousands of years, but hundreds of proofs have been found for it, and even today people enjoy attempting to find new proofs of this theorem. The people who find these proofs are not concerned about whether the statement is true; they are really interested in why it is true.

This interest in the reasons why statements are true or false is one of the things that makes mathematics fascinating. The mathematician is often concerned not only with *what* is true but also with *why* things are true.

As we proceed through this book, we will encounter many statements about mathematics and computer science which will require proof or disproof. With a solid grounding in mathematical logic, we should be able to proceed with some degree of confidence. One warning should be given though. Some things which are " obviously true " are not so obvious when we try to prove them—in fact, some turn out to be false! One of the more bizarre facts of logic tells us that it is even possible that some statements could be true but impossible to prove from the current axioms of mathematics.

In computer science, as in mathematics, it is crucial to *know* what we are doing is correct. In algorithm analysis, for example, we are concerned with knowing that the algorithm does what it is supposed to do and that it does its job efficiently. To correctly address those issues, it is often necessary to use the formalism that mathematics can provide in order to verify that the statements we wish to make about the algorithms in question are correct.

Many mathematical structures are used in computer science. These structures are often best described in terms of the formal theory of mathematics, and in this context proofs play a very important role. Graph theory, for example, provides the foundation for many ideas in data structures, and it forms the basis for some of the theoretical models of a computer which are useful in understanding how computers are able or unable to perform certain tasks.

Some Examples

To begin considering the ideas of mathematical logic and proofs, we look at a few examples of reasoning, some of which represent valid reasoning and others which do not. As we proceed through the chapter, we will be able to discern for certain which arguments are valid and which are not.

EXAMPLE 1.1

Consider the following argument:

If computers are to really find their way into the home of the average person, then the price of some complete systems should be less than $1000. The local computer store has complete systems priced under $1000, so it must follow that computers have found their way into the average home. ■

We need to ask whether the argument convinces us that the conclusion is justified by the evidence. If the conclusion does "follow logically," then we say that the argument *proves* that the conclusion is indeed true. Sometimes we find that the conclusion is true even when the argument is not "logical."

EXAMPLE 1.2

If computers are to find their way into the home of the average person, then the price of some complete systems should be less than $1000. Computers are found in the homes of average people, thus we must conclude that there are complete computer systems which are priced under $1000. ■

EXAMPLE 1.3

All computers have input devices. The Macintosh is a computer. Thus the Macintosh has an input device. ■

EXAMPLE 1.4

All Apple computers can be connected to printing devices. All minicomputers can be connected to printing devices. Thus, some Apple computers are mini-computers. ■

In an argument, the statements used to build the conclusion are called the *premises*, and the final statement is called the *conclusion*. We regard an argument as being valid provided that whenever all the premises are true, the conclusion is guaranteed to be true.

A couple of important facts need to be recognized. First, a valid argument in which some of the premises are false may or may *not* produce a valid conclusion, second an invalid argument (one that is not valid) *can* produce a true conclusion. In the second case, the arriving at a true conclusion is more a chance happening than anything else.

In the problems for this section, we ask you to consider some arguments and decide whether you think they are valid. There obviously cannot be right or wrong answers at this point, since we have no techniques for determining whether an argument is valid. The point is to get you thinking about the problems involved.

Problem Set 1

The following problems should be answered based on your intuitive knowledge of logic. Much of the rest of this chapter is devoted to providing techniques to answer these questions. These problems should get you started thinking about the ideas.

1. Consider the argument in Example 1.1. Do you believe that it is a convincing argument? Why or why not? Do you believe the conclusion?

2. Answer the questions in Problem 1 for Example 1.2.

3. Answer the questions in Problem 1 for Example 1.3.

4. Answer the questions in Problem 1 for Example 1.4.

5. Explain why it is reasonable that a valid argument with false premises produces conclusions which may or may not be true.

6. We said that an invalid argument could lead to a true conclusion. How could this happen? What does this tell us about the process of proof?

7. Analyze the following argument, which is sometimes called the "apple press" argument:

> Apple presses eat apples.
>
> Johnny eats apples.
>
> Thus Johnny must be an apple press.

Obviously this is not a valid argument. Explain why.

In Problems 8 to 15, determine whether the argument given justifies the conclusion stated and explain why you think that this is the case.

8. If I get the job and work hard, then I will be promoted. I was promoted. Thus I got the job.

9. If I get the job and work hard, then I will be promoted. I was not promoted. Thus either I did not get the job or I did not work hard.

10. I will either get an A in this course or I will not graduate. If I don't graduate, I will go into the army. I got an A. Thus, I won't go into the army.

11. I will either get an A in this course or I will not graduate. If I don't graduate, I will go into the army. I got a B. Thus, I will go into the army.

12. If the football game runs late, then *60 Minutes* will be delayed. If *60 Minutes* is delayed, then the local news will not start until after 11:00. The local news started at 11:15, so the football game ran late.

13. If I buy a new car, then I will not be able to go to Florida in December. Since I am going to Florida in December, I will not buy a new car.

14. Either the butler or the maid committed the crime. If the butler did it, he would not have been able to answer the phone at 11:00. Since he did answer the phone at 11:00, the maid must have done it.

15. If the temperature had gone down, Bill would not have gone to the parade. Since Bill did not go to the parade, the temperature must have gone down.

*16. Consider the following pair of statements:
   (a) The statement labeled (b) is false.
   (b) The statement labeled (a) is true.
   Can either statement above be true? Why or why not?

*17. Consider the statement "I am telling a lie." Can this statement be true? Can it be false?

1.2   PROPOSITIONS

Mathematically Useful Statements

As we begin to study the logic used in mathematics, we should note first two commonsense ideas which are built into the system of logic that we use and the reasons why we use these ideas.

First, mathematically interesting statements are those which can be shown to be either true or false—or which at least could be shown to be true or false if some additional information were known regarding variables used in the statement. Statements like "If x is an odd number and y is an odd number, then $x + y$ is an odd number" and "The sum of the first n odd integers is n^2" are examples of statements which are mathematically interesting, because they can be shown to be either true or false with no concern about the values of the variables involved. The statement $x + y = 4$, on the other hand, would be mathematically interesting also, provided that some kind of additional information was given regarding the variables x and y. Do we mean to say this is true for all values of x and y (unlikely)? Or, are we asserting that for all values of x we can find a value of y such that the statement is true? Or, do we mean to assert that this statement is true when $x = 2$ and $y = 3$? In any case, once we make a decision about what we intend to do with x and y, we come up with a statement that will be either true or false.

However, such statements as "Three is a pretty number" or "Fred is tall," though certainly legitimate in English, leave much to be desired mathematically. We have no idea what a pretty number is, and the decision as to whether 3 is pretty is quite subjective. Similarly, unless Fred is 10 feet tall, or 3 feet tall, it will be pretty difficult to come to any general agreement as to whether that statement is true or not.

Second, a mathematical statement is either true or false but never both. It is interesting to consider a statement like "This statement is false," which is true if it is false and false if it is true. But such statements are mathematical curiosities and are not the kind of thing to which we want to devote a lot of time in this book.

There are some rather interesting consequences from these two concepts. The most obvious is usually called the *law of the excluded middle*. If a statement is not true, then it is false (or if it is not false, it is true). This seemingly obvious fact is used very conveniently in some proof techniques, because sometimes it is easier to prove that a statement is not false than it is to prove that the statement is true. The other consequence of this idea is the reverse of what we referred to above: If a statement is true, it is not false; and if a statement is false, it is not true.

Despite the fact that the statements we made above would seem reasonable to most of us, there are people who, for various philosophical reasons, do not accept all that we stated above. In particular, there are mathematicians (known as *intuitionists*) who do not accept the idea that a statement has been proved true if it has been demonstrated that the statement is not false. They feel that there is a difference between not false and true. Actually in everyday life we sometimes use that kind of reasoning as well. People sometimes say, for example, when asked how they feel about a decision that affects them, that they are "not unhappy." A little reflection might suggest that this is a weaker statement than saying that they are happy. The statement that they are happy *should* be true, but maybe it really is in some kind of intermediate state. The mathematical logic of the intuitionists is interesting, but for our purposes we stick with the law of the excluded middle for the things that we do in this book.

Propositional Logic

We begin our study of logic by formally defining the idea of a *proposition*.

> **DEFINITION**
>
> A **proposition** is a statement which, as given, is either true or false.

One of the goals of mathematics is to determine which propositions are true and which are false.

> **DEFINITION**
>
> A **theorem** is a proposition which has been proved to be true about a mathematical system.

Most, but not all, theorems are in the form "if something, then something else." In propositional logic, we learn how to work with statements of that form in order to better understand how to prove theorems. When claiming that a statement is a proposition, we need not *know* whether it is true or false; we only need to know that it is the kind of statement which falls into one or the other of those categories.

The statement "For any integer value n for which $n > 2$, the equation $a^n + b^n = c^n$ has no solutions in which a, b, and c are all nonzero integers" is obviously either true or false. The fact that years of mathematical effort have not resolved the issue as to which category that statement belongs does not alter the fact the statement is a proposition.[1]

EXAMPLE 1.5

The following are all propositions:
(a) There are two solutions to the equation $x^2 + 4 = 20$, and both solutions are integers.
(b) Either this program runs, or there was an error in keying in the data.
(c) It is not the case that 5 is a prime number.
(d) If x is any integer, then x^2 is a positive integer.
(e) Every integer is the sum of four perfect squares.   ∎

EXAMPLE 1.6

The following are examples of statements that are *not* propositions.
(a) $x^2 = 11$.
(b) This is a bad program.
(c) Go for it!   ∎

Our first approach to mathematics is a study of propositional logic. The basic building blocks of propositional logic are, naturally enough, propositions. In propositional logic, the goal is to study the ways of combining propositions to form new propositions and to determine under what circumstances these new "compound" propositions are true. Later we turn to a more inclusive form of logic, which will enable us to use statements like Example 1.6a and in effect turn them into propositions.

Notation

To describe the processes of mathematical logic, we need a way of symbolizing propositions. It has become a tradition of sorts in elementary propositional logic to use lowercase letters starting from p and continuing as needed (q, r, s, etc.) to represent propositions when we wish to study logic in a symbolic fashion. (The reason for the use of p is that p is the first letter in the word "proposition.")

In indicating that we will use a particular letter to stand for a proposition, we will use the following kind of notation:

$$p: \text{If } x \text{ is any integer, then } x^2 \text{ is positive or zero.}$$

[1] This proposition was stated as a theorem by the famous French mathematician Pierre de Fermat (1601–1665) and has been the subject of intense scrutiny ever since. But to this day it remains one of the most puzzling unsolved problems in mathematics.

This means that we would denote the proposition in Example 1.5a by p.

Often these symbols are used as *propositional variables*, that is, as symbols that could represent any proposition, rather than standing for a particular proposition.

In the next section, we begin to explore the process of combining several propositions into one, which is a first step to explaining the process of proof. Remember that our goal in this chapter is to develop the mathematical techniques which we need to prove that a particular proposition is true (or false).

Problem Set 2

1. Classify the following statements as propositions or nonpropositions, and explain your answers:
   (a) The population of the United States is 185 million.
   (b) July 4 occurs in the winter in the northern hemisphere.
   (c) Elephants are smarter.
   (d) X is greater than Y.

2. Classify the following statements as propositions or nonpropositions, and explain your answers:
   (a) Buy bonds!
   (b) The DEC Rainbow is an 8-bit computer.
   (c) $A + B = 17$.
   (d) There is a largest prime number.

3. Construct an example of a proposition which is (a) false, (b) true, and (c) of unknown truth.

4. Construct three statements that are not propositions, and explain why they are not propositions.

Suppose that we have the following:

   p: George Washington was the first president of the United States.

   q: Abraham Lincoln discovered America.

5. Write the proposition which combines the propositions p and q with the word " or." Is this proposition true or false?

6. Repeat Problem 5 with the word "and."

7. Write the proposition which is the "opposite" of p. Is this proposition true or false?

8. Write the proposition "If p, then q" in good English. Is this statement true or false?

Let p: Neil Armstrong walked on the moon.

   q: IBM makes computers.

9. Write the proposition which combines propositions p and q with the word "or." Is this proposition true or false?

10. Repeat Problem 9 with the word "and."

11. Write the proposition which is the opposite of p. Is this proposition true or false?

12. Write the proposition "If p, then q" in good English. Is this statement true or false?

Let p: The capital of Nebraska is Omaha.

  q: North Dakota borders on Canada.

13. Write the proposition which combines propositions p and q with the word "or." Is this proposition true or false?

14. Repeat Problem 13 with the word "and."

15. Write the proposition which is the opposite of p. Is this proposition true or false?

16. Write the proposition "If p, then q" in good English. Is this statement true or false?

17. In everyday English, frequently we combine two propositions with the word "or," and we mean that exactly one of the two propositions is true. Often we signal this by placing the word "either" in front of the compound sentence. Apply this use of "or" to Problem 9. Does it change the results? Why or why not?

*18. We can create tables which indicate the truth or falsity of propositions by T and F, respectively. This is particularly useful in attempting to decide the truth of propositions which are formed as compounds of several other propositions. If we made a table to indicate the truth of the proposition p AND q, it would look like this:

p	q	p AND q
T	T	T
T	F	F
F	T	F
F	F	F

The meaning of this table is that a proposition of the form p AND q is false unless both p and q are true. Construct a table to represent the truth of p OR q.

*19. Construct a table as in Problem 18 for the statement "If p is true, then q is true."

*20. If we wish to prove the statement "No two men are created unequally" to be true, what proposition could be proved false instead, to accomplish the same purpose?

1.3   NEW PROPOSITIONS FROM OLD ONES

Fundamental Connectives

Most, if not all, propositions of interest (both in mathematics and in real life) tend to be rather complicated. The statements in Example 1.5a–c are all propositions which are constructed by combining other propositions. We had some practice in that kind of activity in Problems 5 to 17 in Section 1.2. Just as in arithmetic (in which we combine numbers to make other numbers through the operations of addition, subtraction, multiplication, and division), in logic we find that we can combine propositions to form new propositions through the use of some basic mathematical operations.

Let's go back to the proposition from Example 1.5a. It says, "There are two solutions to the equation $x^2 + 4 = 20$, *and* both solutions are integers." This proposition is composed of two parts connected by the word "and." If we let p stand for the proposition that there are two solutions to the equation, and q be the proposition that all the solutions to the equation are integers, then the statement is one of the form p AND q. In this example, "and" is used to *connect* the two propositions to form a single new proposition whose truth depends on the truth of the propositions p and q.

In a similar manner, the proposition in Example 1.5b, "Either this program runs, or there was an error in keying in the data," is clearly two propositions joined to form a new one, this time by means of the word "or." In Example 1.5c, if we denote by p the statement "5 is a prime number," the statement says that p is not true.

These three statements are examples of the three fundamental *connectives* used to combine propositions and form new propositions from old ones.

The Connective AND

Let's start with the AND connective. The idea of this connective is to take two other propositions and make a new one which says that the two original propositions are *both* true. Since mathematicians are inherently lazy about spelling things out, instead of using the *word* AND in a symbolic description of a compound proposition, we use the symbol $\wedge$. Thus, to combine propositions p and q, using AND as a connective, we write $p \wedge q$.

Truth tables are often used to determine the truth or falsity of compound propositions and to provide formal definitions for the various connectives of propositional logic. The following table defines the connective $\wedge$:

p	q	$p \wedge q$
T	T	T
T	F	F
F	T	F
F	F	F

The two leftmost columns contain all four of the possible combinations of true and false which could occur for the propositions p and q. The rightmost column indicates the end result of $p \wedge q$ under the circumstances described by the preceding two columns. The truth table provides a complete description of the AND operation, because the table includes all possible combinations of true and false which can occur for p and q. Later we discuss how we can construct truth tables to provide the truth values for more complicated compound propositions.

EXAMPLE 1.7

Use the truth table for AND to determine the truth of each of the following propositions:
(a) Jimmy Carter was once president of the United States, and Spiro Agnew was vice president during the time that Carter was president.
(b) Richard Nixon was once president of the United States, and Spiro Agnew was vice president during the time that Nixon was president.
(c) Nelson Rockefeller was president of the United States, and so was Gerald Ford.
(d) Hubert Humphrey was president of the United States, and so was Thomas Dewey.

SOLUTION
(a) The first part of the proposition is true, while the second part is false. Since they are connected with the connective AND, the result is false.
(b) Both parts of the proposition are true; thus, by the defining truth table, the proposition is true.
(c) The first part of the proposition is false, while the second part is true. According to the definition of AND, the resulting compound proposition is false.
(d) Neither part of the proposition is true, and thus the proposition must be false. ■

The Connective OR

In a similar fashion we denote the OR connective by the symbol $\vee$. The truth table for OR is

p	q	$p \vee q$
T	T	T
T	F	T
F	T	T
F	F	F

The intent of the OR connective is to produce a statement which is true when either of its constituent parts is true. In Example 1.5b we said that "either this program runs, or there was an error in keying in the data." What we meant was that at least one of the two things had occurred. We would believe the statement to be true provided that the program ran, if we found an input error, or if both events occurred. This definition of OR may be at least slightly controversial. In normal usage we make statements such as "Either I will pass this course, or I will lose my scholarship," and we mean that one or the other of the parts will be true, but not both. This use of the word OR is referred to as the *exclusive OR*, which is *not* the way in which OR is commonly used in mathematics.

EXAMPLE 1.8

Use the truth table for OR to determine the truth of the following propositions:
(a) Jimmy Carter was president of the United States in 1978, or Spiro Agnew was vice president during the time that Carter was president.
(b) Richard Nixon was president of the United States in 1971, or Spiro Agnew was vice president during the time that Nixon was president.
(c) Nelson Rockefeller was president of the United States, or Gerald Ford was president.
(d) Hubert Humphrey was president of the United States, or Thomas Dewey was president.

SOLUTION
(a) The first part of the proposition is true, while the second part is false. Since they are connected with the connective OR, the result is true.
(b) Both parts of the proposition are true, and thus, by the defining truth table, the proposition is true.
(c) The first part of the proposition is false, while the second part is true. According to the definition of OR, the resulting compound proposition is true.
(d) Neither part of the proposition is true, and thus the proposition must be false. ■

Students first encountering the symbolic renderings of AND and OR often manage to become confused as to which one is which. This confusion is probably precipitated by the fact that the two symbols are identical except that one is upside down from the other. The easiest way to tell them apart is to note that the ∧ looks kind of like a capital A with the horizontal bar removed, and since AND starts with A, the symbol ∧ must stand for AND. Since ∧ stands for AND, the other symbol must be the one that stands for OR. It probably doesn't hurt to note that the symbol ∨ looks like the letter v. But there is no term in logic which starts with the letter v, so the mnemonic device must have something to do with the other symbol.

DEFINITIONS

A proposition constructed by using the OR connective for its main connective is called a **disjunction**, and one whose main connective is AND is called a **conjunction**.

An easy way to keep these two similar-sounding terms straight is to remember that AND requires that the truth of the two parts be connected, while the operation of OR allows for a result of true even when the truth values disagree.

Negation of a Proposition

The final of the three basic operations is called *negation*. The negation of a proposition is the proposition which states that the original proposition is false. Negation is illustrated by the proposition in Example 1.5c, where the proposition says that it is not the case that 5 is a prime number. (In this case the original proposition is true, and that means that the new proposition must be false.) We symbolize the operation of negation by $\sim$, and we write the negation of p as $\sim p$. (We read it as "not p.") The truth table for $\sim p$ is given by

p	$\sim p$
T	F
F	T

EXAMPLE 1.9

Use the truth table for NOT to determine the truth of the following:
(a) Nelson Rockefeller was not a president of the United States.
(b) Richard Nixon was not a president of the United States.

SOLUTION
(a) We have the negation of the statement "Nelson Rockefeller was a president of the United States." Since that is false, the negation is true.
(b) The situation is reversed. In that case, we are negating a statement which is true, and thus proposition (b) is false. ∎

Computer Applications

Most computer languages make use of the logical connectives. In some languages, a data type known as *boolean* (for George Boole, a pioneer in mathematical logic) or *logical* is available in which the values that can be stored are true and false. Within this data type, logical operations are provided, including the operations of AND, OR, and NOT. Typically, suppose we write something like "If $x < 4$ or $y > 5$, then . . . ." The computer will evaluate the logical expression indicated, by first evaluating the results of the comparisons as true or false and then using the definitions of the connectives to determine whether the final logical expression is true or false.

Machine language instructions for most machines include logical operators which work on the binary data stored in the machine.

Binary representation of data uses a sequence of 0s and 1s to represent information. Since most electronic equipment is bimodal (switches are off or on, voltage is high or low, etc.), the binary system is a convenient system to use in the computer. One of the two states of the storage device is used to represent a 0, and the other state is used to represent a 1. Each 0 or 1 thus represented is called a *bit*, and the memory is organized into clusters of bits. The smallest conveniently accessible cluster of bits is often called a *word*, and a word usually consists of one or more clusters of 8 bits, called a *byte*. One byte can contain enough different sequences of 0s and 1s to represent one keystroke from the keyboard or terminal.

Machine language boolean operators act on information "bit by bit," with 0 being regarded as false and 1 being regarded as true. These operations work on information in bytes or words. If a byte contained the data 01010101 and another byte contained 11110000, the result of applying the OR operator would be 11110101, while applying the AND operator to the same bytes would result in 01010000.

Complex Propositions

By using the logical connectives discussed in this section, we can construct some very complex propositions. For example, we can produce (and hopefully interpret) the proposition

$$[(p \wedge q) \vee r] \wedge [\sim(p \wedge r)]$$

This proposition can be read, "p AND q is true or r is true, while at the same time p AND r is false."

A rational question to ask is whether we could find another (hopefully simpler) proposition that has the same logical meaning, that is, one which is true and false under the same circumstances as the original statement. This particular issue is especially important when the logical expression involved is part of a computer program.

EXAMPLE 1.10

Determine whether the expression $(((x = 3)$ AND $(y < 9))$ OR $(y > = 4))$ AND (NOT $((x = 3)$ AND $(y > = 4)))$ is true or false if x is 2 and y is 10. This statement can be seen to be

$$[(p \wedge q) \vee r] \wedge [\sim(p \wedge r)]$$

where p: $x = 3$

q: $y < 9$

r: $y > = 4$

Under these circumstances p is false, q is false, but r is true. Working our way through the parentheses, we see that $p \wedge q$ is false, but $(p \wedge q) \vee r$ is true. And $p \wedge r$ is false, so $\sim(p \wedge r)$ is true. Since both sides of the conjunction are true, the whole proposition is true. ■

If we were writing a computer program and needed to use the logical expression in Example 1.10, obviously it would be extremely helpful to be able to find a simpler logical expression with the same meaning. In the next few sections we begin to discuss the process of finding propositions which are logically equivalent to a given proposition.

Problem Set 3

Let p: A byte has 7 bits.
  q: A word is 2 bytes.
  r: A bit is a 0 or a 1.

p is false, and q and r are true. Write grammatical English statements for the following symbolic statements. In each case determine whether the statement is true or false.

1. $p \wedge q$

2. $p \vee r$

3. $\sim p$

4. $\sim(p \wedge q)$

5. $\sim p \vee \sim q$

6. $[(p \wedge q) \vee r] \wedge [\sim(p \wedge r)]$

Let p: Chicago is in Illinois.

  q: Boston is in New York.

  r: Seattle is in the Pacific northwest.

p and r are true, while r is false. Write the following propositions in symbolic form, and determine in each case whether the proposition is true or false.

7. Chicago is in Illinois, and Boston is in New York.

8. Chicago is in Illinois, or Boston is in New York.

9. Chicago is in Illinois, and Seattle is in the Pacific northwest.

10. Boston is not in New York.

11. It is not the case that either Chicago is in Illinois or Boston is in New York.

12. Either Boston is in New York and Seattle is in the Pacific northwest, or Chicago is not in Illinois.

*13. Define a symbol for the exclusive OR connective and determine the truth table for this connective.

*14. A rule of logic known as *De Morgan's law* says that $\sim(p \vee q)$ is logically the same as $\sim p \wedge \sim q$. Explain why.

*15. Another form of De Morgan's law says that $\sim(p \wedge q)$ is logically the same as $\sim p \vee \sim q$. Explain why.

*16. The connective NOR is defined by giving p NOR q the same result as $\sim p \wedge \sim q$. Define a symbol to represent the connective NOR, and determine the truth table for this connective.

*17. The connective NAND is defined by giving p NAND q the same result as $\sim p \vee \sim q$. Define a symbol to represent the connective NAND, and determine what the truth table for this connective should be.

*18. The connective NEITHER is defined by giving p NEITHER q the same result as $\sim(p \vee q)$. Define a symbol to represent the connective NEITHER, and determine its truth table. Does the truth table for NEITHER look like the result of either Problem 16 or 17? Why?

1.4 ALGORITHMS, TRUTH TABLES, AND TAUTOLOGIES

At this point, we want to look at the problem of analyzing a given proposition.

EXAMPLE 1.11

Symbolize the following propositions:
(a) Either I will wash my car, or it will not rain.
(b) It is humid and cloudy, or it is raining, but at the same time it is false that it is both humid and raining.

SOLUTION
(a) This proposition is constructed by combining two others. The first is obviously "I will wash my car." The second might be "It will not rain," but a

second look reveals that this is really the negation of the proposition "It will rain." Thus we note that we can write

p: I will wash my car.

q: It will rain.

The proposition can then be symbolized as $p \lor \sim q$.
(b) Three propositions are used to construct this one:

p: It is humid.

q: It is cloudy.

r: It is raining.

The first part of the statement (the part before "but at the same time") appears to say $(p \land q) \lor r$. The last part seems to say that $p \land r$ is false, which we can thus symbolize as $\sim(p \land r)$. The part in the middle is a wordy way of saying AND, so we symbolize the statement as

$$[(p \land q) \lor r] \land [\sim(p \land r)] \qquad \blacksquare$$

Having seen how to symbolize propositions, we must now face the issue of determining when a given proposition is true or false.

Truth Tables

One way of analyzing the truth or falsity of propositions like those in Example 1.11 is by means of a truth table. We first encountered truth tables as a means of defining the logical connectives. We can take that idea and expand it to produce the ideas that we want.

A few observations will help us understand the process:

1. The definition of NOT involves a single proposition, and the defining truth table (Table 1.1) consists of two lines.

TABLE 1.1 Defining truth table for NOT

p	$\sim p$
T	F
F	T

2. The definitions of AND and OR involve two propositions, and the defining truth tables (Tables 1.2 and 1.3) each consist of four lines.

TABLE 1.2   Defining truth table for AND

p	q	$p \wedge q$
T	T	T
T	F	F
F	T	F
F	F	F

TABLE 1.3   Defining truth table for OR

p	q	$p \vee q$
T	T	T
T	F	T
F	T	T
F	F	F

3. If we had a compound proposition composed of three basic propositions, we would get eight different combinations of truth values of the three basic propositions. (The four combinations for two of them would be repeated twice, once with the third proposition true and once with the third one false.)

4. Following the same reasoning, we would expect 2^n possible combinations if we had started with n basic propositions.

Using a truth table to analyze the proposition in Example 1.11a will require four lines, and a table for the proposition in Example 1.11b will require eight lines. One important aspect of the process is to find a logical way of constructing all the possibilities. It is very important not to have to resort to a hit-or-miss guessing scheme. We describe one method, but note that this method most definitely is *not* the only way to do it. Indeed, some authors present truth tables which are organized in a quite different fashion from ours.

1. Write the names of the propositions in alphabetical order: p, q.

2. Determine the number of lines that are required. In the column headed by the *last* proposition, write that many lines of alternating T's and F's.

p	q
	T
	F
	T
	F

3. Move to the next column to the left. Instead of alternating T's and F's, alternate *pairs* of T's and F's:

p	q
T	T
T	F
F	T
F	F

4. If it is necessary, repeat the process for each remaining variable, while doubling the number of T's and F's that are alternated with each column that you move to the left.

In a truth table we are setting up, each row represents a different combination of values of the truth of the underlying basic propositions of the proposition which we are analyzing. We take those combinations and determine for which the entire proposition is true and for which the proposition is false. We illustrate with Example 1.11a.

We write the entire proposition to the right of the column headings, and we produce one column underneath each propositional variable and each connective. We do not produce columns underneath parentheses:

p	q	p	$\vee$	$\sim$	q
T	T				
T	F				
F	T				
F	F				

We copy into each column headed by a letter the exact sequence of T's and F's that is found in the column on the left headed by that letter.

p	q	p	$\vee$	$\sim$	q
T	T	T			T
T	F	T			F
F	T	F			T
F	F	F			F

At this point, the process becomes a lot like doing arithmetic calculations—we start evaluating the results of the logical operations. We pick a column and go all the way down the column, evaluating as we go. We use the following rules to determine which columns are to be evaluated.

1. We must go to the farthest inside set of parentheses that has not been done yet.

2. Within the innermost set of parentheses, we do computations in the order:
  (a) ~ are done first.
  (b) ∧ and ∨ are done next, from left to right.
The computations follow the rules for combining T's and F's according to the definition of the connective involved.

EXAMPLE 1.12

Complete the construction of the truth table for Example 1.11a.

We have already set up the truth table for $p \vee \sim q$. The next step is to evaluate the expressions. First we evaluate the ~. Underneath the ~ symbol we write the results of applying that operation *to the column to which it applies*. In this case the ~ is applied to the column under the q, since we need to evaluate $\sim q$. Once we have done this, we will no longer make use of the column q; so we shade it out and use the ~ column to represent $\sim q$. This is done in the same manner in which we replace the $2 + 2$ in the arithmetic computation $(2 + 2) \cdot 3$ by a 4 and thus rewrite it as $4 \cdot 3$.

p	q	p	$\vee$	$\sim$	q
T	T	T		F	T
T	F	T		T	F
F	T	F		F	T
F	F	F		T	F

The only column left is the one under the ∨. We evaluate the results for this column by applying the OR rule to the two unshaded columns nearest the connective (one on the left and one on the right). In this case, this means the column headed by p and the column headed by ~. Again, we shade the two columns so utilized and place the results under the connective. Since this is the *only* unshaded column left, this column will represent the final result for the truth table. Table 1.4 is the completed truth table for $p \vee \sim q$.

TABLE 1.4 Completed truth table

p	q	p	$\vee$	$\sim$	q
T	T	T	T	F	T
T	F	T	T	T	F
F	T	F	F	F	T
F	F	F	T	T	F

This completed table shows that the proposition is false in the case in which p is false and q is true, but true in all other cases. In terms of the proposition in Example 1.11a, this means that the proposition is true *except* when I don't wash my car and it does happen to rain. ∎

EXAMPLE 1.13

21

1.4 ALGORITHMS,
TRUTH TABLES,
AND TAUTOLOGIES

Construct the truth table for the proposition Example 1.11b.

SOLUTION

We find the combinations of possible truth values for the propositional variables by the process described earlier, and we get the following:

p	q	r
T	T	T
T	T	F
T	F	T
T	F	F
F	T	T
F	T	F
F	F	T
F	F	F

We start by writing the proposition and creating the columns.

p	q	r	[(p ∧ q) ∨ r] ∧ [~ (p ∧ r)]
T	T	T	
T	T	F	
T	F	T	
T	F	F	
F	T	T	
F	T	F	
F	F	T	
F	F	F	

Now we copy the appropriate sequences of T's and F's into the columns headed by propositional variables.

p	q	r	[(p	∧	q)	∨	r]	∧	[~	(p	∧	r)]
T	T	T	T		T		T			T		T
T	T	F	T		T		F			T		F
T	F	T	T		F		T			T		T
T	F	F	T		F		F			T		F
F	T	T	F		T		T			F		T
F	T	F	F		T		F			F		F
F	F	T	F		F		T			F		T
F	F	F	F		F		F			F		F

The first connective evaluated is the $\wedge$ in the expression

p	q	r	[(p	$\wedge$	q)	$\vee$ r]	$\wedge$ [$\sim$ (p	$\wedge$	r)]
T	T	T	T	T	T	T	T	T
T	T	F	T	T	T	F	T	F
T	F	T	T	F	F	T	T	T
T	F	F	T	F	F	F	T	F
F	T	T	F	F	T	T	F	T
F	T	F	F	F	T	T	F	F
F	F	T	F	F	F	T	F	T
F	F	F	F	F	F	F	F	F

Again, at each step in the process we shade the columns which have been done. The symbols under the $\wedge$ now stand for the result of $p \wedge q$, so the values under the columns p and q are no longer needed in the computation.

Continuing, we next evaluate the $\wedge$ connecting $p \wedge r$ at the far right to obtain

p	q	r	[(p	$\wedge$	q)	$\vee$ r]	$\wedge$ [$\sim$ (p	$\wedge$	r)]
T	T	T	T	T	T	T	T	T	T
T	T	F	T	T	T	F	T	F	F
T	F	T	T	F	F	T	T	T	T
T	F	F	T	F	F	F	T	F	F
F	T	T	F	F	T	T	F	F	T
F	T	F	F	F	T	T	F	F	F
F	F	T	F	F	F	T	F	F	T
F	F	F	F	F	F	F	F	F	F

Next we must evaluate the $\vee$ in $(p \wedge q) \vee r$. The columns used in the evaluation are the nearest unshaded columns on either side of the $\vee$.

p	q	r	[(p	$\wedge$	q)	$\vee$	r]	$\wedge$ [$\sim$ (p	$\wedge$	r)]
T	T	T	T	T	T	T	T	T	T	T
T	T	F	T	T	T	T	F	T	F	F
T	F	T	T	F	F	T	T	T	T	T
T	F	F	T	F	F	F	F	T	F	F
F	T	T	F	F	T	T	T	F	F	T
F	T	F	F	F	T	F	F	F	F	F
F	F	T	F	F	F	T	T	F	F	T
F	F	F	F	F	F	F	F	F	F	F

Similarly, we evaluate the $\sim$ in $\sim(p \wedge r)$. The $\sim$ operator requires only one column, so we use the unshaded column closest to the $\sim$ on the left. (In this case it is the column under the ∧.) Thus, we get the following:

p	q	r	(	p ∧ q	) ∨ r		∧	∼	(p ∧ r)	
T	T	T	T	T T	T T			F	T T	T
T	T	F	T	T T	T F			T	T F	F
T	F	T	T	F F	T T			F	T T	T
T	F	F	T	F F	F F			T	T F	F
F	T	T	F	F T	T T			T	F F	T
F	T	F	F	F T	F F			T	F F	F
F	F	T	F	F F	T T			T	F F	T
F	F	F	F	F F	F F			T	F F	F

Finally we apply the ∧ rule to the only two remaining unshaded columns and obtain Table 1.5. The last unshaded column now represents the truth value of the proposition for each of the eight possible combinations of T and F

TABLE 1.5 Completed truth table

p	q	r	(	p ∧ q	) ∨ r	∧	∼	(p ∧ r)	
T	T	T	T	T T	T T	F	F	T T	T
T	T	F	T	T T	T F	T	T	T F	F
T	F	T	T	F F	T T	F	F	T T	T
T	F	F	T	F F	F F	F	T	T F	F
F	T	T	F	F T	T T	T	T	F F	T
F	T	F	F	F T	F F	F	T	F F	F
F	F	T	F	F F	T T	T	T	F F	T
F	F	F	F	F F	F F	F	T	F F	F

of the underlying simple propositions. Going back to the proposition in Example 1.11b, we can see that on a humid, cloudy day with rain, the proposition is false, but on a humid cloudy day with no rain, the proposition is true. ■

EXAMPLE 1.14

Determine whether the 'Data out of range' message will be printed if x is 3, y is 1, and w is 15 in a Pascal program which includes the statement

if $(((x < y)$ AND $(y = 1))$ OR $(w = 12))$ AND (NOT $((x < y)$ AND $(w = 12)))$
then write ('Data out of range')

If $p: x < y$, $q: y = 1$, and $r: w = 12$, the proposition is of exactly the same form as the proposition in Example 1.13. From our given values of x, y, z, and w, we can see that p is false, q is true, and w is false. Examining the truth table, we see on the FTF line of the truth table that the compound proposition is false, and thus the 'data out of range' message is not printed by this **if** statement. ■

Algorithms

We can summarize the process of setting up a truth table as an *algorithm*. An algorithm is a sequence of instructions that can be carried out mechanically in order to perform a particular task. There are a number of ways of describing algorithms, and in Chapter 7 we make a more formal definition of an algorithm. But at the moment note that algorithms are the kind of processes that can be given to computers to do. In this book, we describe algorithms by means of a language which looks much like Pascal or Algol. In Appendix 1 we provide a complete description of our language. Here, however, are some of the highlights.

We use an **if**. . .**then**. . .**else**. . .**fi** construction when we need to choose between courses of action. This structure indicates that we are to do the actions following the word **then** when the condition following the word **if** is true, and when the condition is false, we do the actions following the word **else**. The "word" **fi** is used to indicate the end of the structure. (**Fi** is **if** spelled backward.) On occasion we may delete the **else** clause, which indicates that if the condition listed is false, then we wish to do nothing.

We use **while**. . .**do**. . .**od** to indicate a collection of actions that are to be repeated. The actions are repeated as long as the condition following the word **while** is true.

We use braces, { and }, to enclose annotative information for the reader. With the exception of the **if** and **while** control, all instructions are to be done in the order listed, from **begin** until we reach the end of the algorithm (**end**).

Our algorithm to set up a truth table can be listed as follows:

ALGORITHM SETUP

begin
n: = number of basic propositions;
     { : = is used to indicate assigning a value to a variable}
l: = 2^n;
b: = 1;
c: = n;
while c > = 1 **do**
     in column c alternate b T's with b F's, starting with T
     until l lines are filled;
     c: = c − 1;
     b: = 2∗b;
od
end.   {A period is used to indicate the end of the algorithm.}

Tautologies

When we construct the truth tables for propositions, sometimes we find some surprising results.

EXAMPLE 1.15

Construct a truth table for the following symbolic proposition:

$$(p \wedge q) \vee (\sim p \vee \sim q)$$

p	q	(p	$\wedge$	q)	$\vee$	($\sim$	p	$\vee$	$\sim$	q)
T	T	T	T	T	T	F	T	F	F	T
T	F	T	F	F	T	F	T	T	T	F
F	T	F	F	T	T	T	F	T	F	T
F	F	F	F	F	T	T	F	T	T	F
			4		5	1		3	2	

The numbers under the columns represent the order in which the columns were evaluated. ■

The surprising feature of Example 1.15 is that the proposition has the property of always being true, regardless of the truth values of the basic propositions which comprise it. Statements which have this property are called *tautologies*.

Tautologies are interesting propositions. The truth of the statement depends not on the truth of the propositions which comprise it, but on only the fact that the propositions have been combined in a particular fashion. No matter what p and q are, the statement $(p \wedge q) \vee (\sim p \vee \sim q)$ is guaranteed to always be true.

EXAMPLE 1.16

Determine whether the proposition $(p \wedge q) \vee (p \vee q)$ is a tautology.

SOLUTION
This proposition is not a tautology because if both of the underlying basic propositions are false, the entire proposition is false. You can construct a truth table for yourself to verify that this is the case and that the proposition is true in every other case. ■

One final note about truth tables. Understanding them does help us to learn about logic and the way that propositions are combined, but as a tool for proving statements their value is quite limited. Truth tables can be used to show that a proposition is a tautology, but their main value to us will actually be to show us how proofs work.

129, 21

Problem Set 4²

1. Construct the truth table for the proposition $(\sim p \lor q)$.

2. Construct the truth table for the proposition $(p \land \sim q)$.

3. Construct the truth table for the proposition

$$[(p \lor r) \land (q \lor r)] \land [\sim p \lor \sim r]$$

   How does your truth table compare to that of the example which we worked through in this section?

4. Construct the truth table for the proposition

$$[p \land (q \lor r)] \land [q \land (p \lor r)]$$

5. Determine whether or not a proposition of the form $\sim (p \land q) \lor (\sim p \lor \sim q)$ is a tautology.

6. Determine whether a proposition of the form $(p \land \sim q) \lor (\sim p \land q)$ is a tautology.

In each of the following problems, (a) determine the underlying simple propositions, (b) symbolize the proposition, and (c) determine under what circumstances the proposition is true and under what circumstances the proposition is false.

7. Either the Cubs will win the pennant, or the Padres will not win the World Series.

8. Either my car will reach Omaha before it breaks down, or it will break down before it gets to Omaha and I will have to hitchhike to Denver.

9. This company needs a large mainframe computer to do its payroll processing, and the company can either buy a new one or lease a used computer.

10. An employee keeps all financial records on a spreadsheet program and does not use word processing, or the employee keeps all financial records on a spreadsheet and uses a graphics package.

11. This train leaves from Seattle and stops in either Ellensburg or Yakima.

12. This train leaves from Seattle and stops in Ellensburg, or it leaves from Seattle and stops in Yakima.

*13. Explain the reasons for the relationship that you find between the results of Problems 11 and 12.

---

² Problems labeled P are programming exercises—typically in Pascal. If you do not know Pascal, your instructor may modify these problems for another programming language.

*14. Find two symbolic propositions which have identical final results for their truth tables.

*15. Design an algorithm to produce a listing of the possibilities for a truth table with the *first* column alternating T's and F's, the second alternating pairs of T's and F's, etc.

*16. The binary numbers from 0 to 7 are 000, 001, 010, 011, 100, 101, 110, and 111. Modify the algorithm SETUP so that it lists the numbers from 0 to 7 in binary.

*17. Modify the algorithm in Problem 16 so that it lists all the binary numbers from 0 to $2^n - 1$.

P18. The following program in Pascal prints out all combinations of true and false for two basic propositions (albeit upside down from our orientation). Modify it to write the truth table right side up.

```
PROGRAM TABLE (OUTPUT)³
var p,q:boolean;
    function printboolean(p:boolean): char;
    begin
    if p then printboolean:='T'
      else printboolean:='F'
    end;
begin
writeln('p':7,'q':7);
    for p:= false to true do
      for q:= false to true do
              writeln(printboolean(p):7,printboolean(q):7)
end.
```

P19. Modify the program in Problem 18 to write out the truth table for $p \lor q$ (either right side up or upside down, as you prefer).

P20. Modify the program in Problem 18 to write out the truth table for $p \land q$.

P21. Write a program to print out the truth table for the proposition in Problem 3.

P22. Write a program to print out the truth table for the proposition in Problem 4.

P23. Write a program to print out the truth table for the proposition in Problem 5.

P24. Write a program to print out the truth table for the proposition in Problem 6.

---

³ Notice that this program could be used to replace SETUP.

1.5 EQUIVALENT PROPOSITIONS

In Problem 17 of Section 1.3, we described the connective NAND as being defined by $\sim p \lor \sim q$. Other authors define this operator as $\sim(p \land q)$. On the surface, it might appear as though someone must have made a mistake. But when we look at the truth tables for the two propositions, we see that they are identical in terms of the final truth values of the two propositions.

p	q	$\sim$	p	$\lor$	$\sim$	q
T	T	F	T	F	F	T
T	F	F	T	T	T	F
F	T	T	F	T	F	T
F	F	T	F	T	T	T
		1		3		2

p	q	$\sim$	(p	$\land$	g)
T	T	F	T	T	T
T	F	T	T	F	F
F	T	T	F	F	T
F	F	T	F	F	F
		2		1	

In logic, all that really concerns us is the ultimate truth of a proposition, and in those terms, we can make no distinction between the two propositions. They are *logically equivalent* because they have the same truth values, provided that they begin with the same truth values for the underlying basic propositions. Although one proposition may be easier to read than the other, insofar as the truth values of any propositions which depend on one or the other of the two propositions, it makes no difference which of the two is used. Logical equivalence may allow us to replace a complex proposition with a much simpler one.

DEFINITION

Two propositions are **logically equivalent** if they have exactly the same truth values under all circumstances.

EXAMPLE 1.17

In Example 1.11b we considered the proposition "It is humid and cloudy or it is raining, but at the same time it is false that it is humid and raining." Another statement about the same weather conditions is "Either it is not humid and is raining, or it is humid, cloudy, and not raining." We can ask whether this simpler proposition is equivalent to the other, more complicated one. If they are equivalent, then instead of having to deal with the first one, we could consider the second.

The second proposition can be symbolized as $(\sim p \wedge r) \vee [(p \wedge q) \wedge \sim r]$. The truth table for this proposition is as follows:

p	q	r	(~ p ∧ r) ∨ [(p ∧ q) ∧ ~ r]										
T	T	T	F	T	F	T	F	T	T	T	F	F	T
T	T	F	F	T	F	F	T	T	T	T	T	T	F
T	F	T	F	T	F	T	F	T	F	F	F	F	T
T	F	F	F	T	F	F	F	T	F	F	F	T	F
F	T	T	T	F	T	T	T	F	F	T	F	F	T
F	T	F	T	F	F	F	F	F	F	T	F	T	F
F	F	T	T	F	T	T	T	F	F	F	F	F	T
F	F	F	T	F	F	F	F	F	F	F	F	T	F

The final values for this truth table are *exactly* the same as for the truth table for the statement from Example 1.11b (Table 1.5). This means that the two statements in any situation are either both true or both false. ■

Whenever we find logically equivalent statements, we should feel free to substitute one for another as we wish, since this action will not change the truth value of any statement which we are considering.

Some Logically Equivalent Propositions

There are a number of fundamental equivalences which are especially useful in proofs.

Commutative properties:

1. (OR) $p \vee q$ is equivalent to $q \vee p$
2. (AND) $p \wedge q$ is equivalent to $q \wedge p$

Associative properties:

3. (OR) $(p \vee q) \vee r$ is equivalent to $p \vee (q \vee r)$
4. (AND) $(p \wedge q) \wedge r$ is equivalent to $p \wedge (q \wedge r)$

Distributive properties:

5. (AND over OR) $p \wedge (q \vee r)$ is equivalent to $(p \wedge q) \vee (p \wedge r)$
6. (OR over AND) $p \vee (q \wedge r)$ is equivalent to $(p \vee q) \wedge (p \vee r)$

De Morgan's laws:

7. $\sim(p \wedge q)$ is equivalent to $\sim p \vee \sim q$
8. $\sim(p \vee q)$ is equivalent to $\sim p \wedge \sim q$

In view of the associative properties, parentheses are not needed in expressions which involve propositions all joined by only $\lor$'s or only $\land$'s.

EXAMPLE 1.18

Verify the commutative property for AND.

The process is straightforward. We set up truth tables for each proposition, and we note that the final results are identical.

p	q	$p \land q$	$q \land p$
T	T	T T T	T T T
T	F	T F F	F F T
F	T	F F T	T F F
F	F	F F F	F F F

$\blacksquare$

EXAMPLE 1.19

Verify the associative property for AND.

As before, we simply set up the truth tables and compare final results.

p	q	r	$(p \land q) \land r$	$p \land (q \land r)$
T	T	T	T T T T T	T T T T T
T	T	F	T T T F F	T F T F F
T	F	T	T F F F T	T F F F T
T	F	F	T F F F F	T F F F F
F	T	T	F F T F T	F F T T T
F	T	F	F F T F F	F F T F F
F	F	T	F F F F T	F F F F T
F	F	F	F F F F F	F F F F F

$\blacksquare$

When we are writing logical expressions for "if" statements or doing any logical operations within a program, the idea of equivalent expressions can often be quite valuable. For example, if we wish to evaluate the expression ($n = 7$ AND $a > 5$) OR ($n = 7$ AND $y > 0$), we can recognize that this proposition is logically equivalent to the expression $n = 7$ AND ($a > 5$ OR $y > 0$). The first expression is equivalent to the second because of the distributive property of AND over OR.

De Morgan's laws are particularly useful in simplifying logical expressions. If we wanted to do some operation in a program *except* in the case in which $x = 4$ and $y > 5$, we could write if NOT (($x = 4$) AND ($y > 5$)), then . . . . Or by using De Morgan's law (number 7), we could rewrite the logical expression as (NOT ($x = 4$) or NOT ($y > 5$)). This last expression can be further simplified to $((x <> 4)$ OR $(y <= 5))$, which is much easier to follow than the original expression.[4]

In Chapter 3 we explore some ways in which these same ideas are used in the design of electronic circuitry.

Problem Set 5

1. Show that $p \lor q$ is equivalent to $q \lor p$.

2. Show that $p \lor (q \lor r)$ is equivalent to $(p \lor q) \lor r$.

3. Show that $p \land (q \lor r)$ is equivalent to $(p \land q) \lor (p \land r)$.

4. Show that $p \land (q \lor r)$ is *not* equivalent to $(p \land q) \lor r$. Thus expressions involving mixtures of the operations should make use of parentheses to minimize the possibility of confusion.

5. Show that $p \lor (q \land r)$ is not equivalent to $(p \lor q) \land r$.

6. Show that $p \lor (q \land r)$ is equivalent to $(p \lor q) \land (p \lor r)$.

7. Show that $\sim(p \lor q)$ is equivalent to $\sim p \land \sim q$.

8. Show that $p \lor \sim(q \land r)$ is equivalent to $(p \lor \sim q) \lor \sim r$.

Use the examples of equivalent statements found in this section to show that the following pairs of statements (taken from boolean expressions in Pascal programs) are equivalent.

9. $((n = 7)$ OR $(a > 5))$ AND $(x = 0)$; $((n = 7)$ AND $(x = 0))$ OR $((n > 5)$ AND $(x = 0))$

10. NOT $((n = 7)$ OR $(a > 5))$; $((n <> 7)$ AND $(a <= 5))$

11. NOT $((n = 7)$ AND $(a <= 5))$; $((n <> 7)$ OR $(a > 5))$

12. $((n <> 7)$ AND $(a < 5))$ OR $(x = 0)$; $((n <> 7)$ OR $(x = 0))$ AND $((a < 5)$ OR $(x = 0))$

13. $(n = 7)$ OR (NOT $((a <= 5)$ AND $(x = 0)))$; $((n = 7)$ OR $(a > 5))$ OR $(x <> 0)$

14. NOT $(((x = 0)$ AND $(n = 7))$ AND $(a < 5))$; $(((x <> 0)$ OR $(n <> 7))$ OR $(a >= 5))$

Write the following logical expressions as the disjunction of conjunctions of simple expressions. For example, NOT $((x = 0)$ AND $(n = 5))$ could be written as $(x <> 0)$ OR $(n <> 5)$.

15. $(x = 0)$ AND (NOT $(n = 7)$ AND $(a < 5))$

16. $(x = 0)$ OR (NOT $((n > 7)$ OR $(a = 5)))$

---

[4] The symbol $<>$ means "not equal to."

17. NOT $(((x = 0)$ AND $(n = 7))$ AND $(a < 5))$

18. NOT $(((x = 0)$ OR $(n = 7))$ OR $(a < 5))$

19. $(n = 7)$ OR (NOT $((a < = 5)$ AND $(x = 0)))$

20. $((n < > 7)$ OR $(x = 0))$ AND $((a < 5)$ OR $(x = 0))$

*21. Explain De Morgan's laws in words, and explain why they work.

*22. Explain the two distributive properties in words, and explain why they work.

P23. The following Pascal program will test two logical expressions in two variables for equivalence.

PROGRAM EQUIVTESTER (OUTPUT)

```
var x,y,t,p,q: boolean
begin
t:= true
for p:= false to true do
    for q:= false to true do begin
        x:=({insert expression for x});
        y:=({insert expression for y});
        if x < > y then t:= false
    end;
    if t then writeln('the expressions are equivalent')
        else writeln('the expressions are not equivalent');
end.
```

Use this program to show that $\sim(p \wedge q)$ is equivalent to $\sim p \vee \sim q$.

P24. Modify the above program to also print out truth tables for the propositions being tested.

P25. Modify the program in Problem 23 to verify the results of Problems 1 and 7.

P26. Modify the program in Problem 23 to work for logical expressions with three variables. Then use your program to verify the results of Problems 2 to 6 and 8.

1.6 THE CONDITIONAL AND BICONDITIONAL

The Conditional

Suppose that your father offers to give you a new car if you get an A in your discrete mathematics course. You get a B, but your father gives you a car anyway. An important question arises from this: Was he telling the truth? We might first note that the statement: "If you get an A in discrete mathematics, then I will give you a car" is a compound sentence, with underlying basic propositions being "you get an A in discrete mathematics" and "I will give you a car." The construction of this compound sentence involves the use of a new connective.

Common sense should lead us in the right direction for analyzing this proposition. We would be certain that your father was lying if you got an A and the car were not forthcoming. Thus, there is no question about how the truth of our compound proposition should be decided if the first part is true and the second false. Similarly, the proposition is clearly true if both parts are true. When you get both the A and the car, clearly you have not been lied to. The real issue arises when we consider what the truth value should be for this proposition when the first part is false. One way of looking at the problem is to recognize that the statement really was only intended to tell us what would happen when the first part was true, and the statement was never designed to supply any information at all as to what happens when we don't get the A. When the discrete mathematics grade is anything except an A, we *have no right* to expect anything to happen or not happen. No matter what happens, we should be willing to accept the result as *not* false. Hence by the law of the excluded middle, the statement must be *true*.

The kind of considerations involved in this analysis leads us to define a new connective, called the *conditional*. When we want to join propositions p and q by a conditional, we write[5] if p, then q and denote it symbolically as $p \rightarrow q$. We need to observe that $p \rightarrow q$ is not the same proposition as $q \rightarrow p$, because in this compound proposition, the first basic proposition and the second play quite different roles. The truth table which defines $p \rightarrow q$ is given in Table 1.6. In a conditional statement, the proposition p is called the *hypothesis*, and the proposition q is called the *conclusion*. Probably the third line of this truth table causes the most difficulty for students.

TABLE 1.6   Defining truth table for $p \rightarrow q$

p	q	$p \rightarrow q$
T	T	T
T	F	F
F	T	T
F	F	T

Uses of the Conditional

The discussion above should make it more clear why we have chosen to make the definition in the manner we have. Perhaps an example from computer science will help to clarify the situation.

---

[5] Some people read $p \rightarrow q$ as "p implies q," but we prefer to reserve the use of the word "implies" to describe a relationship between propositions, and not to use it as a connective.

In Pascal, as with most other programming languages, we use conditional statements to control program actions. In using the conditional statement "if $y < 3$, then $x := 1$," we observe that if the value of y is less than 3, then a 1 will be stored at x. It is no accident that this observation is a conditional statement. When the statement $y < 3$ is true (and the computer works correctly), we find that a 1 is stored at x. If $y < 3$ is false, we really don't know what to expect at x. A 1 is not beyond the realm of imagination, but not having a 1 there would be no surprise either. The only time we might think that something had gone haywire would be the situation in which $y < 3$ is true but we don't find a 1 stored at x. This situation matches the truth table which is used to define the conditional connective.

Most theorems in mathematics are propositions of the form of a conditional, so it is especially important that we know exactly how this kind of statement works.

EXAMPLE 1.20

(a) If ABC is a right triangle with right angle at C, then $a^2 + b^2 = c^2$.
(b) If $x = y$ and $y = z$, then $x = z$.
(c) If ABC is an isosceles triangle with $a = b$, then angles A and B are congruent.

The above propositions are three examples of theorems and axioms from high school mathematics which are in the form of conditional statements. ∎

One fact must be emphasized: To claim that the statement $p \rightarrow q$ is true does not mean that p *causes* q to be true. In fact, one way to guarantee that the proposition $p \rightarrow q$ is true is to construct p so that it is always false. The common figure of speech "If that's true, then I'm a monkey's uncle" really makes use of the definition of the conditional, because it asserts that the statement is true even though the conclusion (I'm a monkey's uncle) is obviously false. Certainly no one is claiming to really be a relative of any creature of the simian species, but instead the claim is being made (in a colorful fashion) that the hypothesis is false. (It is left as an exercise to explain exactly why this is the case.) The statement $p \rightarrow q$ has a number of equivalents, each of which is valuable in proofs.

THEOREM 1.1

The following propositions are equivalent to $p \rightarrow q$:
(a) $\sim p \vee q$
(b) $\sim q \rightarrow \sim p$
(c) $\sim (p \wedge \sim q)$

SOLUTION
The proof of this theorem is left as an exercise. ∎

The proposition $\sim q \to \sim p$ is called the *contrapositive* of $p \to q$, the proposition $q \to p$ is called the *converse*, and $\sim p \to \sim q$ is called the *inverse*. The converse and inverse of a statement are *not* equivalent to the original statement but are equivalent to each other, because the inverse is the contrapositive of the converse.

Just to make life difficult (actually in order to give us a variety of available language), the proposition $p \to q$ can be expressed in several different ways. The following are the most common:

If p, then q

q if p

p only if q

p is sufficient for q

q is necessary for p

All these reflect the idea that in the proposition under discussion whenever it is known that p is true, it will have to follow that q is also true.

EXAMPLE 1.21

Consider the statement from algebra which says $b^r = b^s$ if $r = s$. The same statement may also be expressed in the following ways:
(a) $r = s$ only if $b^r = b^s$.
(b) $b^r = b^s$ if $r = s$.
(c) $b^r = b^s$ is a necessary condition for $r = s$.
(d) $r = s$ is a sufficient condition for $b^r = b^s$. ∎

EXAMPLE 1.22

Each of the following statements is of the form $p \to q$. Identify p and q.
(a) Lines l and m are not parallel only if they have a point in common.
(b) For two lines to be parallel, it is sufficient that they be perpendicular to the same line.
(c) For two sides of a triangle to be congruent, it is necessary that the angles opposite those sides be congruent.

SOLUTION
(a) In this one, the proposition which is associated with the "only if" phrase is always the conclusion. Thus p must be the proposition that the lines are not parallel, and q is the proposition that the two lines have a point in common.
(b) The sufficient condition is always the hypothesis. Thus p is the proposition that the two lines are perpendicular to the same line, and q is the proposition that the lines are parallel.

(c) The necessary condition is the conclusion. Thus p is the proposition that two sides of a triangle are congruent, and q is the proposition that the angles opposite those sides are congruent. ∎

Note that it is not the order in which the propositions are listed which determines which is the hypothesis and which is the conclusion, but rather how the propositions are used in the conditional statement. We could, for example, rewrite Example 1.22c to say that it is necessary that the angles opposite two sides of a triangle be congruent for the sides to be congruent.

The Biconditional

It is reasonable, in view of the previous discussion in this section, to also wish to consider a connective which is true if both parts have the same truth value and is false if both parts do not have the same truth value. This connective would express the idea that people have in mind when they object to the resolution of the "false true" case for the conditional statement. We use the symbol $\leftrightarrow$ to indicate this connective, and we read it as "if and only if."[6] This connective is called the *biconditional*, and its defining truth table appears as Table 1.7.

TABLE 1.7   Defining truth table for the connective $\leftrightarrow$

p	q	$p \leftrightarrow q$
T	T	T
T	F	F
F	T	F
F	F	T

Most theorems in mathematics which are not in the form of a conditional are in the form of a biconditional. Looking back to high school geometry, we might remember theorems like the following.

EXAMPLE 1.23

(a) Two triangles are congruent if and only if all three sets of corresponding sides are congruent.
(b) Two lines are parallel if and only if they have the same slope.
(c) A line is tangent to a circle if and only if the line is perpendicular to a radius of the circle at the point at which the radius intersects the circle.

   Each of these theorems is well known, and each can be symbolized in the form $p \leftrightarrow q$. ∎

---

[6] Some people read $p \leftrightarrow q$ as "p is equivalent to q," but we prefer to use equivalent only to describe a relationship between propositions rather than to use it as a connective.

The biconditional receives its name from the fact that $p \leftrightarrow q$ is equivalent to $(p \rightarrow q) \wedge (q \rightarrow p)$. See Table 1.8. This means that the biconditional is the conjunction of two conditional statements. This equivalence leads to the terminology "if and only if," because it means that "p if and only if q" is equivalent to the conjunction of the statements "p if q" and "p only if q." Other ways of expressing this connective are "if p, then q and conversely" and "p is necessary and sufficient for q."

TABLE 1.8   Truth tables showing equivalence of $p \leftrightarrow q$ and $(p \rightarrow q) \wedge (q \rightarrow p)$

p	q	(p	$\rightarrow$	q)	$\wedge$	(q	$\rightarrow$	p)	p	$\leftrightarrow$	q
T	T	T	T	T	T	T	T	T	T	T	T
T	F	T	F	F	F	F	T	T	T	F	F
F	T	F	T	T	F	T	F	F	F	F	T
F	F	F	T	F	T	F	T	F	F	T	F

Statements of the form "p if and only if q" are proved by proving that *both* the statements "if p, then q" and "if q, then p" are true.

The propositions P and Q will be equivalent when both are true and false at exactly the same time. This means that $P \leftrightarrow Q$ is a tautology.

EXAMPLE 1.24

Show that $\sim(p \wedge q)$ is equivalent to $\sim p \vee \sim q$.

To do this, we can set up the truth table for the proposition $[\sim(p \wedge q)] \leftrightarrow (\sim p \vee \sim q)$.

p	q	[$\sim$	(p	$\wedge$	q)]	$\leftrightarrow$	($\sim$	p	$\vee$	$\sim$	q)
T	T	F	T	T	T	T	F	T	F	F	T
T	F	T	T	F	F	T	F	T	T	T	F
F	T	T	F	F	T	T	T	F	T	F	T
F	F	T	F	F	F	T	T	F	T	T	T

Since the statement we constructed was a tautology, the two statements with which we started are equivalent. ∎

We have now considered both of the propositional forms which are used to construct most theorems in mathematics. We are now ready to begin considering the ways in which proofs are constructed and the reasons behind such constructions.

Problem Set 6

1. Show that each of the following is equivalent to $p \rightarrow q$:
   (a) $\sim p \vee q$
   (b) $\sim q \rightarrow \sim p$
   (c) $\sim (p \wedge \sim q)$

2. Consider the theorem from calculus which says, "If f is a continuous function on $[a, b]$, then f is integrable." Without worrying about what the theorem means, rewrite the theorem by using (a) a necessary condition, (b) a sufficient condition, and (c) the phrase "only if."

3. Repeat Problem 2, using the statement "If a series is absolutely convergent, it is convergent."

In each of the following, identify the hypothesis and conclusion:

4. To be elected president, it is sufficient to be a politician.

5. To be elected president, it is necessary to be a politician.

6. A person can star on late-night television only if she or he has silver hair.

7. To become rich, it is sufficient to have rich parents.

8. A necessary condition for understanding computer science is a complete knowledge of discrete mathematics.

9. A sufficient condition for getting all A's is being a good student.

10. Only if I wake up will we go to the game.

11. This program will run only if there were no typing errors.

Construct a truth table for the following:

12. $[(p \rightarrow q) \wedge p] \rightarrow q$

13. $(p \leftrightarrow q) \leftrightarrow r$

14. $p \leftrightarrow (q \rightarrow r)$

15. $(p \wedge \sim p) \rightarrow q$

16. Show that $p \rightarrow q$ and $q \rightarrow p$ are *not* equivalent.

Write the inverse, converse, and contrapositive for the following:

17. If the weather is cold, then it will snow.

18. Being able to type is sufficient to learn word processing.

19. If you don't practice, you will never learn how to play your horn.

20. I will keep this job only if I get a raise in salary.

Use the biconditional to determine whether the following pairs are equivalent:

21. $(p \wedge q) \vee r$; $(p \vee r) \wedge (q \vee r)$ **22.** $(p \rightarrow q) \rightarrow r$; $(p \wedge \sim q) \rightarrow r$

23. $p \leftrightarrow q$; $(p \wedge q) \vee (\sim p \wedge \sim q)$ **24.** $p \leftrightarrow q$; $\sim p \rightarrow \sim q$

1.7   ARGUMENTS AND PROOFS

A "Classical" Example

It is now time to consider the means by which we prove statements in mathematics. A good place to start is to examine a proof from high school geometry.

EXAMPLE 1.25

Given: $AB \parallel CD$

        $\overline{AE} \cong \overline{CE}$

Prove: $\overline{BE} \cong \overline{DE}$

Statements	Reasons
1. $AB \parallel CD$	**1.** Given
2. $\angle BAE \cong \angle ECD$	**2.** Alternative interior angles
3. $\angle BEA \cong \angle DEC$	**3.** Vertical angles
4. $\overline{AE} \cong \overline{CE}$	**4.** Given
5. $\triangle AEB \cong \triangle CED$	**5.** ASA
6. $\overline{BE} \cong \overline{DE}$	**6.** Corresponding parts of congruent triangles

Each statement in the left-hand column is supported by a reason listed in the right-hand column. The reasons listed either are "given" or are statements which identify a theorem. When a theorem is referred to as a reason, we mean that the information preceding the statement together with the statement of the theorem *logically implies* the given statement. ∎

DEFINITION

A collection of statements **logically implies** or **implies**[7] another collection of statements if whenever all the statements in the first collection are true, then all the statements in the second collection must be true.

---

[7] This is the use of "implies" we mentioned earlier.

These rules of logic are typically expressed as *rules of inference*. Many of the examples from Section 1.1 are based on the application (or misapplication) of the rules of inference that we develop in this section.

Modus Ponens

One of the best known rules of inference, *modus ponens*, is described symbolically as

$$p \rightarrow q$$

$$p$$

$$\therefore \quad q$$

The notation above means that whenever the statements listed above the line are true, then the statement(s) listed below the line must be true. This rule then says that whenever a statement of the form "if p, then q" is true and at the same time the statement p is true, it must be the case that q is also true. This fact can indeed be seen if we examine the following table:

p	q	p	$p \rightarrow q$	q
T	T	T	T	T
T	F	T	F	F
F	T	F	T	T
F	F	F	T	F

Only on the shaded line (the top line) of the truth table are both propositions true, and on that line we find that q is also true.

Let us recall an example from Section 1.1:

EXAMPLE 1.2 (REPEATED)

If computers are to find their way into the home of the average person, then the price of some complete systems should be less than $1000. Computers are found in the homes of average people, thus we must conclude that there are complete computer systems which are priced under $1000.

ANALYSIS
We can make the following assignments:

p: Computers are found in the home of the average person.

q: There are computers which cost less than $1000.

The argument is exactly the same one we described above, and thus this is a valid argument. The only question which we might raise about this line of reasoning lies not in the logic, but only in whether we accept that the statements which went into it (the *premises*) were themselves true statements. If they are true, we are forced to conclude that the last statement (the *conclusion*) must be true. ∎

This particular form of an argument is usually called *modus ponens*, or the *law of detachment*. The proposition $[p \wedge (p \to q)] \to q$ is a tautology. Whenever the proposition $p_1 \wedge p_2 \wedge \cdots \wedge p_n \to q$ is a tautology, we have that $p_1, p_2, \ldots, p_n$ imply q.

Notation and Terminology

The proof which is demonstrated in Example 1.25 uses a form which is rarely seen outside of high school geometry texts. On the left side we see explicit statements of the propositions which either have been proved true in the course of the proof or were assumed true at the outset. In the right-hand column we see the statements of (or references to) the theorems which, together with the laws of inference from logic, were used to show that the statements on the left should be true. In statement 3, for example, we have stated that angles BEA and DEC are congruent, and the reason given is that vertical angles are congruent. Basically what we have done to get statement 3 is to note that those angles are vertical angles (this becomes proposition p) and to mention the theorem which says that if two angles are vertical angles, then they are congruent (this is, in fact, $p \to q$) to conclude q, which is the proposition listed under statement 3.

In describing the rules of inference, we often use the notation[8] $p \vdash q$ to indicate that q is implied by p. If more than one statement is used to imply q, then all are listed on the left side, separated by commas. The law of *modus ponens* can be written as $p, p \to q \vdash q$.

Most mathematical proofs relax the format which we used for our geometric proof and use a narrative format, which lists the statements to be proved and the reasons that we can give for those statements in a paragraph format. In Example 1.25, the proof would have been written in the following way:

Since AB and CD are parallel and angles BAE and DCE are alternate interior angles to those parallel lines, those angles are congruent. Since angles BEA and DEC are vertical angles, they, too, are congruent. It was given that $AE = CE$. We thus have two triangles with two corresponding angles and the included sides congruent. This implies that the two triangles AEB and CED are congruent; and since BE and DE are corresponding parts of those triangles, they must be congruent.

---

[8] The symbol $\vdash$ can be read "proves."

DEFINITION

A **proof** is a sequence of statements each of which is (a) true by assumption or (b) a previously proven theorem or (c) a previously agreed-upon definition or axiom or (d) a statement which is implied by the statements which precede it.

Testing Rules of Inference

To understand the process of proof, we must understand the rules of inference which we are allowed to use. In propositional logic it is easy to construct and test rules of inference. The basic rule is that

$$p_1, p_2, \ldots, p_n \vdash q$$

is a rule of inference if and only if

$$p_1 \wedge p_2 \wedge \cdots \wedge p_n \rightarrow q$$

is a tautology.

We can list several commonly used rules of inference:

1. $p, p \rightarrow q \vdash q$ (modus ponens)

2. $p \rightarrow q, q \rightarrow r \vdash p \rightarrow r$ (syllogism)

3. $p \rightarrow q, \sim q \vdash \sim p$ (modus tollens)

4. $p \vdash p \vee q$ (addition)

5. $p \wedge q \vdash p$ (specialization)

6. $p, q \vdash p \wedge q$ (conjunction)

Of course, if two statements are equivalent and one of them occurs in a rule of inference, we can make use of the other just as well, so that in any proof, equivalent statements are interchangeable.

EXAMPLE 1.26

Show that modus tollens is a valid rule of inference.

Modus tollens is the rule

$$p \rightarrow q, \sim q \vdash \sim p$$

To verify that this is a valid rule of inference, we need to show that

$$[(p \rightarrow q) \wedge \sim q] \rightarrow \sim p$$

is a tautology. We set up the following table:

p	q	$[(p \rightarrow q) \wedge \sim q] \rightarrow \sim p$
T	T	T T T F F T T F T
T	F	T F F F T F T F T
F	T	F T T F F T T T F
F	F	F T F T T F T T F

Since this is a tautology, the argument is valid.

Another approach to verifying this rule is the following: Modus ponens gives us

$$p,\ p \rightarrow q \vdash q$$

If we apply this rule, clearly

$$\sim q,\ \sim q \rightarrow \sim p \vdash \sim p$$

is valid. The second hypothesis is just the contrapositive of $p \rightarrow q$, and we can substitute $p \rightarrow q$ into the argument. That substitution gives us the statement of the modus tollens rule which is

$$\sim q,\ p \rightarrow q \vdash \sim p \qquad \blacksquare$$

In the problems at the end of this section, you are asked to verify that the other rules of inference listed in this section are also valid.

Fallacies

DEFINITION

A **fallacy** or **invalid argument** is an argument for which in some cases the hypotheses are all true, but the conclusion is false.

Example 1.1 provides an example of one of the most commonly occurring fallacies.

EXAMPLE 1.1 (REPEATED)

Consider the following argument:

If computers are to really find their way into the home of the average person, then the price of some complete systems should be less than $1000. The local computer store has complete systems priced under $1000, so it must follow that computers have found their way into the average home.

We need to ask whether the argument convinces us that the conclusion is justified by the evidence. If it does "follow logically," then we say that the argument *proves* that the conclusion is indeed true. Sometimes we find that the conclusion is true even when the argument is not "logical."

ANALYSIS

We make the following assignments:

p: Computers find their way into the home of the average citizen.

q: The price of some complete systems is less than $1000.

The argument listed above is described by

$$q, p \to q \vdash p$$

The truth table for the proposition associated with this argument is given by

p	q	$[(p \to q) \land q] \to p$
T	T	T T T T T T T
T	F	T F F F F T T
F	T	F T T T T F F
F	F	F T F F F T F

Looking at the third line of the truth table, we see that the argument fails when p is false and q is true, because under those circumstances both hypotheses are true but the conclusion is not. In this example, this says that the conditional statement is true, computers are priced under $1000 but are not found in the homes, contrary to the argument. ∎

Construction of Proofs

The most common kind of theorem which is proved in mathematics is the statement of the form

$$p \to q$$

Suppose we are asked to prove a proposition of this form. What do we do?

To approach the problem, we look once more at the truth table for $p \to q$:

p	q	$p \to q$
T	T	T
T	F	F
F	T	T
F	F	T

We note that as far as the truth of $p \rightarrow q$ is concerned, we couldn't care less what happens when p is false. In that situation, it does not matter whether q is true or false, for in either case $p \rightarrow q$ is true. The important issue arises when p is true. If we can guarantee that the second line of the truth table *never* occurs, we will have that the statement "if p then q" is always true. The best way to make sure of that is to show that whenever p is true, q *must* be true. This guarantees that we will never be on line 2 of the truth table for $p \rightarrow q$. Our strategy for proof should be clear: Assume that p is true and proceed to find a sequence of statements each of which follows from the other (and by already proved theorems) so that the final conclusion that we reach is that q is true.

We write $p \vdash_* q$ if there is a proof in which p is assumed to be true and in which q is demonstrated to be true. This discussion says that to prove "if p, then q," all that we have to do is construct a proof in which $p \vdash_* q$. Similarly, if we have a proof in which

$$p, q, r, \ldots, z \vdash_* a$$

then we have actually shown the statement

$$p \wedge q \wedge r \wedge \cdots \wedge z \rightarrow a$$

is true.

EXAMPLE 1.27

One good practice exercise in the construction of proofs is to attempt to prove some statements symbolically. We might be asked, for example, to prove that $(p \rightarrow q) \wedge (\sim r \rightarrow \sim q) \wedge \sim r \rightarrow \sim p$ is a tautology.

One way to do this would be to construct a truth table, but another way is to construct a proof in which

$$p \rightarrow q, \ \sim r \rightarrow \sim q, \ \sim r \vdash_* \sim p$$

Using the geometry style of proof, we write

Given: $p \rightarrow q$

$\sim r \rightarrow \sim q$

$\sim r$

Prove: $\sim p$

Statements	Reasons
1. $p \rightarrow q$	1. Given
2. $\sim q \rightarrow \sim p$	2. Contrapositive of 1
3. $\sim r \rightarrow \sim q$	3. Given
4. $\sim r \rightarrow \sim p$	4. Syllogism (2 and 3)
5. $\sim r$	5. Given
6. $\sim p$	6. Modus ponens (4 and 5)

This kind of a scheme can be used to verify rules of inference, and sometimes it is easier to verify rules of inference by using this scheme than it is to construct a truth table. The only problem is that this technique will not show that a purported rule of inference is actually *not* a rule of inference, but rather a fallacy. In this example, it seems to be more instructive to construct a proof than to construct an eight-line truth table.

We can also write this proof in paragraph form as follows:

Since $p \rightarrow q$ is given to be true, its contrapositive $\sim q \rightarrow \sim p$ must also be true. We are also given $\sim r \rightarrow \sim q$, and so by the law of syllogism we can conclude that the proposition $\sim r \rightarrow \sim p$ must be true. Since r is false, modus ponens allows us to conclude that p is false. ∎

To prove $p \leftrightarrow q$, we can prove $p \rightarrow q$ and prove $q \rightarrow p$. Since the conjunction of those two propositions is equivalent to p if and only if q, this strategy yields the needed result.

Indirect Proofs

Indirect proof is a method of proving a proposition in which we assume that the conclusion that we wish to reach is false and then we reach a conclusion which requires us to accept the statement that we wished to prove was actually true.

There are actually two forms of indirect proof for a proposition of the form $p \rightarrow q$. The more obvious method is to prove the contrapositive of the original statement by using the usual direct method of proof. To do this, we assume that the conclusion of the original statement is false and show that the hypothesis of that statement is also false.

The second method is assume that the hypothesis is true and the conclusion is false (that is, we assume $p \wedge \sim q$) and show that some statement which we know to be false follows from that assumption. This proves the proposition $(p \wedge \sim q) \rightarrow F$, where F stands for a universally false statement. (And $p \wedge \sim p$ is an example of a proposition which is always false.) By doing this, we have found that $p \wedge \sim q$ cannot be true, for if it were, it would force us to conclude that a statement known to be false is actually true. This proves $\sim(p \wedge \sim q)$. Since $p \wedge \sim q$ is the negation of $p \rightarrow q$, we have proved the negation of $p \rightarrow q$ is false, which means we have shown $p \rightarrow q$ to be true.[9] The second type of the indirect proof is different from the first, in that in the second form we assume both that q is false *and* that p is true, whereas in the proof of the contrapositive the only assumption that we make is that q is false. In the second type of indirect proof, we have completed our proof when we show that some impossible statement must be true; in the first type, we are done when we show that p is false. This second type of indirect proof is the one which probably causes the most confusion among students and as such requires an example.

---

[9] Actually this method works for any proposition, since if we prove $\sim p \rightarrow F$, we must conclude that p is true.

EXAMPLE 1.28

47

1.7 ARGUMENTS
AND PROOFS

Given: $p \vee q$, $\sim q \vee r$.

Prove: $p \vee r$.

PROOF
To show that $p \vee r$ is true by indirect proof, we assume that its negation is true, that is, we assume $\sim p \wedge \sim r$. Now $p \vee q$ is equivalent to $\sim p \rightarrow q$. By modus ponens, the assumption of $\sim p$ yields that q is true. And $\sim q \vee r$ is equivalent to $q \rightarrow r$. By modus tollens, our assumption of $\sim r$ yields $\sim q$. Thus we have proved $q \wedge \sim q$. Since this is a contradiction, we must conclude that $p \vee r$ is true.   ∎

Problem Set 7

Prove that the following rules of inference are valid by using a truth table.

1. $p, p \rightarrow q \vdash q$ (modus ponens)

2. $p \rightarrow q, q \rightarrow r \vdash p \rightarrow r$ (syllogism)

3. $p \vdash p \vee q$ (addition)

4. $p \wedge q \vdash p$ (specialization)

5. $p, q \vdash p \wedge q$ (conjunction)

6. $\sim p \vdash (p \rightarrow q)$ (law of vacuous proof)

7. $q \vdash (p \rightarrow q)$ (law of trivial proof)

8. $p \rightarrow r, q \rightarrow r, p \vee q \vdash r$ (law of reduction to cases)

9. $p \leftrightarrow q \vdash (p \rightarrow q) \wedge (q \rightarrow p)$

10. $(p \rightarrow q) \wedge (q \rightarrow p) \vdash p \leftrightarrow q$

Prove the following *without* using truth tables.

11. Given p, $p \rightarrow q$, and $q \rightarrow r$, prove r.

12. Given $p \vee q$ and $\sim p$, prove q.

13. Given $p \wedge \sim q$, $p \rightarrow r$, and $r \rightarrow (s \vee q)$, prove s.

14. Given $p \leftrightarrow q$, $\sim p \rightarrow r$, and $\sim r$, prove q.

15. Given $p \vee q$ and $\sim q \vee r$, prove $p \vee r$ (use a direct proof).

16. Given $p \leftrightarrow q$ and $q \leftrightarrow r$, prove $p \leftrightarrow r$.

17. Given p and $(p \wedge \sim q) \rightarrow \sim p$, prove $p \rightarrow q$.

18. Given p, q, and $(p \wedge q) \rightarrow r$, prove r.

Consider the following arguments, and analyze them to determine whether they are valid.

19. If I buy a new car, then I will not be able to go to Florida in December. Since I am going to Florida in December, I will not buy a new car.

20. Either the butler or the maid committed the crime. If the butler did it, he would not have been able to answer the phone at 11:00. Since he did answer the phone at 11:00, the maid must have done it.

21. If the temperature had gone down, Bill would not have gone to the parade. Since Bill did not go to the parade, the temperature must have gone down.

22. If the football game runs late, then they will delay the start of *60 Minutes*. If *60 Minutes* runs late, the local news will start after 11:00. The local news started at 11:15 tonight, thus the football game ran late.

23. If I get the job and work hard, then I will be promoted. I was promoted. Thus I got the job.

24. If I get the job and work hard, then I will be promoted. I was not promoted. Thus either I did not get the job or I did not work hard.

25. Either I will get an A in this course, or I will not graduate. If I don't graduate, I will go into the army. I got an A. Thus, I won't go into the army.

26. Either I will get an A in this course, or I will not graduate. If I don't graduate, I will go into the army. I got a B. Thus, I will go into the army.

1.8 PREDICATE LOGIC

Predicates

Many statements in mathematics are *not* simply propositions. In many instances we must make statements whose truth depends on the value of a variable which occurs in the statement.

> **DEFINITION**
>
> A statement which would be a proposition except for the fact that it includes variables whose values are not specified is called a **predicate**.

EXAMPLE 1.29

49

1.8   PREDICATE LOGIC

The following mathematical sentences are all predicates.
(a) $x + 4 = 3$.
(b) The sum of the first n odd integers is n^2.
(c) If $x < 3$, then $x^2 = -1$.
(d) $x + y = 4$.
(e) If $x < y$, then $x^2 < y^2$.
(f) $x + y = 4$ if and only if $y = 4 - x$.   ■

A predicate which involves one variable is called a **one-place predicate**. A predicate which involves two variables is called a **two-place predicate**, and in general a predicate which involves n variables is called an ***n*-place predicate**.

DEFINITIONS

The collection of variables which can replace a variable in a predicate is called the **universe** of the variable. In a one-place predicate, if a value from the universe of the variable can be substituted for the variable to make it true, then we say that the value **satisfies** the predicate.

If an n-tuple of values exists which can be substituted for the variables in an n-place predicate and make the predicate become a true proposition, then we say that that n-tuple satisfies the predicate.

We generally denote predicates by capital letters such as P, Q, R, etc., and following the name of the predicate we list in parentheses the variables which are used by the predicate. Thus in Example 1.29 we might describe the predicates as

(a) $P(x)$ (b) $Q(n)$

(c) $R(x)$ (d) $S(x, y)$

(e) $T(x, y)$ (f) $U(x, y)$

It is interesting to note in this case that the predicates $R(x)$, $T(x, y)$, and $U(x, y)$ are actually compound predicates consisting of predicates which are combined by the same kind of logical connectives that are used in propositional logic. This means that we can write $R(x)$ as $A(x) \to B(x)$, where $A(x)$ is the predicate $x < 3$ and $B(x)$ is the predicate $x^2 = -1$. Similarly, we write $T(x, y)$ as $C(x, y) \to D(x, y)$ and write $U(x, y)$ as $F(x, y) \leftrightarrow G(x, y)$.

Terminology

DEFINITIONS

An n-place predicate is said to be **satisfiable** provided that there is an n-tuple which satisfies it; and if all n-tuples satisfy it, it is said to be **valid**.

Two predicates will be **equivalent** if they have the same truth value for all possible values of their variables. Another way to phrase this would be to say that $X(x)$ and $Y(x)$ are equivalent predicates provided that $X(x) \leftrightarrow Y(x)$ is valid. The idea of equivalent predicates plays the same role in predicate logic as the idea of equivalent propositions does in propositional logic.

Some of the terminology relating to predicates may seem quite familiar. You may recall that in algebra we said that the value -1 *satisfies* the equation $x + 4 = 3$. This equation actually is a predicate, and solving the equation actually means that we are trying to find all values of the variable which satisfy the predicate.

There are also many examples of predicates which occur in common usage of our everyday language.

EXAMPLE 1.30

(a) She runs the mile faster than any of them.
(b) It is a good idea.
(c) If he wins the election, then they will lose their jobs. ∎

Each of the above statements contains pronouns. In English, pronouns act much as variables do in mathematics, and hence English sentences which involve pronouns with unknown referents are predicates. In English, of course, we often use pronouns for variety in our conversation and writing when it is clear to whom the pronoun refers. In cases like that, the pronoun is clearly not a variable, in the same sense that using e in a discussion of natural logarithms really is not using the letter e as a variable.

Quantifiers

If we take the time to examine some of the theorems and axioms that we find in mathematics books, we might note some interesting features of many of those statements.

EXAMPLE 1.31

We consider some of the statements of the basic properties of algebra as typically presented in an algebra textbook. Let a, b, c denote real numbers.
(a) $a + b = b + a$
(b) $a + (b + c) = (a + b) + c$
(c) $a + 0 = 0 + a = a$
(d) For every a there is a b such that $a + b = 0$ and $b + a = 0$.
(e) If $a < b$, then there is a c such that c is positive and $a + c = b$.
(f) If $a < b$ and $b < c$, then $a < c$.

The most striking fact about these statements is that each one involves predicates, yet we claim that they are all true statements. ∎

The fundamental question is, How can a predicate like $P(a, b)$: $a + b = b + a$ be regarded as a proposition? Of course, we know what the authors of the algebra book had in mind when they wrote that predicate as an axiom. They meant to say that *no matter what* values we substitute for a and b in the predicate, the resulting proposition is a true statement. In this instance, then, the symbols a and b act as though they are in some sense not really variables at all, since the truth of what we mean depends not on particular values of a and b but rather on the fact that the statement is true for *all* values of a and b predicates.[10] In this instance we say that the variables a and b have been *quantified universally*. To indicate this fact, we would write symbolically $\forall a \, \forall b P(a, b)$, which is read "for all a and b, $P(a, b)$."

One way of effectively removing variables from a predicate is to quantify them in this fashion. Again when we have used the symbol $\forall$ to quantify the symbolic representation of a predicate, or the phrase "for all x" to precede the statement of a predicate given in natural language, the intention is to claim that the predicate is true for all values of x in the universe.

We note that the statement in Example 1.31c is really $\forall a R(a)$, and the predicate will be true no matter what the value of the variable happens to be.

Example 1.31b is $\forall a \, \forall b \, \forall c Q(a, b, c)$. We should note that in many cases when we state theorems or axioms, it is customary not to explicitly state universal quantifiers for all the variables, except when it is necessary to avoid confusion.

In Example 1.31d, however, we have a predicate of the form $S(a, b)$, and in this case the a has been explicitly quantified. What about the b? We have said that given the value of a, we can find *some* value of b to make the predicate $(a + b = 0)$ true. Once again, we do not really care what value of b makes the predicate true, but only that there exists one that does. Thus, in a different fashion, we have removed the explicit concern about what b is from the statement of the predicate. In this case we say that b has been *existentially quantified*. When we quantify a variable existentially in a predicate, we are essentially saying that the predicate can be made true by proper selection of a value of the variable. The symbol used to indicate that a variable has been quantified existentially is $\exists$, and thus we can write Example 1.31d as $\forall a \, \exists b S(a, b)$. We read this, "for every a there is a value of b such that $S(a, b)$ is true." This means that if we are given a value for a, we can find a value for b to make $S(a, b)$ true.

Changing the order of the quantifiers does change the meaning considerably. If we were to write $\exists b \, \forall a S(a, b)$, we would be saying (as we read from left to right) that there is some value of b which can be chosen so as to make S true no matter what value of a is chosen. Notice that in the first arrangement we say no matter what a is, we can find a b that works, whereas here we say there is one value of b which, once chosen, makes the predicate true regardless of the value of a.

---

[10] Logicians regard a and b as still being variables but assign them to a special class of variables called *bound variables* which are allowed to occur in propositions.

Moving on to Example 1.31e, we see that without using any universal quantifiers the statement is

$$A(a, b) \to \exists c[B(c) \wedge C(a, bc)]$$

The brackets have been added to the last phrase to indicate that the existential quantifier applies to the whole phrase enclosed in brackets. Once again we must assume, for this to become a proposition, that all variables which have not been quantified must be universally quantified. And their values should be quantified through the entire predicate, so we write $\forall a \; \forall b\{A(a, b) \to \exists c[B(c) \wedge C(a, bc)]\}$. We use parentheses, brackets, and braces to indicate the *scope* of quantifiers. If no parentheses follow a quantifier or sequence of quantifiers, we assume that it (or they) applies only to the predicate immediately following. This means that $\forall a[P(a) \to Q(a)]$ is a different statement from $\forall a P(a) \to Q(a)$.

Some terminology is certainly useful here. A variable which has been either quantified or substituted for in a predicate is said to be *bound*, while variables which have not been bound are said to be *free*. Bound variables can be included in a proposition because with the "binding" of the variable, it now becomes possible to determine the truth of a predicate without regard to any unknown values of variables.

We should summarize the meaning of the use of quantifiers.

1. $\forall x P(x)$ means that for all values in the universe of the variable x, the predicate $P(x)$ is true.

2. $\exists x P(x)$ means that there is at least one value in the universe of the variable x which satisfies the predicate $P(x)$.

A statement which includes free variables is *not* a proposition, but only a predicate, and a predicate in which all the variables have been bound is, in fact, a proposition.

EXAMPLE 1.32

Let $P(x, y)$: x is older than y.

$Q(x, y)$: x is wiser than y.

$R(x)$: x is bald.

Consider the following symbolic statements and translate them into English.
(a) $\forall x \; \exists y[P(x, y) \to (Q(x, y) \wedge R(y))]$
(b) $\forall x \; \forall y[Q(x, y) \to R(y)]$
(c) $\forall x[\forall y \sim P(x, y) \to R(x)]$

(a) In this statement, the universal quantifier applies to the entire statement. So in our convention of usage, we may leave out the phrase "for all x" if the resulting statement is not ambiguous. This means that we could write this as "There is some y such that if x is older than y, then x is wiser than y and y is bald." This does seem ambiguous when we read it, so we backtrack and must write instead, "For all people x, there is a y such that if x is older than y, then x is wiser than y and y is bald." Note also that we inserted at the beginning of this statement an indication of the universe of the variable x.

(b) Again in this example, the universal quantifiers both apply to the entire statement, so we can attempt to write it by not explicitly including the phrase "for all." Our proposition reads, "If x is wiser than y, then y is bald."

(c) We use the same conventions as before and write, "If the statement x is older than y is false for all y, then x is bald." ∎

Negations of Quantified Predicates

The last part of this example is hard to deal with because of the presence of something new—the negation of a predicate. It is important to consider the interaction between the negation sign and quantifiers.

   If we claim that a universally quantified predicate is false, then we must be saying that there is some value of its variable which makes the predicate false. (After all, if that weren't the case, the universally quantified statement would be true.) Similarly, if we claim that an existentially quantified statement is false, the only way that can occur is if the predicate is *never* true. This line of reasoning leads us to the following two rules for negation of quantified predicates.

NEGATION OF QUANTIFIED PREDICATES

(a) $\sim \forall x P(x)$ is equivalent to $\exists x \sim P(x)$.
(b) $\sim \exists x P(x)$ is equivalent to $\forall x \sim P(x)$.

EXAMPLE 1.33

If $P(x)$ is the predicate "x is taller than 1 meter" with the universe of x being adult males, the universally quantified statement is "All adult males are taller than 1 meter." For that statement to be false, it is necessary to exhibit only one adult male who is shorter than 1 meter. Thus the negation of the statement is "There is an adult male who is not taller than 1 meter." The original statement is $\forall x P(x)$, and its negation is $\exists x \sim P(x)$.

In a similar vein, let $Q(x)$ represent the predicate that x is taller than 2.5 meters. To quantify that statement existentially is to say that there is an adult male who is taller than 2.5 meters, and to claim that in fact the proposition is false is to say that the predicate is not satisfied. Another way to put it is that all people make that predicate false. In this case the original statement is $\exists x Q(x)$, and its negation must be $\forall x \sim Q(x)$. ∎

Quantifiers and Finite Universes

It is illustrative to consider what happens to the process of quantification when the universe of discourse of a predicate is finite. Suppose we consider the situation in which the universe of discourse for P consists of two elements a and b. It should seem reasonable under those circumstances to note that $\forall x P(x)$ is saying that *both* $P(a)$ and $P(b)$ are true, and in these circumstances, the universal quantifier acts just as a conjunction. In a similar fashion, the existential quantifier acts as a disjunction. The laws of negation discussed above, and the equivalences and rules of inference which we discuss in the next section, are easily understood in that light: In particular, the laws of negation are then seen to be a kind of De Morgan's law situation.

Predicates and Computer Science

A postscript might be added to our introduction of predicates. Modern computing is beginning to make use of predicates at both the programming and applications level.

The language PROLOG (*pro*gramming in *log*ic) allows a programmer to specify predicates and a universe of discourse and then to do logical manipulations with the predicates. The language PROLOG has been extensively used in new research into artificial intelligence.

At the application level, relational database query languages permit the storing of information by use of records which contain information about various aspects of an item. A car dealer might, for example, store data about cars on the lot in terms of year of manufacture, maker, model, body type, and mileage. The user can pose queries to the database which are actually in the form of predicates, such as body type is a wagon, mileage is less than 50,000. The database could then print all data items for which the given predicate is true.

Problem Set 8

Consider the following predicates:

$P(x, d)$: $|x - 3| < d$

$Q(x, e)$: $|x^2 - 9| < e$

$R(x)$: $x^2 = 9$

$S(x, y)$: $|x - y| < 0.01$

Write out the following statements:

1. $P(3.01, 0.02)$

2. $Q(4, 10)$

3. $R(5)$

4. $S(e, 2.78)$

5. $P(x, d) \rightarrow Q(x, d)$

Consider the following predicates:

$P(x, y)$: x is faster than y.

$Q(x, y)$: y is taller than x.

$R(x)$: x is more than 200 pounds.

Write good English sentences for the following:

6. $P(x, \text{Fred})$

7. $Q(\text{Sam}, \text{Mike}) \wedge R(\text{Sam})$

8. $P(x, y) \rightarrow Q(x, y)$

9. $Q(x, y) \rightarrow \sim R(x)$

10. $P(\text{Sam}, \text{Fred}) \vee [Q(\text{Sam}, \text{Fred}) \wedge R(\text{Fred})]$

Using the list of predicates from Problems 1 to 5, translate the following into English:

11. $\forall x \, \forall y [S(x, y) \rightarrow P(x, 3)]$

12. $\forall x [S(x, 0.01) \leftrightarrow Q(x, 0.0001)]$

13. $\forall x [P(x, 0.01) \wedge S(y, x)]$

14. $\forall e \, \exists d [P(x, d) \rightarrow Q(x, e)]$

15. $\exists x [R(x) \wedge \forall y P(x, y)]$

Identify the predicates used in the following statements, and write them in symbolic form:

16. Every planet moves in a plane, and each planet has an elliptical orbit with the sun at one focus.

17. For some people in the $100,000 income bracket, there are no taxes.

18. Some horses can run faster than some cars.

19. All cars can run faster than some horses.

20. Either there is a number which is greater than every known solution to the problem, or there is no solution.

21–25. Write the negations for the propositions in Problems 11 to 15 symbolically.

26–30. Write good English statements to negate the statements in Problems 16 to 20.

*31. If the universe of the variable x is finite, then it can be argued that the proposition $\forall x P(x)$ is really a conjunction of propositions, while $\exists x P(x)$ is a disjunction of propositions. Explain why.

*32. Given the fact of Problem 31, explain why the two rules of negation for quantified predicates should be regarded as forms of De Morgan's laws.

1.9 PROVING QUANTIFIED STATEMENTS AND MATHEMATICAL INDUCTION

Many mathematical statements are in the form of quantified statements. A good example is the statement from algebra which says that for all n the sum of the first n odd integers is n^2. Since so many mathematical theorems are of this form, it is important to understand how such statements are to be proven.

Rules of Inference

First note that all the rules of inference which we discussed earlier still work in these circumstances. Propositional logic is just a special case of predicate logic in which all the predicates have no variables. The fact is that all we need to do is to add rules of inference which relate to quantifiers—the rest of our logic works the same way as before.

There are four very important rules of inference which enable us to move from statements with constants to quantified predicates and back under some circumstances. A statement with a constant in it is easier to deal with than a statement which involves quantifiers. In the following, x represents a variable, a is a constant, and P a one-place predicate.

> **RULES OF INFERENCE FOR SIMPLE PREDICATES**
>
> (a) $P(a) \vdash \exists x P(x)$ (Existential generalization)
> (b) If a is arbitrary (an "average" member of the universe of discourse), $P(a) \vdash \forall x P(x)$ (Universal generalization).
>
> If, for an arbitrary member of the universe of x, we can show that P is true without making use of any properties of a other than the fact that it is in that universe, then we have shown that $P(x)$ is true regardless of the value of x.

(c) $\exists x P(x) \vdash P(a)$, where a is chosen to be a special element of the universe of x in the sense that it is an element which satisfies P. (Existential instantiation).

(d) $\forall x P(x) \vdash P(a)$, where a is *any* element of the universe of x. (Universal instantiation).

The first two rules of inference are important because they demonstrate the means by which we can prove quantified predicates. The last two show us how we can use quantified predicates to make valid conclusions involving constants.

To prove a statement of the form $\forall x P(x)$, we have to take an arbitrary element from the universe of x and show that P is true for that element. The only restriction is that in the process, the only things we are allowed to use about x are facts known to be true for *any* element of the universe of x. Similarly, to prove $\exists x P(x)$, all we need to do is to find *one* element of the universe of x which satisfies P. When that is done, we have the proposition we sought.

For more complicated expressions, we use those rules in concert with the other rules of inference that we have already considered.

EXAMPLE 1.34

Determine the steps needed to prove a statement of the form

$$\forall x \ \forall e \ \exists d [P(x, d) \to Q(f(x), e)]$$

According to the rules of inference, we must take x_1 and e_1 to be arbitrary elements of the universes of x and e, respectively. Our task then becomes one of finding some value for d, say d_1, so that the statement

$$P(x_1, d_1) \to Q(f(x_1), e_1)$$

can be proved to be true. If we determine such a value for d_1, then we can assume that $P(x_1, d_1)$ is true. We then proceed to find a proof for the statement $Q(f(x_1, e_1))$, using the usual rules of inference. ■

Note that the whole process involves using *all* the rules of inference in concert with one another. This technique is used over and over in the construction of proofs in mathematics. We should observe that all the techniques of proof still apply here and that we can use the techniques of negating quantified statements discussed in Section 1.8 to produce the negations as needed.

Another concern in dealing with predicate logic is the problem of dealing with quantifiers which apply to conjunctions or disjunctions of predicates, each of which may (or may not) include the variable referred to by the quantifier. We list several rules of inference which can be applied here:

RELATIONSHIP BETWEEN QUANTIFIERS
AND LOGICAL CONNECTIVES

(a) $\forall x[P(x) \land Q(x)]$ is equivalent to $\forall x P(x) \land \forall x Q(x)$.

(b) $\exists x[P(x) \lor Q(x)]$ is equivalent to $\exists x P(x) \lor \exists x Q(x)$.

(c) $\forall x P(x) \lor \forall x Q(x) \vdash \forall x[P(x) \lor Q(x)]$. (These two are *not* equivalent.)

(d) $\exists x[P(x) \land Q(x)] \vdash \exists x P(x) \land \exists x Q(x)$. (These two are *not* equivalent.)

The statement $\forall x[P(x) \land Q(x)]$ means that for all values of x *both* $P(x)$ and $Q(x)$ are true, which means that for all values of x, each must be true individually. The rule of inference (a) illustrates again the close relationship between universal quantification and the operation of conjunction.

Statement (c), on the other hand, shows that universal quantification does not work so well with disjunction. To verify that the two statements are not equivalent, we use the universe of discourse as the set of real numbers; when $P(x)$ is the predicate, $x = 2$ and $Q(x)$ is $\sim P(x)$. Then we have a situation in which $\forall x[P(x) \lor Q(x)]$ is true, but $\forall x P(x) \lor \forall x Q(x)$ is false.

Using these rules of inference, we can produce several others as well. For example, we can prove that the proposition $\forall x[P(x) \land Q]$ is equivalent to $\forall x P(x) \land Q$, where Q is a proposition or a predicate which does not involve the variable x.

EXAMPLE 1.35

Prove that $\forall x[P(x) \land Q] \vdash \forall x P(x) \land Q$.

PROOF

To prove this, we assume $\forall x[P(x) \land Q]$ and show $\forall x P(x) \land Q$. To prove the latter, we can show that Q is true and that $P(a)$ is true for an *arbitrary* element of the universe of x, so we let a be arbitrary. By universal instantiation and our hypothesis, we know that $P(a) \land Q$ is true. In particular this means that $P(a)$ is true; but since a was arbitrary, universal instantiation means that $\forall x P(x)$ is true. Since $P(a) \land Q$ is true, it also means that Q is true. We have proved both $\forall x P(x)$ and Q, which means that $\forall x P(x) \land Q$ is true. ∎

In the problems you are asked to provide the proof of the other half of the equivalence we mentioned above and to verify some other facts of a similar nature.

Mathematical Induction

One very special, but very important, kind of quantified predicate to consider is the case in which the statement to be proved is of the form $\forall n P(n)$ and the universe of n is the set of positive integers. A statement like the one in Example 1.29b falls into that category.

The *axiom of mathematical induction* provides the rule of inference that we need to prove statements like this one. Stated as a rule of inference, where P is a predicate whose variable has as a universe of discourse the positive integers, the axiom looks like this:

$$P(1), \forall k[P(k) \rightarrow P(k+1)] \vdash \forall n P(n)$$

Obviously, we need to back up and see exactly what this means, and how it can be applied to proving statements.

The axiom says we can prove $\forall n P(n)$ provided that we can prove two things:

1. $P(1)$ is true.

2. Whenever $P(k)$ is true, $P(k+1)$ is also true.

It seems reasonable that these two facts would prove $\forall n P(n)$, since if we were concerned, say, about $P(4)$, we would need only note that statement 1 tells us that $P(1)$ is true, statement 2 tells us that $P(1) \rightarrow P(2)$ and $P(2) \rightarrow P(3)$ and $P(3) \rightarrow P(4)$.

The law of syllogism yields that $P(1) \rightarrow P(4)$ is true, and by modus ponens it must follow that $P(4)$ is true. Similar reasoning could be followed for any integer. Thus the predicate must be true for any integer.

It has been said that showing that $P(n)$ is true for a particular n is like climbing a ladder, once the two parts of an induction proof have been completed. Part 1 tells us we can get on the ladder at rung 1. Part 2 tells us that if we find ourselves on any rung of the ladder, we can climb up to the next one. Thus, to prove that we can reach rung 4, we note that according to part 1 we can get on at rung 1, and part 2 tells us that from rung 1 we can step up to rung 2 and then to 3 and 4.

Part 1 of a proof by mathematical induction is usually called the *basis step*, or the basis for induction, and part 2 is called the *induction step*.

The induction step is the part of the process which usually creates all kinds of difficulties for students. In part 2 we have to prove $\forall k[P(k) \rightarrow P(k+1)]$, which means that we must assume $P(k)$ to be true and then conclude that $P(k+1)$ is also true. In doing this we take k to be "any old integer" and proceed from there. The assumption that $P(k)$ is true is called the *induction hypothesis*, and it causes the question to be asked, Aren't we assuming that the statement we are trying to prove is true? The answer is most definitely *no*. We have only assumed that k was a number such that $P(k)$ is true (using the ladder analogy, we have assumed we find ourselves on rung k), and we use that to try to show that $P(k+1)$ must also be true. (That is, finding ourselves on rung k, we can climb to rung $k+1$.)

Many of you may have seen mathematical induction before, but some examples are needed to clarify the process.

EXAMPLE 1.36

Prove that the sum of the first n integers is $n(n + 1)/2$ for all n.

SOLUTION

Basis Step   Clearly $1 = 1(1 + 1)/2$ is true.

Induction Step   We must prove $\forall k[P(k) \to P(k + 1)]$, by assuming

$$P(k): 1 + 2 + 3 + \cdots + k = \frac{k(k + 1)}{2}$$

We must get to

$$P(k + 1): 1 + 2 + 3 + \cdots + k + 1 = \frac{(k + 1)(k + 2)}{2}$$

To get the statement of $P(k + 1)$ we replaced all occurrences of k in $P(k)$ with $k + 1$ (in the sum on the left this was actually accomplished by increasing the number of summands by 1). We proceed by starting with the left-hand side of $P(k + 1)$ and working to show that by use of $P(k)$ this is actually equal to the right-hand side of $P(k + 1)$:

$$1 + 2 + 3 + \cdots + k + 1 = (1 + 2 + 3 + \cdots + k) + k + 1$$

By the induction hypothesis,

$$1 + 2 + 3 + \cdots + k + 1 = \frac{k(k + 1)}{2} + k + 1$$

By factoring out a common factor of $k + 1$, we then get

$$1 + 2 + 3 + \cdots + k + 1 = (k + 1)\left(\frac{k}{2} + 1\right)$$

Elementary   algebra   yields   that $k/2 + 1 = (k + 2)/2$ and   in   turn $1 + 2 + 3 + \cdots + k + 1 = (k + 1)(k + 2)/2$ as required.   ∎

EXAMPLE 1.37

Prove that the sum of the first n odd integers is n^2 for all integers n.

SOLUTION

Basis Step   Here $P(1)$ says that the sum of the first 1 odd integers is 1^2. That is quite obviously true.

Induction Step We assume that the sum of the first k odd integers is k^2, and we attempt to prove that the sum of the first $k + 1$ odd integers is $(k + 1)^2$. This time, we use the summation notation $\sum_{i=1}^{n} f(i)$, which means add all values of the function f, where i takes on the integer values from 1 to n. In computer terms, this means the result of the following loop:

```
sum := 0;
for i := 1 to n do
    sum := sum + f(i)
od;
```

Thus the sum of the first k odd integers becomes $\sum_{i=1}^{k} (2i - 1)$. A couple of other facts about the summation notation are handy to note:

(a) $\displaystyle\sum_{i=1}^{n} f(i) + f(n + 1) = \sum_{i=1}^{n+1} f(i)$

(b) $\displaystyle\sum_{i=1}^{n} f(i) = \sum_{i=1}^{n-1} f(i) + f(n)$

Returning to the problem at hand, we need to show that

$$\forall k \left(\left[\sum_{i=1}^{k} (2i - 1) = k^2 \right] \rightarrow \left[\sum_{i=1}^{k+1} (2i - 1) = (k + 1)^2 \right] \right)$$

We assume for a *fixed value* of k that the statement $\sum_{i=1}^{k} (2i - 1) = k^2$ is true. This is the induction hypothesis. We need to show that $\sum_{i=1}^{k+1} (2i - 1) = (k + 1)^2$. We work with the left side of this equation and attempt to produce a string of equalities which leads to the desired right-hand side:

$$\sum_{i=1}^{k+1} (2i - 1) = \sum_{i=1}^{k} (2i - 1) + 2(k + 1) - 1 \qquad \text{[we used (a) above]}$$

$$= k^2 + 2k + 1 \qquad \text{(simplification and use of the induction hypothesis)}$$

$$= (k + 1)^2 \qquad \text{(factoring)} \quad \blacksquare$$

One of the big problems with making use of mathematical induction is the fact that many students don't really understand what is happening in an induction proof. For that reason we have been very explicit in describing the parts of an induction proof. And throughout most of this book we will generally be much more explicit about induction proofs than is common practice, although in the next example we demonstrate a more informal version of a proof by induction.

Sometimes we will state that a predicate is true for all integers greater than or equal to some integer (such as 3). In that case, all we need to do to prove that kind of statement is to revise the basis step to start at the specified point [in our hypothetical case we would make our basis step the proof of $P(3)$ rather than the proof of $P(1)$]. It is also possible to prove statements for all nonnegative integers by using $P(0)$ as the basis step. (In terms of the ladder analogy, we simply show that we can get on the ladder at a different rung.)

Sometimes it is extremely useful to use induction to make a definition of something which may depend on integer values. The definition of $n!$ (n factorial), for example, is given sometimes as

1. $0! = 1$.
2. $(n + 1)! = (n + 1)(n!)$ for $n \geq 1$.

Mathematical induction can be used to show that this definition does provide a value of $n!$ for all n and that that value is the usual one of $(n)(n - 1) \cdots (3)(2)(1)$. In Problems 11 and 12 you are asked to do exactly this. The definition does most of the work for us; the main problem is actually the formulation of P. A definition of this type is called an *inductive*, or *recursive*, definition.

Informal Induction

Many authors are not as explicit about the use of mathematical induction as we have been. It is a good idea to note how a more informal proof by induction would proceed. This example and the one that follows also illustrate the fact that induction is *not* always used to verify theorems that are in the form of an equation.

EXAMPLE 1.38

Prove that $2^n \geq 2n + 1$ for all $n \geq 3$.

PROOF
We proceed by induction on n. We note that 2^3 is greater than $2(3) + 1 = 7$. If k is such that $2^k \geq 2k + 1$, then $2^{k+1} = 2^k + 2^k$.

By the induction hypothesis, we know that $2^k \geq 2k + 1$, and since $k \geq 3$, it follows that $k \geq 2$. As a consequence, $2^{k+1} \geq (2k + 1) + 2 = 2k + 3$, which is what we needed to show. ∎

Strong Induction

On occasion, it is convenient to make use of a stronger induction hypothesis than the standard axiom of mathematical induction allows us. We would sometimes like to make the assumption that $P(n)$ is true for all $n \leq k$ in order to prove $P(k + 1)$. It turns out that this is perfectly permissible and does not allow us to prove anything that would not be true under the usual axiom of mathematical induction. The axiom of strong induction is as follows:

$$P(1), \forall k \; \forall n\{[(k \le n) \to P(k)] \to P(n + 1)\} \vdash \forall n P(n)$$

Strong induction is related to ordinary induction because of the fact that it actually proves that $Q(n)$ is true for all n, where $Q(n)$ is the predicate that $\forall x[(x \le n) \to P(x)]$, and the truth of $Q(n)$ clearly implies the truth of $P(n)$ for all n.

EXAMPLE 1.39

Prove that every integer which is greater than or equal to 2 either is a prime number or can be written as a product of primes.

PROOF
We use strong induction for this proof. We let $P(n)$ be the statement that n is either a prime or a product of primes.

Basis Step   Clearly 2 is a prime number, so $P(2)$ holds.

Induction Step   This time we use strong induction and assume that $P(x)$ is true for all x less than or equal to k. If we consider $k + 1$, we note that there are two possibilities: either $k + 1$ is a prime, in which case $P(k + 1)$ is true, or $k + 1$ is not a prime, in which case we can write $k + 1 = ab$, where both a and b are greater than 1 and less than $k + 1$. By the induction hypothesis, a and b each can be written as products of prime numbers, so ab can also be written as a product of primes, namely the product of the primes whose product is a with those primes whose product is b. This verifies $P(k + 1)$ as required.   ■

A "False Proof" by Induction

It is probably advisable at this point to see what can go wrong in a proof by induction and to note the reasons for the problem. To that end, we present (and hope it doesn't destroy all faith in induction in the process) the following example.

EXAMPLE 1.40

Show that the following proof is not valid.

THEOREM (?)   All horses are the same color.

PROOF (?)
We let $P(n)$ be the predicate that in all collections of n horses, all are the same color. Clearly $P(1)$ is true. For the induction step we assume that $P(k)$ is true

and show that $P(k + 1)$ is true. Take a collection of $k + 1$ horses and send one away—we now have k horses, and thus all k horses are the same color. Call the exiled horse back and send a different one away. Again we have a collection of k horses, and all are the same color. Since horses never change color, this must be the same color as the first group, and thus the horse we brought back was the same color as the horse we sent away, and so all $k + 1$ horses are the same color. ■

The problem lies in the interaction of the induction step and the basis step. The proof of the induction step is actually correct *as long as* $k \geq 2$. (It fails at $k = 1$ because there are *no* horses which belong to both groups of horses.)

The basis step, on the other hand, certainly is not true when $n = 2$. We can "get on the ladder" at rung 1, but we can't climb to rung 2. If someone were to "place" us at rung 2 or higher on the ladder, we could get as far as we like (but, of course, we can't).

Problem Set 9

In the following we consider arguments which involve quantified statements. You should symbolize the arguments and determine whether the argument is valid. One hint is that a statement of the form all A's are B's should be rendered as $\forall x[A(x) \rightarrow B(x)]$, where A is the predicate x is an A and B is the predicate x is a B. The expression "some A's are B's" can be given as $\exists x[A(x) \wedge B(x)]$.

1. All computers have input devices. The Macintosh is a computer. Thus the Macintosh has an input device. (This is Example 1.3 from Section 1.1.)

2. All Apple computers can be connected to printing devices. All minicomputers can be connected to printing devices. Some Apple computers are minicomputers. (This is Example 1.4 from Section 1.1.)

3. Some cats have claws. All animals with claws scratch. Thus some cats scratch.

4. All integers are terminating decimals. All terminating decimals are rational numbers. No rational number is an imaginary number. Thus no integer is an imaginary number.

5. Two is an integer. The square root of 2 is an irrational number. Thus some integers have irrational numbers as their square roots.

6. Explain why $\exists x[P(x) \vee Q(x)]$ is equivalent to $\exists x P(x) \vee \exists x Q(x)$.

7. Explain why $\exists x[P(x) \wedge Q(x)]$ is *not* equivalent to $\exists x P(x) \vee \exists x Q(x)$ by finding an example in which the second is true but the first is false.

8. Complete the proof that $\forall x[P(x) \wedge Q]$ is equivalent to $\forall x P(x) \wedge Q$ by showing that $\forall x P(x) \wedge Q \vdash \forall x[P(x) \wedge Q]$.

*9. Prove that $\exists x[P(x) \wedge Q]$ is equivalent to $\exists x P(x) \wedge Q$.

*10. Prove that the following equivalences are also valid:
  (a) $\forall x[P(x) \vee Q]$ is equivalent to $\forall x P(x) \vee Q$.
  (b) $\exists x[P(x) \vee Q]$ is equivalent to $\exists x P(x) \vee Q$.

*11. Prove that our definition of $n!$ given above provides a value of $n!$ for all integers greater than or equal to zero. [*Hint*: Let $P(n)$ be the predicate that $n!$ is defined.] Use the definition to compute $4!$.

*12. A commonly used definition of $n!$ is that $n! = (n)(n-1) \cdots 1$ and $0! = 1$. Prove that the inductive definition and this definition provide exactly the same results. [*Hint*: Let $P(n)$ be the predicate that both definitions provide the same value for n.]

Prove the following statements by mathematical induction:

13. $\displaystyle\sum_{i=1}^{n} 2i = n(n+1)$

14. $\displaystyle\sum_{i=1}^{n} (3i - 2) = \frac{1}{2}n(3n - 1)$

15. $\displaystyle\sum_{i=1}^{n} i^2 = \frac{1}{6}n(n+1)(2n+1)$

16. $1/(1 \cdot 3) + 1/(3 \cdot 5) + 1/(5 \cdot 7) + \cdots + 1/[(2n-1)(2n+1)] = n/(2n+1)$

17. $\sim(p_1 \wedge p_2 \wedge \cdots \wedge p_n)$ is equivalent to $(\sim p_1 \vee \sim p_2 \vee \cdots \vee p_n)$ for all n. (Note: De Morgan's law is only stated for a *pair* of propositions.)

18. $p \wedge (q_1 \vee q_2 \vee \cdots \vee q_n)$ is equivalent to $(p \wedge q_1) \vee (p \wedge q_2) \vee \cdots \vee (p \wedge q_n)$ for all n.

*19. $n! > 2^n$ for $n \geq 4$.

*P20. Write a program to compute $n!$, using the definition $n! = (n)(n-1) \cdots (3)(2)(1)$. This program can be written in any programming language. Then write a program which computes $n!$ by making use of the definition of $n!$ supplied in the text. This second program will be an example of a *recursive* program. This *can* be done in both BASIC and Pascal but not in FORTRAN. You might want to compare the two programs in terms of size, ease of understanding, and execution time.

CHAPTER SUMMARY

In this chapter, we have presented the basic ideas of mathematical logic, beginning with the propositional logic and ending with the predicate logic. In the process, we have introduced some of the important techniques involved in the construction of sound mathematical proofs.

As a particular example of a technique of proof, we considered the process of proof by induction, and we were able to relate this process to the ideas and techniques of predicate logic.

Another thread of the development in this chapter is the relationship between the basic connectives of propositional logic and the boolean (or logical) operations which are available in many computing languages, and which also occur as part of the very architecture of most modern computers.

Many of the ideas that we have discussed in this chapter are used extensively in the remaining chapters of the book. In particular, a large number of the concepts from propositional logic next appear, in only slightly modified form, in the chapter on set theory and in the chapter on boolean algebras. This chapter is the foundation upon which most of the remainder of this book is built.

KEY TERMS

Upon completion of this chapter, you should be familiar with the following terms. (Section numbers are in parentheses.)

algorithm (1.4)
and (1.3)
argument (1.7)
basis step (1.9)
biconditional (1.6)
conditional (1.6)
conjunction (1.3)
connective (1.3)
contradiction (1.7)
contrapositive (1.6)
converse (1.6)
disjunction (1.3)
exclusive OR (1.3)
fallacy (1.7)
hypothesis (1.6)
indirect proof (1.7)
induction step (1.9)
inverse (1.6)
logically implies (1.7)

mathematical induction (1.9)
modus ponens (1.7)
modus tollens (1.7)
negation (1.3)
not (1.3)
predicate (1.8)
proof (1.7)
proposition (1.2)
rules of inference (1.7)
satisfiable (1.8)
satisfies (1.8)
strong induction (1.9)
summation notation (1.9)
syllogism (1.7)
tautology (1.4)
truth table (1.4)
valid argument (1.7)
valid predicate (1.8)

2

Sets and Counting

2.1 INTRODUCTION

In Chapter 1 many readers may have noted a certain reluctance on the part of the author to use the word "set" in the places in which we have become accustomed to finding that word used. We did this because we wished to make precise the ideas we had in mind for dealing with sets *before* we used that kind of terminology. In this section we present some examples of problems (without solutions) in which the terminology and techniques of set theory provide a convenient language and methodology for solution.

EXAMPLE 2.1

United Computer Technologies has written a new package which integrates a word processing program with a spreadsheet program, and they wish to make the program so that it will run on the Admiral 64, which is a 64K machine. They wish to do this in such a way as to minimize the need to overlay code and fetch information from the disk. The word processor requires 40K for program and data, and the spreadsheet requires 32K for program and data. If 16K must be reserved for the code which integrates the two packages, and that code *must* remain resident in the memory, what is the minimum amount of overlay space which will be necessary? ■

EXAMPLE 2.2

Consider the collection of all numbers which are divisors of 24 and the collection of all numbers which are divisors of 36.
(a) Find all numbers which are divisors of both.
(b) Find the largest number which is a divisor of both.
(c) Find all numbers which are divisors of either number.
(d) Find all numbers which are divisors of 36 but not of 24. ∎

EXAMPLE 2.3

Consider the collection of all numbers which are multiples of 6 and the collection of all numbers which are multiples of 4.
(a) Find the collection of all numbers which are multiples of both 6 and 4.
(b) Find the smallest number which is a multiple of both 4 and 6.
(c) Find the collection of all numbers which are multiples of either 6 or 4.
(d) Find the collection of all numbers which are multiples of neither 6 nor 4.
(e) Find the collection of all numbers which are multiples of 6 and not multiples of 4. ∎

Each of these questions can be answered in terms of the basic ideas and terminology of set theory. In Chapter 2, as we did in Chapter 1, we answer these questions in the course of developing the fundamental concepts of the topic of the chapter. In Chapter 2, we introduce the basic concepts of set theory, the methods of using set theory, and the connections which exist between set theory and the topics of mathematical logic discussed in Chapter 1.

2.2 BASIC DEFINITIONS

Probably almost everyone is familiar with the concept of a set from earlier courses in mathematics. The purpose of this section is to make precise the basic terms of set theory.

First, set is used as an *undefined term* (similar to the way that point and line are used as undefined terms in geometry) but with the sense that a set refers to any collection of objects. There are two reasons for this. First, something has to be left as an undefined term, and we might as well stop at "set," rather than make "collection" the undefined term. Second, under certain technical conditions not everything that we might view as a collection can actually be a set. The main determining factor as to whether a collection is a set is whether it is well defined, i.e., whether when given an object which might belong to the collection, it is possible to determine whether that object belongs in the set. To be more precise, we should say that a collection is well defined if for every potential member of the collection, a statement that that object belongs to the collection is a proposition. Just as with logic, there are sets for which we have no effective way of determining whether an element belongs to that set. Later in this book we will meet some sets which have the property that no computer program can answer the membership question for all possible candidates.

EXAMPLE 2.4

The following are all examples of sets:
(a) The collection of all even integers
(b) The collection of all real solutions to the equation $x^2 + 2x - 3 = 0$
(c) The collection of all real solutions to the equation $x^2 + x + 1 = 0$ ■

EXAMPLE 2.5

The following collections are not sets:
(a) The collection of all tall people
(b) The collection of all pretty numbers
(c) The collection of all sets which are not elements of themselves ■

In each case in Example 2.4, given a candidate for membership in the collection, the question as to whether the object belongs to the collection is clearly a proposition. Example 2.4c points out a rather interesting case. The equation is one which has *no* real solutions, so the collection which consists of all real solutions should contain no elements. This set is like a box which has nothing in it.

DEFINITIONS

An **empty** set is a set which has nothing in it. The empty set is denoted by ∅.

The **elements** of a set are those objects which belong to the collection.

We usually denote sets with capital, or uppercase, letters and the possible elements of sets with lowercase letters. To indicate that a is an element of the set A, we write $a \in A$.

Logic and Set Theory—Connections

In most cases, we are interested in studying sets which have all their elements coming from some given set, called the *universal set*. If A is a set and x is an element of the universal set, then $x \in A$ is a proposition. This means that if x is used as a variable, $x \in A$ must be a predicate in which the universe of x is the universal set.

The collections in Example 2.5 are not sets. The first two collections lead to the same kinds of problems that we had with similar statements about tall people and pretty numbers in Chapter 1. The last collection creates real headaches. If we name that collection A, we would want to ask, Is $A \in A$ true? If that is true, by the very definition of A, it is impossible for A to belong to A. On the other hand, if $A \in A$ is false, then by the definition of A we must have $A \in A$. This is closely related to the problem of dealing with the statement "this statement is false." These are examples of what is known as *Russell's paradox*. Russell's paradox was named after the great English logician and philosopher Bertrand Russell (1872–1970) who first pointed out this paradox and who also proposed some technical revisions to set theory which makes it impossible for such a problem to arise in modern set theory.

We must note the close connections between logic and set theory. Already we have seen the connection between predicates and sets. The set of all values from the universe of discourse which satisfy a particular predicate form a set, and this particular idea is very often used to describe sets in mathematics. In logic we have predicates and a universe of discourse, which corresponds directly to sets and the universal set. We find that the concepts of predicate logic and mathematical set theory are inextricably tied, and collectively the two topics are referred to as the *foundations* of mathematics. Almost everything that is done in mathematics can ultimately be traced to the topics of logic and set theory.

Notation

To indicate the elements that comprise a set, we usually use one of two methods. The first, and simplest, is simply to list the elements between braces, such as $A = \{1, 2, 3, 4\}$ or $B = \{1, 2, \ldots, 99\}$. In this notation, if we wish to include a large number of elements which continue a pattern, we use the ... notation. The ... notation can also be used to indicate the continuation of a pattern indefinitely, as in the set $C = \{2, 4, 6, \ldots\}$.

The other commonly used notation is to specify a set by means of a predicate. We can write $D = \{x \mid x \text{ is odd}\}$ to indicate the set of all odd numbers. This is read, "the set of all x such that x is odd." Since every set can be defined by a predicate, this notation can be used to specify any set. The general form is $\{x \mid P(x)\}$, where P is a predicate whose universe is the universal set.

EXAMPLE 2.6

Let $E = \{x \mid 3 < x < 75, x \text{ an integer}\}$. This set can also be written as $E = \{4, 5, 6, \ldots, 75\}$. Let $F = \{x \mid x \text{ is a perfect square}, x \text{ an integer}\}$. This set can also be written as $F = \{1, 4, 9, 16, 25, \ldots\}$. Although the second notation is more graphic and points out some of the set's elements explicitly, the first is much more precise and is especially useful when we deal with proving facts about some sets. ■

If A and B are sets, then we say that A and B are **equal** (and write $A = B$) if A and B contain exactly the same elements.

We say that A is a **subset** of B and we write $A \subset B$ if every element of A is also an element of B. (Note that if $A = B$, then *both* $A \subset B$ and $B \subset A$ are true.)

If A is a subset of B but A and B are not equal, then we say that A is a **proper subset** of B.

We say that A and B are **disjoint** if A and B have no elements in common.

EXAMPLE 2.7

Let U = set of all integers

$A = \{1, 2, 3, 4, 5, 6, 7, 8, 9, 10\}$

$B = \{1, 2, 3, 4\}$

$C = \{3, 4, 5, 6\}$

$D = \{x \mid 1 \leq x \leq 4\}$

$E = \{5, 6, 7\}$

We then have the following relations holding among A, B, C, and D: Because every element of B is also an element of A, we can write that $B \subset A$. Similarly $C \subset A$, $D \subset A$, and $E \subset A$. In each case, we have proper subsets. The only equality which holds is $B = D$. The fact that $B = D$ does not prevent us from noting that both $B \subset D$ and $D \subset A$, since the definition of subset is satisfied by the elements of the two sets. Finally we note that E and B, and E and D, are disjoint, because they have no elements in common. ■

In terms of the language of logic, we can write the following:

$A = B$ means that $\forall x (x \in A \leftrightarrow x \in B)$.

$A \subset B$ means that $\forall x (x \in A \rightarrow x \in B)$.

A and B disjoint means $\forall x [\sim (x \in A \wedge x \in B)]$.

THEOREM 2.1

$A = B$ if and only if $A \subset B$ and $B \subset A$.

PROOF
This is left to Problem 26.

Cardinal Numbers; Finite and Infinite Sets

We leave the concepts of finite and infinite sets as undefined terms, although it is possible to be very precise about them.

> ### DEFINITION
>
> The **cardinal number** of a finite set is the whole number which represents the number of elements that belong to the set. For any set A, we denote its cardinal number by $|A|$.

The concept of cardinal number for infinite sets is a very important topic in courses in advanced set theory and logic, but we choose not to discuss this issue in a course in discrete mathematics. In determining the cardinal number of a finite set, we are not concerned with the number of times the same element is written in the description of the set, but only with how many distinct elements belong to the set. The cardinal number of the set $\{a, a, a, b\}$ is 2 because the set contains only two distinct elements, a and b.

The point made above about not being concerned with the number of times that an element is listed as belonging to a set also applies to determining if two sets are equal or if one is a subset of the other. The crucial issue is whether a particular element does or does not belong to a set, not how many times it is listed in the set description.

EXAMPLE 2.8

The following sets are all equal because they contain exactly the same elements:

$$A = \{a, a, a, b, b, a, c\}$$
$$B = \{c, b, a, c\}$$
$$C = \{a, b, c\}$$

Each set contains the elements a, b, and c. Thus they are all equal. The cardinal number of each set is 3. The set $D = \{a, a, a\}$ is a subset of $E = \{a, b, c\}$, because every element of D (only a) is also an element of E. ∎

Problem Set 2

Write descriptions for the following sets by using predicates:

1. $A = \{a, e, i, o, u\}$
2. $B = \{2, 3, 5, 7, 11, 13, 17, 19, 23, 29, \ldots\}$
3. $C = \{1, 3, 5, 7, 9, \ldots\}$

4. $D = \{1, 8, 27, 64, \ldots\}$

5. $E = \{$George Washington, John Adams, $\ldots$, Ronald Reagan$\}$

List the elements which belong to the following sets. Use the $\ldots$ notation where necessary.

6. $F = \{x \mid x$ is a positive even integer$\}$

7. $G = \{x \mid x$ is a multiple of 3 and $0 < x < 15\}$

8. $H = \{x \mid x^2 = 9\}$

9. $I = \{x \mid x$ is a multiple of 5$\}$

10. $J = \{x \mid x$ is a state whose first letter is A$\}$

11. $K = \{x \mid x \in F \wedge x \in G\}$

12. $L = \{x \mid x \in H \wedge x \in I\}$

13. $M = \{x \mid x \in F \vee x \in H\}$

14. $N = \{x \mid \sim(x \in F) \wedge x \in G\}$

15. $O = \{x \mid (x$ is a letter$) \wedge \sim(x \in A)\}$

16. $P = \{\sim(x \in C) \wedge x \in B\}$

17. For each of the sets in Problems 1 to 15 (odd) which are finite, determine the cardinal number of the set.

18. For each of the sets in Problems 2 to 16 (even) which are finite, determine the cardinal number of the set.

19. Find a set whose cardinal number is 4.

20. Find a set whose cardinal number is 1.

21. Find a set whose cardinal number is 0.

22. Find an infinite set.

Suppose

$$U = \{1, 2, 3, 4, 5, 6, 7, 8, 9\}$$
$$A = \{1, 4, 9\}$$
$$B = \{x \mid x \in U \text{ and } x \text{ is a perfect square}\}$$
$$C = \{1, 2, 3, 5, 7, 9\}$$
$$D = \{2, 3, 5, 7\}$$
$$E = \{x \mid x \in U \text{ and } x \text{ is a prime number}\} \qquad (1 \text{ is } not \text{ prime})$$

23. Indicate which sets are subsets of others.

24. Indicate which sets are *proper* subsets of others.

25. Find the pairs of sets which are disjoint.

26. Prove Theorem 2.1.

*27. Prove the following: If $A \subset B$ but A and B are disjoint, then $A = \varnothing$.

*28. Prove the following: If $A \subset B$ and $C = \{x \mid x \in A \wedge x \in B\}$, then $C = A$.

*29. Prove the following: If $A \subset B$ and $D = \{x \mid x \in A \vee x \in B\}$, then $D = B$.

*30. Prove the following: If A and B are sets and $C = \{x \mid x \in A \wedge x \in B\}$, then $C \subset A$.

2.3   OPERATIONS ON SETS

One of the more interesting facts about set theory is that the basic operations of set theory correspond very closely with the basic connectives of logic. Since we have already seen that the basic definitions of sets correspond so closely with the basic ideas of predicate logic, perhaps that should not come as quite so great a surprise after all.

Looking back at Example 2.2 from Section 2.1, we will discover the motivation for the definition of the three basic operations of set theory.

EXAMPLE 2.2 (REPEATED)

Consider the collection of all numbers which are divisors of 24 and the collection of all numbers which are divisors of 36.
(a) Find all numbers which are divisors of both.
(b) Find the largest number which is a divisor of both.
(c) Find all numbers which are divisors of either number.
(d) Find all numbers which are divisors of 36, but not of 24.   ■

Intersection

To answer Example 2.2a, we need to look at two sets: $A = \{x \mid x$ is a divisor of 24$\}$ and $B = \{x \mid x$ is a divisor of 36$\}$.

It is clear that both A and B are sets and that what we need to do is construct a new set from these two which consists of all elements belonging to both sets. The new set which we form in this way will contain the solution to Example 2.2a; furthermore, the largest member of the set will be the solution to Example 2.2b. Since $A = \{1, 2, 3, 4, 6, 8, 12, 24\}$ and $B = \{1, 2, 3, 4, 6, 9, 12, 18, 36\}$, the elements common to both sets are $\{1, 2, 3, 4, 6, 8, 12\}$. This operation of finding all the elements common to a pair of sets occurs frequently enough that it is one of the basic operations of set theory, called the *intersection* of the two sets.

DEFINITION

The **intersection** of the sets A and B is the set of all elements which are elements of both sets. Symbolically we denote this operation by $A \cap B$, and we write

$$A \cap B = \{x \mid x \in A \land x \in B\}$$

The operation of intersection of sets shares many properties with the connective $\land$. The following theorem lists some of those common properties.

THEOREM 2.2

The following properties hold for the intersection of two sets (U represents the universal set):

(a) $A \cap B = B \cap A$ (Commutative)

(b) $(A \cap B) \cap C = A \cap (B \cap C)$ (Associative)

(c) $A \cap U = A$ (Identity)

PROOF

We prove Theorem 2.2a and leave the other two parts as exercises.

To prove that two sets are equal, we must show that $X \subset Y$ and $Y \subset X$. Thus in this case we must show that both $A \cap B \subset B \cap A$ and $B \cap A \subset A \cap B$. We recall that this means we must show that

$$\forall x(x \in A \cap B \to x \in B \cap A) \quad \text{and} \quad \forall x(x \in B \cap A \to x \in A \cap B)$$

This means we must prove that $\forall x(x \in A \cap B \leftrightarrow x \in B \cap A)$. When we consider the definitions, we recall that $x \in A \cap B$ means that $x \in A \land x \in B$ and $x \in B \cap A$ means that $x \in B \land x \in A$. But because the conjunction operation in logic is commutative, the statements in question must be equivalent. Hence connecting them by an $\leftrightarrow$ must yield a statement which is always true. ∎

A Note about Proofs

One note should be added about the manner in which proofs are constructed. In this particular case we worked our way backward from the statement that we wished to prove until we reached a statement which we were sure was true. This is a very legitimate technique in constructing proofs. This process is, in fact, quite similar to the "working backward" technique of problem solving which is often used in the design of computer algorithms. In that process, as with the method we demonstrated of constructing proofs, there are three phases:

1. Determining exactly what the goal is for the problem (in this case it was restating what needed to be proved).

2. Working back to the beginning of the problem by determining what is needed to reach the goal and then repeating the process with the newly determined need as the new goal. This process is continued until the solution becomes obvious.

3. Writing the solution in a "frontward manner."

Thus far, we have dispensed with step 3, but we present that proof shortly. In most cases, people who wish to write proofs which seem polished will turn their thought processes around when actually presenting the proof. This does describe better the final logic in the proof, but it does not demonstrate the thought processes involved in the construction of the proof. We write out the proof in the usual fashion, but we encourage you to try to do it yourself before reading further.

"FRONTWARD PROOF" OF THEOREM 2.2a

We will show that $A \cap B$ and $B \cap A$ are subsets of each other and thus, by Theorem 2.1, they must be equal. Because $p \wedge q$ is equivalent to $q \wedge p$, it follows that the predicates $x \in A \wedge x \in B$ and $x \in B \wedge x \in A$ are equivalent predicates. Since that is the case, by the definition of intersection it follows that for all x, $x \in (A \cap B)$ if and only if $x \in (B \cap A)$. But this last statement says that both $A \cap B \subset B \cap A$ and $B \cap A \subset A \cap B$, as required. ∎

Union

Example 2.2c leads us to consider the process of forming a set from A and B which consists of all elements belonging to either of the sets. In this case, since $A = \{1, 2, 3, 4, 6, 8, 12, 24\}$ and $B = \{1, 2, 3, 4, 6, 9, 12, 18, 36\}$, the new set will become $\{1, 2, 3, 4, 6, 8, 9, 12, 18, 24, 36\}$. This new set is called the *union* of A and B.

DEFINITIONS

The **union** of the sets A and B consists of all elements which belong to either set. We denote this formally as

$$A \cup B = \{x \mid x \in A \vee x \in B\}$$

The symbol $\cup$ is used for the union operation for two reasons. First, it looks kind of like a U, which is the first letter in "union." Second, it looks kind of like a $\vee$, which is the logical connective used to define it. Those two facts should help keep you from confusing the symbols for union and intersection. The operation of set union shares many properties with the operation of disjunction. We have a theorem which is quite similar to Theorem 2.2.

The following properties hold for the union of sets:
(a) $A \cup B = B \cup A$ (Commutative)
(b) $(A \cup B) \cup C = A \cup (B \cup C)$ (Associative)
(c) $A \cup \varnothing = A$ (Identity)

PROOF
These properties follow directly from the properties of the disjunction operation from logic, and we leave the details to the reader as an exercise. ∎

One problem that students sometimes have when they are asked to define union and intersection on an examination is to use colloquial rather than technical language. Some students will write for the union of two sets "all elements in both sets," while for the intersection they will write "all elements in both sets at the same time." The difficulty is that while they probably mean in their definition of union to "lump" the two sets together, they actually use nontechnical language which does not carry the proper meaning in terms of logic, and so they finish by making the definitions of union and intersection identical. It is very important to be precise in the use of language in mathematics.

Complements and Differences

Moving ahead to the final part of Example 2.2, we are asked to find all numbers which are divisors of 36 and not divisors of 24. Since $A = \{1, 2, 3, 4, 6, 8, 12, 24\}$ and $B = \{1, 2, 3, 4, 6, 9, 12, 18, 36\}$, what we are after is the set of all elements which are in B but are not in A. This set is referred to as the *difference* between B and A, and it is denoted by $B - A$. In this case $B - A$ is $\{9, 18, 36\}$.

DEFINITIONS

The **difference** between B and A (or A removed from B) is the set consisting of all elements which belong to B and do *not* belong to A. Formally we write

$$B - A = \{x \mid x \in B \wedge \sim(x \in A)\}$$

The **complement** of a set A, denoted by A', is the set $U - A$, which can be described as $\{x \mid \sim(x \in A)\}$.

The operation of set complementation behaves quite similarly to the operation of negation in logic.

THEOREM 2.4

The following facts are true about a set and its complement:
(a) $A \cap A' = \emptyset$.
(b) $A \cup A' = U$.

PROOF
Both parts of this theorem can be translated to basic facts from logic. Any proposition is either true or false. Thus when for any element in U we make the statement $x \in A$, that statement will either be true, putting x in A, or false, which puts x into A'. All elements of U fall in one set or the other and thus must be in the union of the two sets. The other part is just as easy, since no element can have both $x \in A$ and $\sim (x \in A)$ true at the same time. Thus the defining property for $x \in (A \cap A')$ cannot be satisfied, and the set must be empty. ∎

Properties Common to Logic and Set Theory

Every one of the properties which we have discussed about sets corresponds exactly to a related property from logic. One more such property is the following.

THEOREM 2.5

The following properties hold for the operations of set union and intersection:
(a) $A \cup (B \cap C) = (A \cup B) \cap (A \cup C)$
(b) $A \cap (B \cup C) = (A \cap B) \cup (A \cap C)$
These properties are called the *distributive* properties, and they correspond exactly to the distributive properties in logic. The proof is left to the problems. ∎

We can summarize the properties of sets that we have found thus far:

1. Associative properties:
   Union: $A \cup (B \cup C) = (A \cup B) \cup C$
   Intersection: $A \cap (B \cap C) = (A \cap B) \cap C$

2. Commutative properties:
   Union: $A \cup B = B \cup A$
   Intersection: $A \cap B = B \cap A$

3. Identity:
   Union: $A \cup \emptyset = A$
   Intersection: $A \cap U = A$

4. Complement:
Union: $A \cup A' = U$
Intersection: $A \cap A' = \emptyset$

5. Distributive:
Union: $A \cup (B \cap C) = (A \cup B) \cap (A \cup C)$
Intersection: $A \cap (B \cup C) = (A \cap B) \cup (A \cap C)$

Among the properties listed above are the defining properties for a more general mathematical structure known as a *boolean algebra*. Both the properties of propositional logic and the properties of the subsets of a given universal set satisfy all the conditions for being a boolean algebra. In Chapter 3 we discuss boolean algebras in more detail.

One final theorem is quite important for doing operations on sets, and once again the connection to the properties of mathematical logic should be obvious.

THEOREM 2.6   DE MORGAN'S LAWS

If A and B are subsets of a universal set U, then the following two properties hold:
(a) $(A \cup B)' = A' \cap B'$.
(b) $(A \cap B)' = A' \cup B'$.
The proof is actually a simple matter of translating the statements from set theory to logic. ∎

One final fact might be noted about all this: there is a principle known as *duality* which is convenient for dealing with theorems about sets. In every theorem we have proved about sets, we have obtained corresponding statements for union and intersection. In fact, to get from one of these statements to the other, all that is needed is to interchange $\cup$ and $\cap$ and to interchange $\emptyset$ and U. We see later that this property of duality holds for all boolean algebras, and this fact means that every time we prove a theorem about a boolean algebra, we have really proved two theorems.

Problem Set 3

Let $U = \{a, b, c, d, e, f, g, h, i\}$ and suppose that

$A = \{a, b, c, d, e\}$ $C = \{e, f, g, h, i\}$ $E = \{b, d, f, h\}$

$B = \{d, e, f, g\}$ $D = \{a, c, e, g, i\}$ $F = \{a, e, i\}$

1. Find the following subsets of U:
(a) $A \cup B$ (b) $A \cap B$ (c) $C \cup D$
(d) $C \cap D$ (e) $E \cup F$ (f) $E \cap F$
(g) $A \cup C$ (h) $A \cap C$

2. Find the following subsets of U:
   (a) A' (b) B' (c) $A - B$
   (d) $B - A$ (e) $E' \cap F'$ (f) $(E \cup F)'$

3. Find the following subsets of U:
   (a) $A \cap (B \cup C)$ (b) $(A \cap B) \cup (A \cup C)$
   (c) $(A \cap D) - B$ (d) $(A - E)'$

Let U be the set of real numbers given by $\{x \mid -11 < x < 11\}$. Consider the following subsets of U:

$$A = \{x \mid 0 < x < 3\} \qquad C = \{x \mid -1 < x < 1\} \qquad E = \{x \mid 5 < x < 10\}$$

$$B = \{x \mid 2 < x < 6\} \qquad D = \{x \mid -10 < x < 10\} \qquad F = \{x \mid 0 < x < 6\}$$

4. Find the following subsets of U:
   (a) $A \cup B$ (b) $A \cap B$ (c) $C \cup D$
   (d) $C \cap D$ (e) $E \cup F$ (f) $E \cap F$
   (g) $A \cup C$ (h) $A \cap C$

5. Find the following subsets of U:
   (a) A' (b) B' (c) $A - B$
   (d) $B - A$ (e) $E' \cap F'$ (f) $(E \cup F)'$

6. Find the following subsets of U:
   (a) $A \cap (B \cup C)$ (b) $(A \cap B) \cup (A \cup C)$
   (c) $(A \cap D) - B$ (d) $(A - E)'$

7. Solve the problems posed in Example 2.3: Consider the collection of all numbers which are multiples of 6 and the collection of all numbers which are multiples of 4.
   (a) Find the collection of all numbers which are multiples of both 6 and 4.
   (b) Find the smallest number which is a multiple of both 4 and 6.
   (c) Find the collection of all numbers which are multiples of either 6 or 4.
   (d) Find the collection of all numbers which are multiples of neither 6 nor 4.
   (e) Find the collection of all numbers which are multiples of 6 and not multiples of 4.
   Note that since all these will be infinite sets, you will have to use the ... notation to list the elements which belong to each of the sets.

Using the properties of the set operations which we have discussed in this section, prove the following:

8. $(\emptyset \cup A) \cap (B \cup A) = A$

9. $(A \cup B) \cap (A \cup B') = A$

10. $A \cap (A \cup B) = A$

*11. If A is any set and $\emptyset$ is empty, prove that $\emptyset \subset A$. (*Hint*: Remember the logic involved, especially the definition of $\rightarrow$.)

*12. Prove that if $\emptyset_1$ and $\emptyset_2$ are both empty sets, then they must be equal to each other. (*Hint*: Use the result of Problem 11.)

*13. Prove that $(A')' = A$ for any set A. (*Hint*: Remember your logic.)

14. Complete the proof of Theorem 2.2.

15. Prove Theorem 2.3.

16. Prove Theorem 2.5.

17. Prove Theorem 2.6.

P18. In Pascal, provision is made for the use of a set as a data type. The universal set (in Pascalese this is called the *base type*) must be declared as a type, and the type of a universal set must be a finite subrange of an ordinal type. Typically there is a limit of 16 or 32 elements for the base type, but this is implementation-dependent. Although sets cannot be directly written, by use of a procedure we can output the elements of a set. The following program produces the set $\{a, b, c\}$ and prints its elements:

```
PROGRAM SETS (OUTPUT)
type universal = 'a' ... 'z';
     subset = set of universal;
var A:subset;
    procedure writeset(Q:subset);
    var x: char;
    begin
    write('{');
    for x := 'a' to 'z' do
        if x in A then write(x, ' ');
    writeln('}');
    end;
begin
A := ['a','b','c'];
writeset(A)
end.
```

The operation of union is indicated in Pascal by a plus, intersection by an asterisk, and set difference by a minus.

Write a program to print out each of the results from Problem 1.

P19. Write a program to compute and print the results for all the parts of Problem 2. (*Hint*: $A' = U - A$.)

P20. Write a program to compute and print the results for all the parts of Problem 3.

2.4 VENN DIAGRAMS

One of the more useful devices in understanding the relationships between sets, and for visualizing how the set operations work, is the Venn diagram. The Venn diagram represents sets as regions in a plane. Usually a set is represented in a Venn diagram by the region contained inside a closed curve, such as a circle or a rectangle. This representation can be used to illustrate the relationships between the set operations and to demonstrate physically why certain of the theorems of set theory should be true. Although results from a Venn diagram usually are not considered to constitute a proof, they are often suggestive of the reasons behind the truth of a theorem; and with a certain amount of care, the ideas in the Venn diagram can be converted into a proof.

Illustrating Set Operations with Venn Diagrams

A typical method of dealing with Venn diagrams is to use shadings to indicate which regions represent which sets and then to interpret the results of the operations by ascertaining which regions in the diagram are shaded in which fashion. Typically we use a rectangle to indicate the universal set, and most of the other sets are represented by circles. The complement of a set is represented by that portion of the universal set which is *not* in the set. We usually shade the portion in which we are interested, whether that set is a given set (Fig. 2.1a) or its complement (Fig. 2.1b). This arrangement makes it quite easy to illustrate the set operations.

The important thing in illustrating operations on sets is that care should be taken that the sets are arranged in the most general fashion. With two sets, for example, part of the region in the universal set should not be in either set, part should be in one set but not in the other, and part should be in both sets. With three sets the situation becomes more complicated, and with four it is worse yet. The number of different regions which must be formed by sets in "general position" in a Venn diagram is 2^n, where n is the number of sets used. It is *not* an accident that this is the same number as the number of lines required for a truth table which has n basic propositions in the statement. This number allows one to consider all the cases. As long as only two or three sets are involved, the diagram can be standardized as illustrated in Fig. 2.2. Usually, Venn diagrams are not used when the number of sets involved exceeds three, unless some special relationships are assumed about the sets.

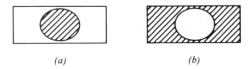

(a)   *(b)*

Figure 2.1 (a) The set A. (b) The set A'.

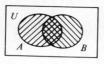 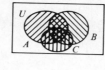

(a)   (b)

Figure 2.2 (a) Two sets. (b) Three sets.

Venn diagrams can be used to illustrate the basic set operations. In Fig. 2.2a the region shaded in both directions represents $A \cap B$, while the union of the two sets is represented by that portion of the diagram which has any shading in either direction.

EXAMPLE 2.9

Use a Venn diagram to illustrate De Morgan's law:

$$(A \cup B)' = A' \cap B'$$

We can do this by first drawing A and B in general position, and finding $A \cup B$. Then $(A \cup B)'$ is everything in the universal set which was not shaded.

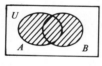   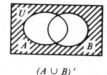

$A \cup B$ $(A \cup B)'$

Next we form the other set by taking A' and shading it and B' and shading it in a different fashion. Then $A' \cap B'$ is the area in the diagram which has been shaded in both ways.

$A' \cap B'$

We can then note that this figure is identical to the one for $(A \cup B)'$. ■

EXAMPLE 2.10

Illustrate the distributive property:

$$A \cap (B \cup C) = (A \cap B) \cup (A \cap C)$$

First we draw the sets in general position (Fig. 2.3). The next step is to find $A \cap B$ and $A \cap C$ (Fig. 2.4). The result of $(A \cap B) \cup (A \cap C)$ is the region that is shaded in either fashion.

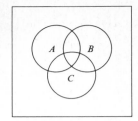

Figure 2.3 Three sets
in general position.

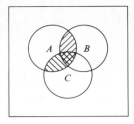

Figure 2.4 $A \cap B$ ///;
$A \cap C$ \\\.

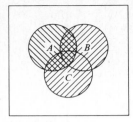

Figure 2.5 $B \cup C$ ///;
A \\\.

To find the other set, we first find $B \cup C$ and note its intersection with A. In Fig. 2.5 the intersection is the area which is shaded both ways. This is exactly the same region as was shaded in either direction in Fig. 2.4, which confirms the distributive property. ■

Illustrating Specific Relationships between Sets

Venn diagrams can also be used when the relationships between the sets are specified. For example, we can use Venn diagrams to illustrate a subset relation holding between two sets or two sets being disjoint (see Fig. 2.6). We can use these to illustrate theorems from set theory which hold under special circumstances.

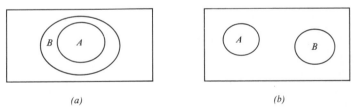

(a)   (b)

Figure 2.6 (a) $A \subset B$. (b) A, B disjoint.

EXAMPLE 2.11

Illustrate the fact that if $A \subset B$, then $A \cup B = B$.

A /// B \\\

The figure shows A as a subset of B, and the union of A and B is the area shaded in either direction, which is quite clearly the same as B. ■

Venn diagrams are often quite helpful in lending us understanding as to exactly what some theorems in set theory actually say. In the problems for this section you are asked to find some regions by using Venn diagrams and to illustrate some theorems from set theory.

Problem Set 4

Illustrate the following sets with Venn diagrams:

1. $A \cup B$

2. $A \cap B'$

3. $A \cup (B \cap C')$

4. $(A \cup B) \cap (A' \cup C)$

5. $(A' \cap B') \cap C'$

6. $A' \cap B$

7. $B' \cup A'$

8. $A' \cap B'$

9. $(A \cup B) \cup C$

10. $(A \cup B) \cap (A \cup C)$

11. $((A \cup B) \cup C)'$

12. $A \cap (B \cup C)'$

13. $(A \cup B)' \cup C$

14. $(A \cup B)' \cap (A \cup C)'$

15. $(A' \cap B') \cap C'$

16. $(A \cap B)' \cup (A \cap C)'$

17. Draw $A \cup B$ and explain why it illustrates that $A \subset A \cup B$ and $B \subset A \cup B$.

18. Draw $A \cap B$ and explain why it illustrates that $A \cap B \subset A$ and $A \cap B \subset B$.

19. Use a Venn diagram to illustrate the fact that if $A \subset B$, then $A \cap B = A$.

20. Region 1 in Fig. 2.7 can be expressed as $A' \cap B'$. Provide similar expressions for the other three regions illustrated in the figure.

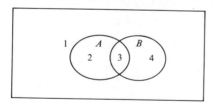

Figure 2.7

21. Region 1 in Fig. 2.8 can be expressed as $A' \cap B' \cap C'$. Provide similar expressions for the other seven regions in Fig. 2.8.

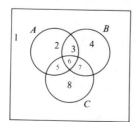

Figure 2.8

22. Illustrate the associative property for intersection with a Venn diagram of three sets.

23. Illustrate the distributive property of union over intersection $[A \cup (B \cap C) = (A \cup B) \cap (A \cup C)]$.

24. Illustrate De Morgan's law: $(A \cap B)' = A' \cup B'$.

25. Illustrate De Morgan's law: $(A \cup B)' = A' \cap B'$.

26. Suppose that $A \subset B$. Use a Venn diagram to show that $B' \subset A'$.

27. Show that $(A' \cup B) \cap B' \subset A'$.

28. Show that $(A \cup B)' \subset A' \cup B'$.

2.5 PRODUCTS OF SETS; RELATIONS AND FUNCTIONS

In discussing sets, we have stated that the order in which elements are listed is not important and that the only concern is whether a particular element is contained in a set. Thus $\{a, b\}$ is the same set as $\{b, a\}$. With regard to sets, first and second have no meaning, but in some cases in mathematics we are very much concerned with the order in which things happen. What is first, second, third, etc., can be very important.

Ordered Pairs

A first step toward considering this issue is the *ordered pair*. Ordered pairs are indicated by listing the elements of the ordered pair in parentheses. In an ordered pair, the order in which the entries are written is important. Two ordered pairs are the same provided that they contain the same entries in the same order. The ordered pair (a, b) is *equal to* the ordered pair (c, d) provided that both $a = c$ and $b = d$. In ordered pairs, unlike sets, (a, b) and (b, a) *are different.*

EXAMPLE 2.12

If $(a, b) = (1, 3)$, then $a = 1$ and $b = 3$.
$(2, 3) \neq (3, 2)$ because $2 \neq 3$.
$(2, 3) \neq (2, 4)$ because although $2 = 2$, $3 \neq 4$. ■

In many books on set theory, a formal definition of ordered pairs is made in the following way:

DEFINITION

The **ordered pair** (a, b) is the set $\{a, \{a, b\}\}$.

This clever definition makes an ordered pair into a set. And this set does meet our needs for ordered pairs, because if $(a, b) = (c, d)$ according to this definition, then it has to follow that $a = c$ and $b = d$. Verification of this fact is left as an exercise. We will not make a big point of making use of this definition, and our main point in including this definition is to emphasize the fact that ordered pairs can be described in terms of sets if necessary.

EXAMPLE 2.13

$(1, 2) = \{1, \{1, 2\}\}$
$(2, 3) = \{2, \{2, 3\}\}$
$(3, 2) = \{3, \{3, 2\}\} = \{3, \{2, 3\}\}$
$(2, 2) = \{2, \{2, 2\}\} = \{2, \{2\}\}$ ∎

Products of Sets

Often it is necessary to form a set consisting of all ordered pairs in which the first entry comes from one set and the second entry comes from a second set. This new set is known as the *product* of the two sets, and it forms the basis for two very important mathematical ideas, those of the relation and the function.

> **DEFINITION**
>
> The **product** $A \times B$ of sets A and B is the set
>
> $$\{(x, y) \mid x \in A \land y \in B\}.$$

EXAMPLE 2.14

If $A = \{1, 2\}$ and $B = \{a, b\}$, then $A \times B = \{(1, a), (1, b), (2, a), (2, b)\}$. Sometimes we can visualize the product of two sets by drawing a picture which represents the first set on a horizontal axis and the second set on a vertical axis, with dots located at the "grid points" at the intersection of the horizontal and vertical lines corresponding to the points of the individual sets of the product. See Fig. 2.9.

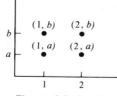

Figure 2.9 $A \times B$.

This technique for illustrating the product of sets is the one used in courses in algebra and calculus to represent graphs of the mathematical functions studied in those courses. ∎

We can extend the idea of ordered pairs to ordered triples in the following way: An ordered triple is a list of three numbers (a, b, c), such that $(a, b, c) = (x, y, z)$ if and only if $a = x$ and $b = y$ and $c = z$. Ordered 4-tuples, and in general ordered n-tuples, can be defined in a similar fashion. An interesting application of the use of induction to provide a definition is seen in the following:

DEFINITION

An **ordered n-tuple** is defined for all n in the following way:

An ordered 1-tuple is a single element.

An ordered $(n + 1)$-tuple is an ordered pair whose first entry is an ordered n-tuple and second entry a single element.

Thus, an *ordered triple* is an ordered pair in which the first entry is an ordered pair and the second entry is a single element. An *ordered 4-tuple* similarly is an ordered pair in which the first entry is an ordered triple and the second entry is a single element.

It is possible to show that the definition above provides exactly the same condition for equality of ordered n-tuples as described above. You are asked to carry this out in Problem 5 for ordered triples and to show that it is true for ordered n-tuples in Problem 6.

EXAMPLE 2.15

Form ordered pairs corresponding to the ordered triple $(1, 2, 3)$, the ordered 4-tuple (x, y, z, w), and the ordered 5-tuple (a, b, c, d, e).

SOLUTION

The ordered triple $(1, 2, 3)$ is the ordered pair $((1, 2), 3)$. The ordered 4-tuple (x, y, z, w) is the ordered pair $((x, y, z), w)$. We can further decompose the ordered triple (x, y, z) as $((x, y), z)$, so we could write this 4-tuple as $(((x, y), z), w)$. The ordered 5-tuple (a, b, c, d, e) is $((a, b, c, d), e)$, which can ultimately be seen to be $((((a, b), c), d), e)$. ∎

If necessary, it is possible to take an ordered n-tuple all the way back to the technical definition of an ordered pair, and the problems provide that opportunity, but it is by no means crucial to the use of these items. The important ingredient is the fact that two ordered n-tuples are equal if and only if they have exactly the same entries in the same positions.

Relations and Functions

The product of two sets is a set, and thus it is possible to talk about a subset of the product of two sets. We call such a subset a *relation* on the two sets.

DEFINITION

If A and B are sets, a **relation between A and B** is a subset of $A \times B$. If a relation is a subset of $A \times A$, we call it a relation on A.

Relations are often used to describe relationships between objects.

EXAMPLE 2.16

The following are relations:
(a) $M = \{(x, y) \mid x$ is married to $y\}$ is a relation on the set of people in the human race.
(b) $R = \{(x, y) \mid$ language x is available on computer $y\}$ is a relation between the set of languages and the set of computers.
(c) $C = \{(x, y) \mid x$ is a cousin to $y\}$ is a relation on the set of people in the human race.
(d) $F = \{(x, y) \mid$ airline x flies to city $y\}$ is a relation between the set of airlines and the set of cities with airports. ∎

EXAMPLE 2.17

The following are examples of relations on the set of integers:
(a) $\{(x, y) \mid x < y\}$
(b) $\{(x, y) \mid y = x^2\}$
(c) $\{(x, y) \mid x^2 + y^2 = 9\}$
(d) $\{(x, y) \mid x + y = 9\}$ ∎

DEFINITIONS

The **domain** of a relation is the set of all first entries (or first coordinates) which occur in the relation.[1]

The **range** of a relation is the set of all second coordinates.[2]

A **function** is a relation in which each element of the domain belongs to only one ordered pair in the relation.

Example 2.17b and Example 2.17d are functions, because for each value of x, there is only one value of y such that $(x, y) \in R$. On the other hand, Example 2.17a and Example 2.17c are not functions because in each case we find more than one ordered pair with the same x coordinate. And $(0, 1)$, $(0, 2)$,

---

[1] Some authors refer to A as the domain of a relation on A and B, even if not all elements of A occur as first elements.

[2] Some authors call B the range of a relation on A and B and use the term *image* for what we have called the range. Others call the set B the *codomain*. We prefer to avoid these "theological" disputes.

... are all examples of ordered pairs belonging to the relation in Example 2.17a, and (0, 3) and (0, −3) both belong to the relation in Example 2.17c.

It is often convenient to use the usual notation for functions, by writing $f(x) = y$ to mean that (x, y) is one of the ordered pairs in the relation f. Usually when we write this, we are thinking of the relation as a rule which supplies us with a value of y whenever we specify a value of x. The function in Example 2.17b could be described by the notation $f(x) = x^2$ if we follow that convention.

Functions have been a focus of mathematics courses for a long time, and are usually discussed in depth in high school mathematics courses, so we do not take a great deal of time to consider functions. Instead we focus on the properties of relations, since the properties of relations may not be as well known to the reader. Relations form the foundation for much of what is known as discrete mathematics.

Use of Relations

If R is a relation on A and B, then given an ordered pair (a, b) from $A \times B$, the ordered pair may or may not belong to R. If (a, b) does belong to R, we could write $(a, b) \in R$, but we often write aRb to indicate that membership. This notation emphasizes the idea that when (a, b) belongs to a relation, there is a relationship which exists between a and b. Thus in the relation M of Example 2.16, if John and Mary were married, we would have (John, Mary) $\in M$, and could also write John M Mary. This would be read perhaps as John "is married to" Mary. Similarly, if Pascal were available on a PDP 11/70, we could write Pascal R PDP 11/70 and read that as Pascal "runs on the" PDP 11/70. Similar descriptions and terminology can be provided for the relations C and F of Example 2.16.

Relations on a Single Set

Most of the relations with which we are most familiar are defined on a single set. Relations on sets of numbers such as $<$, $>$, $=$, etc., are relations which we have used many times in the past. The notation that we discussed above comes from our usage of these relations. Under most normal circumstances, if x and y have the same value, we write $x = y$, and not $(x, y) \in =$. These relations also exhibit some important properties. For example, we are familiar with the fact that to write $a = b$ means that we can also write $b = a$; that a number is always equal to itself; and that if $a = b$ and $b = c$, we can conclude that $a = c$. These three properties of equality are called the *symmetric*, *reflexive*, and *transitive properties* of equality, respectively. For a relation on a single set, we can attempt to determine which, if any, of those properties hold for that relation. There are also a couple of other properties that a relation may have which are illustrated by the inequality relations. If $a < b$, then $b < a$ must be false. This property is called the *asymmetric property*. If $a \leq b$ and $b \leq a$, then $a = b$. This property is called the *antisymmetric property*. We formalize the definitions of all these properties as follows:

> **DEFINITION**
>
> Let R be a relation on the set A. The relation R is said to be
> (a) **Reflexive** if $\forall a(aRa)$.
> (b) **Symmetric** if $\forall a\ \forall b(aRb \rightarrow bRa)$.
> (c) **Transitive** if $\forall a\ \forall b[(aRb \wedge bRc) \rightarrow aRc]$.
> (d) **Asymmetric** if $\forall a\ \forall b[(aRb) \rightarrow \sim(bRa)]$.
> (e) **Antisymmetric** if $\forall a\ \forall b[(aRb \wedge bRa) \rightarrow a = b]$.
>
> An **equivalence relation** is any relation which satisfies (a), (b), and (c).
>
> A **partial order** is any relation which satisfies (a), (e), and (c).

The concept of an equivalence relation was inspired by equality (in fact, the term "equivalent" is closely related to the term "equals"), which is indeed an equivalence relation, and the idea of a partial order was inspired by the relation $\leq$, which is a partial order.

Note that the relation M described in Example 2.16 is not reflexive, but it is symmetric. The transitive property does not hold for this relation. (If a were married to b, since b is married to a, the transitive property would force us to conclude that a is married to a, which is nonsense.) The relation M is not asymmetric (no symmetric relation can be asymmetric), and it is not anti-symmetric. The properties discussed above are considered only for relations defined on a single set, so the relations R and F from that example cannot have any of the properties. We leave it as an exercise for the reader to show why the relation C in Example 2.16 has the same properties as M.

EXAMPLE 2.18

Let R be the relation defined by xRy if and only if $|x - y|$ is even. Show that R is an equivalence relation.

SOLUTION
We must show that R satisfies the three properties for an equivalence relation.

Reflexive Since $|x - x| = 0$ and 0 is even, it follows that xRx for all x.

Symmetric Since $|x - y| = |y - x|$ for all x and y, it follows that if $|x - y|$ is even, so is $|y - x|$. And this means that for all x and y if xRy, then yRx.

Transitive We suppose that xRy and yRz and show that xRz. Since xRy, $|x - y|$ is even, and, of course, so is $x - y$. Similarly since yRz, $y - z$ is even.

$$x - z = (x - y) + (y - z)$$

Since $x - y$ and $y - z$ are both even, it follows that $x - z$ (and, of course, $|x - z|$) is even. ∎

Partitions and Equivalence Relations

DEFINITIONS

A **partition** of the set S is a collection of subsets of S such that each element of S is in one of the subsets and no two of the subsets have any elements in common.

The **cells** of a partition are the subsets of the original set which comprise the partition.

It is interesting to note that every equivalence relation produces a partition of a set. A partition of the set A is illustrated in Fig. 2.10. If R is an equivalence relation on the set A, then $[a]$, the *equivalence class of a*, is $\{x \in A \mid xRa\}$. The collection of all equivalence classes formed by an equivalence relation forms a partition of a set.

Figure 2.10 A partition of the set A.

THEOREM 2.7

If R is an equivalence relation on the set A, then the collection of all equivalence classes of R is a partition of A.

PROOF
We must prove that (a) every element of A is in one of the cells of the partition and (b) no two cells have any elements in common.
(a) If $a \in A$, then obviously $a \in [a]$.
(b) To show that no two cells have any elements in common, it is necessary to show that if $[a] \cap [b]$ is not empty, then $[a] = [b]$. If $c \in [a] \cap [b]$, then we have cRa and cRb. By the symmetric property we have aRc, and thus by the transitive property we have aRb. Since aRb, it is easy to show that xRa if and only if xRb, and thus $[a] = [b]$. ∎

EXAMPLE 2.19

Find the equivalence classes for the relation R from Example 2.18.

SOLUTION
We first note that $[0]$ consists of all integers which are an even distance from 0. That means that $[0]$ consists of all even integers.

Since 1 does not belong to $[0]$, we can look for $[1]$, which consists of all integers which are an even distance from 1. That means that $[1]$ must be the set of all odd integers. Since every integer is either even or odd, these two equivalence classes form the complete partition of the integers induced by the equivalence relation R.

THEOREM 2.8

If $\{C_1, C_2, \ldots\}$ is a partition of set A and if R is defined by xRy if and only if x and y are in the same cell of the partition, then R is an equivalence relation on A. Furthermore, the equivalence classes of R are the cells of the partition.

PROOF
The proof of this is left as an exercise.

Theorem 2.8 is remarkable in the sense that it tells us that equivalence relations and partitions are "equivalent" ideas because each leads to the other, and the cells of the partition are the equivalence classes of the equivalence relation. It proves to be very convenient to be able to go back and forth between the two ideas.

EXAMPLE 2.20

The following relation on the set of all Pascal (actually the language is irrelevant, any language will do) programs is an equivalence relation. We let P_1RP_2 if and only if P_1 and P_2 produce the same output when they are given the same input.

We leave it as an exercise for the student to prove that this is an equivalence relation.

The equivalence classes for this relation are particularly interesting. They consist of programs each of which does the same thing. For many purposes and many problems, it is enough to know that the equivalence class of all programs which solve a particular problem is not empty. In other ways, however, it becomes useful and important to be able to distinguish between elements of the same equivalence class.

Some programs may do the same things as other programs, only much more efficiently. In our discussion of algorithm complexity later on, this is the primary issue. We must decide what we mean by efficient and determine which algorithms best meet this definition. When we tackle this issue, regardless of what definition we use, we are attempting to make choices between elements of the same equivalence class.

Problem Set 5

1. Let $A = \{1, 2\}$ and $B = \{a, b, c\}$.
   (a) Find the elements of $A \times B$.
   (b) Represent the set $A \times B$ by a diagram.

2. Let $C = \{x, y\}$ and $D = \{1, 2\}$. Repeat Problem 1 for $C \times D$.

3. Consider the ordered triple (1, 3, 5).
   (a) Represent the ordered triple as an ordered pair.
   *(b) Represent the ordered pair in (a) as a set.

*4. Prove that according to the set definition of an ordered pair, if $(a, b) = (c, d)$, then $a = c$ and $b = d$.

5. Show that, according to the definition of the ordered triple (a, b, c) as $((a, b), c)$, the ordered triples (a, b, c) and (x, y, z) are equal if and only if $a = x, b = y$, and $z = c$.

*6. Prove that the ordered n-tuples $(x_1, x_2, \ldots, x_n)$ and $(y_1, y_2, \ldots, y_n)$ are equal if and only if $x_1 = y_1, x_2 = y_2, \ldots, x_n = y_n$. (*Hint*: The definition is given inductively, so use induction in your proof.)

For each of the following relations on the set of integers, determine which of the five properties discussed in this section are satisfied. Also indicate whether the relation is an equivalence relation or a partial order.

7. aRb if and only if $a - b$ is divisible by 3.

8. aRb if and only if $a > b$.

9. aRb if and only if $a + b$ is odd or $a = b$.

10. aRb if and only if $a + b = 8$.

11. aRb if and only if $a + b < 8$.

12. aRb if and only if $|a - b| = 8$.

13. aRb if and only if b is a multiple of a.

14. aRb if and only if $b = 3a$.

15. If A is the set of lines in a plane, define the following relation on A: lRm if and only if l is perpendicular to m. Determine what properties of a relation are satisfied by R.

16. Repeat Problem 15 with the relation "is parallel to." (A line can be considered parallel to itself.)

17. If B is the set of all triangles in the plane, define the relation R on B by aRb if and only if a is congruent to b. Determine what properties of a relation are satisfied by R.

18. Repeat Problem 17 with the relation "is similar to."

19. Let $A = \{p, q, r, s, t\}$ be a set of propositions. Define the relation R on the set A to be xRy if and only if $x \leftrightarrow y$ is a tautology. Show that R is an equivalence relation. (This means that for propositions to be equivalent is an equivalence relation. The symbol $\Rightarrow$ is sometimes used to denote equivalence.)

20. Repeat Problem 19 with the relation M being xMy if and only if $x \rightarrow y$ is a tautology. Show that M is a partial order.

21. Determine which of the relations in Problems 7 to 13 (odd) are functions.

22. Determine which of the relations in Problems 8 to 14 (even) are functions.

23. Determine which properties of a relation are satisfied by the relations given in Example 2.17.

24. Determine which relations in Example 2.17 are functions.

25. Let $A = \{1, 2, 3, 4\}$ and let $R = \{(1, 1), (1, 2), (1, 3), (2, 1), (2, 2), (3, 1), (2, 3), (3, 2), (3, 3), (4, 4)\}$. Show that R is an equivalence relation, and determine the equivalence classes.

26. Let R be the relation defined on the integers by aRb if $a - b$ is even. Show that R is an equivalence relation, and determine the equivalence classes.

27. Partition the set $A = \{1, 2, 3, 4, 5\}$ by $\{\{1, 2\}, \{3\}, \{4, 5\}\}$. List the equivalence relation determined by this partition.

*28. Partition the set $A = \{1, 2, 3\}$ by $\{\{1\}, \{2\}, \{3\}\}$. What is the equivalence relation determined by this partition? What do we usually call this equivalence relation?

29. Complete the proof of Theorem 2.6 by showing that if R is an equivalence relation and aRb, then xRa if and only if xRb.

*30. Prove Theorem 2.8.

In Pascal, relations on a set can be conveniently represented by arrays if each element of the set is given a number and in the array R, $R[i, j]$ is given the value 1 if the ordered pair consisting of element number i and element number j is in the relation. (Unfortunately it is not convenient to define a set of ordered pairs.) Consider the following relations on the set $\{1, 2, 3, 4\}$:

$$R = \{(1, 1), (1, 2), (1, 3), (2, 2), (2, 3), (4, 4)\}$$
$$S = \{(1, 2), (2, 1), (2, 2), (3, 4), (3, 3), (4, 4)\}$$
$$T = \{(1, 1), (1, 2), (2, 1), (2, 2), (3, 3), (4, 4)\}$$

P31. Write a program to input (in any fashion) information describing a relation and test it for the reflexive property. Use your program on the relations R, S, and T.

P32. Repeat Problem 31, only now use the symmetric property.

*P33. Repeat Problem 31 with the transitive property.

*P34. Put your programs from Problems 31, 32, and 33 together to test a relation and determine whether it is an equivalence relation. Also, make your program so that it prints out the equivalence classes for each relation that is an equivalence relation.

2.6   COUNTING TECHNIQUES—UNIONS OF SETS

One of the most important ideas in discrete mathematics relates to the process of counting. Often we need to count the number of elements in a set, the number of ways that a process can happen, the number of steps required for the completion of an algorithm, and other similar quantities. In this section we begin the process of considering such questions by looking at the number of elements in the union of sets.

Disjoint Pairs of Sets

On the surface it would seem quite reasonable to expect that $|A \cup B|$ would be equal to $|A| + |B|$, because the union of A and B is made up of all elements belonging to either set. This is the case when A and B are disjoint, because when we go to count $|A \cup B|$, we find that each element comes from either A or B but never from both sets. This leads us to state the following:

COUNTING PRINCIPLE 1

If A and B are disjoint sets, then $|A \cup B| = |A| + |B|$. ∎

EXAMPLE 2.21

Let $A = \{1, 2\}$ and $B = \{a, b, c\}$. Find $|A \cup B|$.

It is quite clear that $A \cup B = \{1, 2, a, b, c\}$ and that the cardinal number of this set is 5. When we count the elements in the union, each element in the union is found to be from one of the two sets, but not both. So each element of A and each element of B are counted exactly one time when we add $|A| + |B|$. ∎

Nondisjoint Pairs of Sets

If A and B in Example 2.21 were *not* disjoint, the problem of determining the cardinal number of $A \cup B$ would be harder. If we were to count the elements of A and then the elements of B and then add the numbers to find $|A \cup B|$, we would find that we had counted some of the elements two times. In the Venn diagram of Fig. 2.11, we represent the situation which occurs when $|A|$ is 5, $|B|$ is 6, and $|A \cap B|$ is 2. The area shaded $///$ is counted when we find

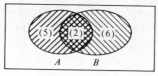

Figure 2.11

$|A|$, and the area shaded \\\ is counted when we find $|B|$. The region shaded twice has been counted twice; and if we add $|A|$ to $|B|$ to find $|A \cup B|$, we need to subtract it from the total. The double-shaded region is $A \cap B$.

This reasoning leads us to the following theorem.

THEOREM 2.9

Let A and B be any two sets. Then $|A \cup B| = |A| + |B| - |A \cap B|$.

PROOF

Although the discussion above is convincing, we use counting principle 1 to illustrate another approach to the same theorem. From Fig. 2.12, $A \cup B$ can be seen to be the union of the three disjoint sets $A - B$, $B - A$, and $A \cap B$.

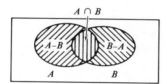

Figure 2.12

Counting principle 1 tells the following facts:

1. $|A \cup B| = |A - B| + |B - A| + |A \cap B|$
2. $|A| = |A - B| + |A \cap B|$
3. $|B| = |B - A| + |A \cap B|$

Using 2 and 3, we note that $|A| + |B| = |A - B| + |B - A| + 2|A \cap B|$. This is just $|A \cup B| + |A \cap B|$, and from this fact we can conclude that our result holds. ∎

EXAMPLE 2.22

Let $A = \{1, 2, 3\}$ and $B = \{3, 4, 5\}$. Then $A \cup B = \{1, 2, 3, 4, 5\}$ and $A \cap B = \{3\}$.

In this case, $|A| = 3$, $|B| = 3$, $|A \cap B| = 1$, and $|A \cup B| = 5$. Thus

$$|A| + |B| - |A \cap B| = 3 + 3 - 1 = 5$$

as expected. ∎

Theorem 2.9 can be used in a number of ways to determine the sizes of various sets. Any time that we know the cardinal number for three of the four sets in the equation, we can solve for the cardinal number of the remaining one.

EXAMPLE 2.23

Suppose that A and B are sets for which $|A| = 5$ $|B| = 8$, and $|A \cup B| = 11$. Find $|A \cap B|$.

SOLUTION
Since $|A \cup B| = 11$, from Theorem 2.9 we have $11 = |A| + |B| - |A \cap B| = 5 + 8 - x$. Solving this equation for x, we get $x = 2$, which says that $|A \cap B| = 2$. ∎

Theorem 2.9 also provides us with the tools needed to solve the problem posed in Example 2.1 from Section 2.1.

EXAMPLE 2.1 (REPEATED)

United Computer Technologies has written a new package which integrates a word processing program with a spreadsheet program, and they wish to make the program so that it will run on the Admiral 64, which is a 64K machine. They wish to do this in such a way as to minimize the need to overlay code and fetch information from the disk. The word processor requires 40K for program and data, and the spreadsheet requires 32K for program and data. If 16K must be reserved for the code which integrates the two packages, and that code *must* remain resident in the memory, what is the minimum amount of overlay space which will be necessary?

SOLUTION
If we let $A = \{$bytes occupied by word processor$\}$ and $B = \{$bytes occupied by spreadsheet$\}$, then we must have $|A \cup B| \leq 48K$, because 16K of the 64K is already occupied. Then $|A \cup B| = |A| + |B| - |A \cap B|$, which means that $|A| + |B| - |A \cap B| \leq 48K$, so we get $40K + 32K - x \leq 48K$, or $72K - x \leq 48K$. From this we conclude that $x \geq 24K$. Thus the smallest possible space for the intersection of the two sets is 24K. ∎

Union of Three Sets

Let us now move to the problem of determining the cardinal number of the union of three sets. If we had sets A, B, and C, we would want to determine the cardinal number of $A \cup B \cup C$. By using Theorem 2.9, we proceed:

$$|A \cup (B \cup C)| = |A| + |B \cup C| - |A \cap (B \cup C)|$$

$$= |A| + |B| + |C| - |B \cap C| - |A \cap (B \cup C)|$$

(This is true from the use of Theorem 2.9 on $|B \cup C|$.) Now, since $A \cap (B \cup C) = (A \cap B) \cup (A \cap C)$, it follows that

$$|A \cap (B \cup C)| = |A \cap B| + |A \cap C| - |(A \cap B) \cap (A \cap C)|$$

This last quantity can be rewritten as

$$|A \cap B| + |A \cap C| - |A \cap B \cap C|$$

Putting all this together, we finally get the following for the final result of the cardinal number of the union:

$$|A| + |B| + |C| - |A \cap B| - |A \cap C| - |B \cap C| + |A \cap B \cap C|$$

The above discussion has actually provided us with the proof of the following theorem.

THEOREM 2.10

Let A, B, and C be sets. The cardinal number of the union of the three is given by

$$|A| + |B| + |C| - |A \cap B| - |A \cap C| - |B \cap C| + |A \cap B \cap C|$$

We have really shown that in this case the following process works:

1. Add the cardinalities of the three sets.

2. Subtract the cardinalities of the three possible intersections of pairs of sets. (Points in these intersections will be "overcounted" otherwise.)

3. Add the cardinality of the intersection of the three sets. (The points in all three sets were subtracted once too many times, so add them again.)

The Venn diagram of Fig. 2.13 helps to illustrate what has happened. In adding $|A| + |B| + |C|$, we have counted all the elements of all three sets once—and some of them 2 or 3 times. The double-shaded areas were counted 2 times, and the region shaded all three directions was counted 3 times. To adjust, we subtract the three "pairwise" intersections. The section shaded in

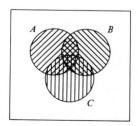

Figure 2.13

all three directions was included in all three of these intersections, and although it was originally counted 3 times, its cardinal number has now been subtracted 3 times. To resolve this, we add its cardinal number back into the total.

We leave the problem of determining the cardinal number of the union of four sets as an exercise. ∎

Often in determining the relationships and cardinal numbers involved when three sets are used, the use of a Venn diagram can simplify the process.

EXAMPLE 2.24

A researcher studying snakes finds the following population of snakes in a swamp:

32 were green.

20 were water snakes.

45 were poisonous.

15 were green and poisonous.

7 were green water snakes.

10 were poisonous water snakes.

5 were poisonous green water snakes.

(a) How many snakes were there all together?
(b) How many snakes fit in only one category?
(c) How many snakes were green water snakes but were not poisonous?

SOLUTION
The answer to (a) is straightforward from Theorem 2.10. We let A be the set of green snakes, B be the set of water snakes, and C be the set of poisonous snakes. We have

$$|A \cup B \cup C| = 32 + 20 + 45 - 15 - 7 - 10 + 5 = 70$$

To answer (b) and (c), a Venn diagram can be filled in to indicate the cardinal number of each of the eight sections (actually, in this case, seven sections, since we have no elements in the complement of all three sets).

In a Venn diagram of sets A, B, and C in standard position, we can fill in the numbers in the regions, working from the "inside" and proceeding outward. Our given information tells us that $|A \cap B \cap C| = 5$, and thus our diagram can reflect that information, as in Fig. 2.14a. We also know, however, that $|A \cap B| = 7$. And since the intersection of all three sets accounts for five of the elements in $A \cap B$, the region corresponding to $(A \cap B) - C$ must be

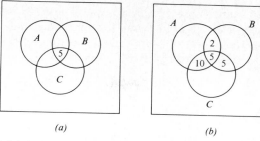

(a)   (b)

Figure 2.14 (a) $A \cap B \cap C$ filled in. (b) All pairwise intersections filled in.

labeled with a 2, to indicate two elements in the intersection of A and B which are not in C. This leaves both $A \cap B$ and the intersection of all three sets with the right cardinal number. Following the same reasoning, we can fill in the region in A and C but not in B with 10, and the region in B and C but not in A with a 5 (Fig. 2.14b).

Finally we see from the diagram that we have accounted for 17 of the elements of A. Since $|A| = 32$, we know that 15 elements must be in A only. Filling in the rest of the diagram leads to Fig. 2.15.

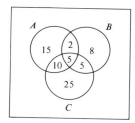

Figure 2.15 A completed Venn diagram.

Now, reading from the diagram, we can answer (b) as $15 + 8 + 25 = 48$ and (c) as 2. ∎

Students who first see a problem like the one above often become confused. They think that the description which says that 32 snakes were green means "32 snakes were green and not poisonous and not water snakes." This is *not* what is intended in this problem. The category of green is intended to include all the green snakes—even those who have some of the other characteristics as well! The technique of using Venn diagrams is very useful in making sense out of problems like this one.

Problem Set 6

1. A recent survey of 100 rock and roll musicians showed that 40 wore gloves on the left hand and 39 wore gloves on the right hand. If 60 wore no gloves at all, how many wore a glove on only the right hand? Only the left hand? On both hands?

2. Two hundred students enrolled in a physical fitness class were asked to select at least one of the activities of jogging and swimming. There were 85 who chose to jog, 60 of whom also chose to swim. Altogether, how many chose to swim? How many chose to swim and not jog?

3. A file of size 96K must be written on a mini disk of capacity 143K. If 100K of the disk is already occupied with files, what is the minimum amount of the disk that must be erased in order to store the new file?

4. A poll of 30 people found that 12 ate fish at least once a week and 5 ate both potatoes and fish at least once a week. If everyone in the survey ate either fish or potatoes at least once a week, how many ate potatoes?

5. Of 30 students enrolled in a discrete mathematics course at a small university, 26 passed the first examination and 21 passed the second. If two students failed both examinations, how many passed both examinations?

6. Of 1200 first-year students at a large university,

   582 took physical education.

   627 took English.

   543 took mathematics.

   217 took both physical education and English.

   307 took both physical education and mathematics.

   250 took both mathematics and English.

   222 took all three courses.

   How many took none of the three?

7. During a certain period, it rained 12 times, the sun shone 15 times, and on 5 instances both happened. If on each day of the period at least one of the two occurred, how many days long was the period?

8. A total of 35 programmers interviewed for a job; 25 knew FORTRAN, 28 knew Pascal, and 2 knew neither language. How many knew both languages?

9. Suppose that $|B| = 12$, $|C| = 11$, $|D| = 8$, $|B \cup C| = 20$, $|B \cup D| = 20$, and $|D \cap C| = 3$. Find
   (a) $|B \cap C|$ (b) $|B \cap D|$
   (c) $|B - D|$ (d) $|B - C|$
   (e) $D \cup C|$

10. Suppose that

$$|A| = 65 \quad\quad |A \cap B| = 20 \quad\quad |A \cup B \cap C| = 100$$
$$|B| = 45 \quad\quad |A \cap C| = 25$$
$$|C| = 42 \quad\quad |B \cap C| = 15$$

Find the following:
(a) $|A \cap B \cap C|$
(b) $|(A \cap B) - C|$
(c) $|(A \cap C) - B|$
(d) $|(B \cap C) - A|$

11. Suppose that

$$|A| = 35 \qquad |A \cap B| = 15 \qquad |A \cup B \cup C| = 52$$
$$|B| = 23 \qquad |A \cap C| = 13$$
$$|C| = 28 \qquad |B \cap C| = 11$$

Find the following:
(a) $|A \cap B \cap C|$
(b) $|(A \cap B) - C|$
(c) $|(A \cap C) - B|$
(d) $|(B \cap C) - A|$

12. A total of 60 potential customers came into a computer supply store. Of these, 52 made purchases: 20 bought paper, 36 bought diskettes, and 12 bought ribbons. If 6 bought both paper and diskettes, 9 bought both diskettes and ribbons, and 5 bought paper and ribbons, how many bought all three items?

13. Find a formula for $|A \cup B \cup C \cup D|$.

*14. Determine a general formula for the cardinal number of the union of any number of sets.

15. (a) How many positive integers less than 100 are multiples of 2?
(b) How many positive integers less than 100 are multiples of 3?
(c) How many positive integers less than 100 are multiples of 5?
(d) Using this information and the principle for counting unions of three sets, determine how many positive integers less than 100 are not multiples of any of these three numbers.

16. Repeat Problem 15, only now consider all integers less than 1000.

17. Use the ideas in Problems 15 and 16 to determine how many positive integers less than 100 are not multiples of any of the numbers 2, 3, 5, and 7.

18. In the early 1950s, 120 residents of New York City were surveyed about their interest in the three baseball teams from that area. Of these, 40 followed the Giants, 28 followed the Yankees, and 31 followed the Dodgers; 23 followed both the Giants and the Yankees, 19 followed the Yankees and Dodgers, and 25 followed the Giants and Dodgers. There were 17 people who followed the fortunes of all three teams. How many people followed none of the teams? How many followed exactly one team? How many followed the Giants and Dodgers but not the Yankees?

P19. Write a function which computes the cardinal number of a set of characters, and use your function to show that the cardinal number of $\{a, b, c, d, e, f, g, h, i, j\}$ is 10.

P20. Use your function from Problem 19 to verify Theorem 2.10 with the sets $A = \{a, b, c, d, e, f\}$, $B = \{d, e, f, g, h, i, j\}$, and $C = \{a, e, i, o, u\}$.

2.7 COUNTING TECHNIQUES—PRODUCTS OF SETS AND SEQUENCES OF EVENTS

Products of Sets

The most common situation in which counting techniques are needed arises when it is necessary to determine the number of ways that a sequence of events can occur. The best place to start considering this problem is to count the number of elements in the product of two sets. The name "product" is suggestive of the result, which we state as follows:

COUNTING PRINCIPLE 2

If A and B are sets, then $|A \times B| = |A| \cdot |B|$.

That this should be the case is made quite clear by examining a pictorial representation of the product of two sets:

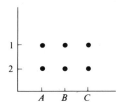

The points which represent the product form a rectangular region which is $|A| \times |B|$. The number of points represented is clearly $|A| \cdot |B|$. ∎

EXAMPLE 2.25

Suppose that a computer algorithm contains the following code:

```
for i := 1 to 5 do
    ( a sequence of 10 statements)
    od;
```

How many total statements are executed in the loop?

SOLUTION
Each statement executed in the process of completing the loop can be represented by an ordered pair (i, j), in which the first element represents the iteration number and the second represents the position of the statement in the list of statements in the loop. Since each statement which is executed by the loop can be described in this way, and each ordered pair gives exactly one statement execution, the number of possible ordered pairs will count the

number of statements executed. The set of ordered pairs is $A \times B$, where $A = \{1, 2, 3, 4, 5\}$ and $B = \{1, \ldots, 10\}$. From counting principle 2 we conclude that $|A \times B| = 5 \cdot 10 = 50$. ∎

EXAMPLE 2.26

Suppose that an algorithm contains the following code sequence:

```
if A = B then (statement 1) else
    if A < B then (statement 2) else
        (statement 3)
    fi
fi;
if X < Y then (statement 4) else
    (statement 5)
fi:
```

How many different sequences of statements can be executed by this algorithm?

SOLUTION

Once again, we can use ordered pairs to describe the possibilities. In this case, we use an ordered pair (i, j), where i represents the statement done in the first **if** structure and j represents the statement done in the second **if** structure. This ordered pair comes from {statement 1, statement 2, statement 3} × {statement 4, statement 5}. The cardinal number of this product set is 6, and thus the number of possible sequences is 6. ∎

Examples 2.25 and 2.26 illustrate in a very elementary fashion how counting principles can be applied to the *analysis of algorithms*, which is the field of computer science concerned with the study of the efficiency of algorithms. This use of counting (or combinatorial) techniques is a very important use of discrete mathematics in computer science.

Trees

Many problems in computer science can be analyzed by means of the "tree." The *tree* is a structure which is represented by vertices connected to other vertices by lines called *edges* in such a way that there is exactly one path between the top, or "root," vertex and any other vertex. In Chapters 5 and 6, which deal with graph theory, we make this idea much more precise, but for now Fig. 2.16 should help to clarify the situation. In this figure, it is intended that the edges be considered as directed downward, so that each vertex has only one entrance but several exits. A path is a sequence of edges leading from one vertex to another which follows the downward orientation. (Again this

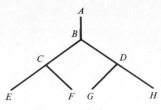

Figure 2.16 A tree.

will all be made precise later.) The number of edges leading out of a vertex is called its *outdegree*. The top or initial vertex is called the *root*. We use the idea of a tree here in order to prove a lemma about paths from the root of a tree which can be used in our analysis of counting. (A *lemma* is a theorem which is proved mainly for the purpose of helping to prove another theorem.)

LEMMA 2.1

If the root of a tree has outdegree m and all the vertices which are connected to the root by a single edge have outdegree p, then there are mp paths which start at the root of the tree and follow two edges.

PROOF
We can translate this problem into one which deals with ordered pairs. We can let $A = \{1, \ldots, m\}$ and $B = \{1, \ldots, p\}$. An element of $A \times B$ will be taken to represent a path in the following way. We number the edges at each vertex from left to right. The ordered pair (i, j) represents a path chosen by taking edge i from the root and edge j from the end of the first edge. Because each vertex connected to the root has outdegree p, each path can be described by an ordered pair in $A \times B$, and each ordered pair describes a path. The number of paths is then $|A \times B|$, which is mp. ∎

This lemma is all that we need to prove one of the most important theorems in combinatorial analysis:

The Fundamental Theorem

THEOREM 2.11   The Multiplication Principle

If a process consists of two stages, the first of which can be accomplished in m ways and the second of which can be accomplished in p ways after the first stage has been done, then the entire process can be done in mp ways.

PROOF
Once again we use the technique of transforming the problem into one which we have already solved. In this case we can describe the process by a tree. The start of the process is represented by the root, and each edge leading from the

root corresponds to a way of completing stage 1 of the process. From the end of each of these edges, we can construct an edge for each way that the second stage can be accomplished, given that the first stage was done by the given method. Each path through the tree corresponds to a way of completing the process in two steps, and each way of completing the process corresponds to one path through the tree. From the lemma, we know there are mp paths through the tree, and thus there are mp ways of completing the process. ∎

By using mathematical induction it is possible to extend this theorem to any finite number of stages.

EXAMPLE 2.27

Use Theorem 2.11 to show that a truth table for a proposition involving four simple propositions will have 16 lines.

SOLUTION

The number of lines in the truth table is equal to the number of possible combinations of T and F which can be assigned to the four basic propositions. We can choose for each of the four propositions one of two values, and by Theorem 2.11, the total number of possibilities is thus $2 \cdot 2 \cdot 2 \cdot 2 = 2^4 = 16$. ∎

EXAMPLE 2.28

If a computer has a 16-bit word, then, since each bit can contain either a 0 or a 1, the same kind of reasoning as above leads us to conclude that one word can contain any of 2^{16} different sequences of 0s and 1s. When a word is used to represent an integer, the first bit is often used as a sign bit, with 1 representing a negative value. It follows that there must be $2^{15} = 32,768$ possible nonnegative integers represented by a single word. Since zero is a nonnegative value, the largest possible integer that can be represented in one word is 32,767. ∎

The multiplication principle can be used just as easily to solve problems in which the number of choices available varies from step to step.

EXAMPLE 2.29

If repetitions are not permitted, how many different three-digit numbers can be formed from the digits, 2, 3, 5, and 7?

SOLUTION

One way to approach problems like this one is to make a "box diagram." We use a box to represent each step in the process, and we fill in each box with the number of ways the step represented by that box may be accomplished. The multiplication principle then tells us that the final solution can be found by multiplying the numbers in the boxes.

Using that idea in this case, we have

First digit	Second digit	Third digit

The first step is the choice of the first digit. This can be accomplished in four ways; once that is done, there are three choices for the second digit and two choices for the third. Our completed box diagram is

First digit	Second digit	Third digit
4	3	2

Multiplying, we see that there are 24 possibilities. ■

A final observation about counting is that sometimes it is better to count the number of ways that some process can *fail* to happen. If we have counted the total number of elements in the universal set, then we can subtract from that number the number of elements in the complement, to obtain the number of ways that the event we want *can* happen. This is true because $U = A \cup A'$ and since A and A' are disjoint, we know that $|U| = |A| + |A'|$. Subtracting $|A'|$ from both sides leads to the equation $|A| = |U| - |A'|$.

EXAMPLE 2.30

In Example 2.29 how many of the numbers do not end in a 3?

SOLUTION
The easiest thing is to count the number that do end in a 3, which is $3 \cdot 2 \cdot 1$, and subtract that number from 24, which was the total number possible. Thus 18 of the numbers do not end in a 3. ■

Problem Set 7

1. A street vendor sells five kinds of sandwiches (beef, ham, salami, turkey, and pastrami) and three beverages (coffee, tea, and milk). How many different menu combinations are available to the person who wishes to purchase a sandwich and drink for lunch?

2. A jogger has six different T-shirts and three pairs of shorts. How many different combinations can she wear?

3. Suppose that a state has car license plates consisting of a letter followed by four digits.
   (a) How many different plate numbers are possible?
   (b) How many are possible if no digits can be repeated?
   (c) How many are possible if no digits are repeated and if the letter must be a vowel?
   *(d) How many plates either have a digit repeated or include a consonant?

4. It is possible to get from Montpelier, Vermont, to Boston by bus or by plane. It is possible to get from Boston to Seattle by bus, train, or plane. It is possible to get from Seattle to Victoria, British Columbia, by plane, bus, or steamship.
   (a) How many different itineraries are possible which take one from Montpelier to Victoria by way of Boston and Seattle?
   (b) Of those in (a), how many itineraries use a different mode of transportation on each leg of the trip?

5. Suppose that a computer is to process a list of five numbers. In how many different orders can the data be input?

6. A baseball manager wants to make out his batting order for a game. He has already chosen the nine players and their positions on the field.
   (a) How many different batting orders are possible?
   (b) Pete Pitcher is a *terrible* hitter. He must bat ninth. How many batting orders are now possible?
   (c) Willie Walker is the best leadoff man, and he should bat first. Given the fact in (b), now how many lineups are possible?
   (d) Steve Slugger is the team's leading homerun hitter, and he should bat third or fourth. Again, given the facts in (b) and (c), how many lineups are now possible?

7. A basketball coach has a roster of 11 players. There are five different positions to be played (center, strong forward, shooting forward, point guard, and off guard).
   (a) How many different lineups can be formed if every player can play every position?
   (b) Tall Paul is 7 feet tall and can play only center. Now how many different lineups are possible?
   (c) In addition to the restrictions in (b), Gary Gimp has been injured and cannot play. Now how many lineups are possible?
   (d) Sam Shrimp is 5 feet 11 inches and can play only the two guard positions. If we still respect the information in (b) and (c), how many lineups are now possible?

8. (a) How many three-digit numbers can be formed from the digits 0, 1, 4, 5, 7, 8, 9?
   (b) If repetitions are not allowed, how many numbers are possible?
   (c) If repetitions are not allowed, how many of the numbers are even?

(d) If repetitions are not allowed, how many of the numbers are less than 500?

(e) If repetitions are not allowed, how many of the numbers will be multiples of 5?

9. A television news program has five segments: local news, regional news, national news, sports, and weather. How many different orders of presentation are possible? How many are possible if local news must be first and weather may *not* be last?

10. What would be printed by a computer program based on the following algorithm?

ALGORITHM SILLY

```
begin
    n:=0;
    for i:=1 to 5 do
        for j:=1 to 5 do
            n:=n+1
        od
    od;
print (n)
end.
```

11. Determine what would be printed by a computer program based on the following algorithm:

ALGORITHM SILLY2

```
begin
    n:=0;
for i:=1 to 3 do
        for j:=1 to 4 do
            for k:=1 to 2 do
                n:=n+1
            od
        od
    od;
print (n)
end.
```

*12. Use the multiplication principle to show that a set with three elements must have eight subsets.

*13. Prove that a set with n elements has 2^n subsets.

*14. Prove that a truth table for a compound proposition with n basic propositions must have 2^n lines.

*15. State and prove the extension of Theorem 2.11 to *n*-step processes. (*Hint*: The proof will require the use of induction.)

*16. The following algorithm sorts a list of *n* numbers into increasing order. It is a technique known as the *selection sort*.

ALGORITHM SORT

```
begin
for i:= 1 to n−1 do
    min:= x[i];
    minspot:= i;
    for j:= i to n do
        if x[j] < min then
                    min := x[j];
                    minspot := j
        fi;
    od
    x[minspot] := x[i];
    x[i] := min
od
end.
```

Determine the number of times that the **if** statement will be executed.

P17. Write a program to print out all the subsets of the set {a, b, c, d, e}.

P18. Write a program to produce all the numbers described in Problem 8.

2.8 COUNTING TECHNIQUES—SUBSETS, COMBINATIONS, AND PERMUTATIONS

In the process of counting, often we need to count subsets of a set or to count the number of ways that subsets of a set can be listed if the order in which the elements occur in the list is significant. These are the basic ideas behind the concepts of *combination* and *permutation*. We present two examples which relate to these kinds of problems, and we present solutions later.

EXAMPLE 2.31

A programmer must write a program to sort a list of 10 numbers into alphabetical order. In how many different orders can the data be presented? ■

EXAMPLE 2.32

A programmer has a list of 10 names and needs to write a program which randomly selects 4 of those names. How many different outputs can the program provide? ■

These two examples provide us with samples of the kinds of problems that can be asked regarding permutations and combinations.

> **DEFINITION**
>
> A **permutation** of a set is a list of the elements of the set in a particular order. A permutation of *n* **items taken** *m* **at a time** is a listing of *m* elements from a set with *n* elements which are arranged in order.

Example 2.31 asks us to count the number of permutations of 10 items taken 10 a time.

> **DEFINITION**
>
> A **combination** is a subset with a specified number of items selected from a given set. A combination of *n* **items taken** *m* **at a time** is a subset with *m* elements chosen from a set which has *n* elements.

Since order does *not* matter in sets, combinations are not concerned with order. Example 2.32 can be answered as the number of combinations of 10 items chosen 4 at a time.

Note: The terminology of permutations and combinations is quite standard, but the notations for counting permutations and combinations are quite diverse. It seems that every author has a different way of denoting the same idea. When you read a mathematics book which deals with combinations and permutations, it is important to look carefully at the notation being used. The notation that we use here is simply one author's choice.

Power Sets

Before we consider how to count permutations and combinations of sets, we should look at the problem of determining how many subsets a set has.

> **DEFINITION**
>
> The **power set** of the set A is the collection of all subsets of the set A. The power set of A is denoted by $P(A)$.

EXAMPLE 2.33

Determine the power sets of the following sets:
(a) $A = \{1, 2, 3\}$
(b) $B = \{1, 2\}$
(c) $C = \varnothing$

SOLUTION

(a) The power set of A consists of all subsets of A. One way to find them is to try to find first all subsets of A with zero elements, then all subsets with one element, then all subsets with two elements, and finally all subsets with three elements.

0 : The only subset of A with zero elements is, of course, $\varnothing$.

1 : The sets $\{1\}$, $\{2\}$, and $\{3\}$ are obviously all the one-element subsets of A.

2 : One way to find two-element subsets of a set with three elements is to look at the problem negatively. We see that to form a two-element subset of A, we must choose one element to leave out for each subset. Leaving out 1 gives $\{2, 3\}$, and the others are $\{1, 3\}$, and $\{1, 2\}$.

3 : The only subset of A with three elements is A itself, namely, $\{1, 2, 3\}$.

Hence, $P(A) = \{\varnothing, \{1\}, \{2\}, \{3\}, \{2, 3\}, \{1, 3\}, \{1, 2\}, \{1, 2, 3\}\}$.

(b) $P(B) = \{\varnothing, \{1\}, \{2\}, \{1, 2\}\}$

(c) $P(C) = \{\varnothing\}$

■

We see in Example 2.33 that $|A| = 3$ and $|P(A)| = 8$, that $|B| = 2$ and $|P(B)| = 4$, and that $|C| = 0$ and $|P(C)| = 1$. The examples above were chosen because the power sets were easy to list. If we had chosen a set such as $\{1, 2, 3, 4, 5, 6, 7, 8\}$, our job would have been much tougher (especially when we tried to find the four- or five-element subsets); but from even our simple examples, a nice pattern emerges, namely, that if $|A| = n$, then $|P(A)| = 2^n$. This is a theorem, and we formally state and prove this theorem now.

THEOREM 2.12

If A is any set and $|A| = n$, then $|P(A)| = 2^n$.

PROOF

There are actually two very good proofs for this theorem—one is based on the multiplication property of counting, and the other uses mathematical induction. We supply the proof which uses the multiplication property of counting, and we leave as an exercise the proof which uses induction.

If we consider the process of constructing a subset from A, we observe that we must make a decision for each element in A as to whether it should be placed in the subset or left out. For each of n elements, there are two choices. The multiplication property of counting states that the total number of ways that the task can be accomplished is found by multiplying the number of ways each step can be performed. The total number possible is 2^n. ■

From Example 2.33, we observe that the set A has one subset with no elements, three subsets with one element, three subsets with two elements, and one subset with three elements. The number of combinations of n items chosen r at a time (denoted by $_nC_r$ and read as "n choose r") is the number of subsets with r elements that a set which has n elements possesses. The first statement in the remark above says that $_3C_0 = 1$. The others say that $_3C_1 = 3$, $_3C_2 = 3$, and $_3C_3 = 1$. We will develop a general formula for dealing with combinations, but even before we do that we can state a theorem.

THEOREM 2.13 $\displaystyle\sum_{k=0}^{n} {}_nC_k = 2^n.$

PROOF
This is left as an exercise.   ∎

To approach the problem of counting combinations, it is best to start with permutations. The reason is that the multiplication property of counting deals with the situation in which the process being counted must occur in some order. Subsets have *no* order, and this makes it impossible to make direct use of the multiplication property in order to count them. We turn to Example 2.31.

In solving that example, we can construct a box diagram. We observe that the first value could occur in 10 ways, the second in 9 (the first item cannot also be the second), the third in 8, etc.

10	9	8	7	6	5	4	3	2	1

In this problem there are $10 \cdot 9 \cdots 1 = 3,628,800$ possible arrangements of the data. This is, of course, 10! Factorials occur with great regularity in counting permutations and combinations, and so it is worth repeating the definitions of factorial which we discussed quite briefly in another context in Chapter 1. The notation of factorials is simple and makes it much easier to write some products. For example, 125! is considerably easier to write and comprehend than a product with 125 factors.

DEFINITION

$$0! = 1$$
$$n! = n[(n-1)!]$$

ALTERNATE DEFINITION

$$n! = n \cdot n - 1 \cdots 3 \cdot 2 \cdot 1$$

The following theorem is a simple consequence of the factorial notation.

THEOREM 2.14

115

2.8 COUNTING
TECHNIQUES—SUBSETS,
COMBINATIONS, AND
PERMUTATIONS

$$\frac{N!}{M!} = N \cdot (N - 1) \cdots (M + 1) \qquad \text{provided that } M < N$$

PROOF
This is left as an exercise. ∎

Now $N!$ is a very rapidly growing quantity. We have seen that 10! is about 3.7 million, but 20! is 2.4×10^{18}, so the values of $N!$ become too large for computers to handle for quite small values of N. A machine using 4 bytes to represent real numbers cannot even take care of 35!, so a lot of care is needed in dealing with factorials to avoid problems with overflow. If we needed the value of 45! divided by 40!, for example, it would be important to use Theorem 2.14 to do the calculation as $45 \cdot 44 \cdots 41$, rather than computing 45! and dividing by 40!. We would obtain a result of 146,611,080 from that computation, but we might get only an error message from our computer if we attempted to calculate this result without using the theorem.

We return to the discussion of Example 2.32. Our discussion of this example will enable us to find formulas for both permutations and combinations. We need to find the value of $_{10}C_4$. If we were to attempt to make a box diagram to solve this problem, we would come up with this:

1st element	2d element	3d element	4th element
10	9	8	7

The solution would appear to be $10 \cdot 9 \cdot 8 \cdot 7 = 5040$. This number is *not* $_{10}C_4$. The use of a box diagram in this manner requires that the order of selection of the elements be significant to the problem, and it is *not*. Actually 5040 represents the number of *permutations* of 10 items chosen 4 at a time, rather than the number of combinations. We denote by $_{10}P_4$ the number of permutations of 10 items chosen 4 at a time, and in general we use $_nP_r$ to denote the number of permutations of n items chosen r at a time. So $_{10}P_4$ is 10!/6!. We can conclude that in general $_nP_r = n!/[(n - r)!]$, since in making a box diagram for it we would have r positions to fill and would start with n and work our way down to $n - r + 1$. Theorem 2.14 supplies the correct result.

THEOREM 2.15

$$_nP_r = \frac{n!}{(n - r)!} \qquad \text{provided that } n \text{ and } r \text{ are integers with } n \geq r$$

PROOF
It is given above. ∎

This does not answer our question about combinations. We are now close to a solution, if we rearrange our thoughts. We could construct a permutation of 10 items taken 4 at a time by *first* selecting a subset of four elements and *then* arranging them in order *after* they have been selected. Viewing the process of constructing permutations in this fashion leads us to construct the box diagram below:

Select 4 items.	Arrange items in order.

That leads us to conclude that $_{10}P_4 = {}_{10}C_4(4!)$, which means that $_{10}C_4 = {}_{10}P_4/(4!)$.

We can generalize this to a theorem about combinations in general:

THEOREM 2.16

If $n \geq r$, then $_nC_r$ is given by the expression

$$\frac{_nP_r}{r!} = \frac{n!}{(n-r)!\,(r!)}$$

PROOF
This is left as an exercise. ∎

Students often become confused when attempting to choose between permutations and combinations in solving a problem. For a permutation to be used, the problem needs to require that the order in which events occur, or the order in which the items are selected, be significant. If, on the other hand, the concept of a *set* (in which order is irrelevant) seems to apply to the problem, then the use of a combination is required.

Applications

EXAMPLE 2.34

(a) In how many different ways can seven people be seated in a row?
(b) In how many ways can seven people be seated at a circular table?
(c) In how many ways can a committee of four be chosen from a group of seven people?
(d) In how many ways can an executive committee of president, vice president, parliamentarian, and treasurer be selected from a group of seven people?

SOLUTIONS
(a) This is simply a question of forming a permutation of the seven people and using all of them. Order gets into this problem by means of left and right, which produces an order to the arrangement. The solution to (a) is then $_7P_7 = 7! = 5040$.

(b) This is also a permutation problem, once we recognize the effect of the circular table. If an arrangement of people is rotated around the table by one or more positions, the arrangement of people is really still the same. For example, the arrangements

are identical. To solve this problem, we recognize that if each position at the table were labeled with a position, there could be 7! arrangements, as in part (a). This actually counts every distinct arrangement 7 times, since position 1 could be *any* of the seven positions at the table. The solution is then $7!/7 = 6! = 720$.

(c) In the case of this committee, it does not matter who is selected first, second, third, or fourth. A committee is simply a subset of the entire group. We must have $_7C_4 = 7!/[(3!)(4!)] = 35$.

(d) This appears to be another committee problem, but in this case the order of selection *does* matter: There is a different committee formed from the same set of people if a different person is chosen to be president. The solution must be $_7P_4$, or 840. ■

The idea of *overcounting* which occurred in Example 2.34b is a recurring theme in the counting process. Use of this idea will enable us to solve many problems in which the simple application of formulas does not provide a solution.

EXAMPLE 2.35

In how many ways can a set with nine elements be partitioned into three subsets, each with three elements?

SOLUTION
We can start by making a box diagram:

Choose set 1.	Choose set 2.	Choose set 3.

Set 1 can be chosen in $_9C_3$ ways, set 2 can be chosen in $_6C_3$ ways, and set 3 can be chosen in $_3C_3$ (or 1) way. Using the box diagram directly gives the product of the three numbers for a result, but it does not matter whether the subsets in the partition are called 1, 2, and 3 or 3, 2, and 1, or any other labeling. Thus our current solution overcounts by a factor of 3! The final solution is found by dividing by 3!, so the result is $(_9C_3 \cdot _6C_3 \cdot 1)/(3!) = 280$. ■

EXAMPLE 2.36

An instructor makes a test by writing 10 questions and then selects at least 3 at random to appear on the test. If the 10 questions have been written, how many different tests are possible?

SOLUTION

This is an example of using combinations and indirect counting. If subsets of any size were allowed for the test, there would be 2^{10} possible tests; but by saying that at least 3 would be chosen, we have ruled out tests with 0, 1, and 2 questions. The number possible is $2^{10} - {}_{10}C_0 - {}_{10}C_1 - {}_{10}C_2 = 1024 - 1 - 10 - 45 = 968$. ■

Problem Set 8

Compute the following:

1. $P(A)$ if $A = \{a, b, c\}$
2. $P(A)$ if $A = \{a, b, c, d\}$
3. $|P(A)|$ if $|A| = 12$
4. $|P(A)|$ if $|A| = 11$
5. (a) $_{10}P_2$ (b) $_{10}P_8$ (c) $_7P_3$ (d) $_7P_4$
6. (a) $_6P_5$ (b) $_6P_1$ (c) $_8P_3$ (d) $_8P_5$
7. (a) $_{10}C_2$ (b) $_{10}C_8$ (c) $_7C_3$ (d) $_7C_4$
8. (a) $_6C_5$ (b) $_6C_1$ (c) $_8C_3$ (d) $_8C_5$

*9. In Problems 7 and 8, the results found in (a) and (b) were equal, as were the results found in (c) and (d), whereas in Problems 5 and 6 this was not the case. Explain why this happened as it did. Your explanation should be in terms of the definitions involved, and you should not need to make any reference at all to the formulas for computing the quantities in question.

10. Prove that $_nC_r = {}_nC_{n-r}$ *without* making reference to the formulas for counting combinations.

*11. Prove that $_nC_r + {}_{n-1}C_{r-1} = {}_{n-1}C_r$.

12. A program is needed to construct all 16-bit binary numbers with exactly two 1s. How many such numbers can be constructed?

13. X Airline flies to 50 cities and needs to test all pairs of cities to determine whether there are connecting flights joining them. How many pairs of cities need to be tested?

14. How many committees with 4 ordinary members and a specified chairperson can be selected from a pool of 11 applicants?

15. How many committees with 3 ordinary members and a specified chairperson can be selected from a pool of 8 applicants?

16. NBA basketball teams have 11-man rosters. If there are five players in the game from one team at any given time, and if the positions are not relevant (a poor assumption, perhaps), how many different lineups could be used by a team in the course of a game?

17. Major league baseball teams have a roster of 25 players; typically 10 of these players are pitchers, and 3 are catchers. If a team must have a pitcher, a catcher, and 7 other players in the game at one time (regard the seven other positions as interchangeable), how many lineups are possible? Suppose that a team is allowed to use a designated hitter who may be any of the players who is not otherwise in the game. Then how many lineups are possible?

18. Eight people are to be seated in a row at the head table of a banquet. How many arrangements are possible? Suppose that A and B are feuding and refuse to sit next to each other. Now how many arrangements are possible?

19. Suppose the table in Problem 18 is circular and answer the questions posed.

20. With a standard deck of cards (52 cards in four suits), how many different 13-card bridge hands can be dealt?

*21. Given a standard deck of cards, how many of the following possible five-card poker hands are possible?
(a) A flush (all the same suit)
(b) Four of a kind
(c) Three of a kind
(d) Full house (three of one kind, two of another)
(e) Two pairs

*22. Suppose that a joker is added to a standard deck of cards. The joker cannot be used in a flush, but it can be used as any card to complete a pair, three of a kind, or four of a kind. A natural four of a kind plus a joker becomes a five-of-a-kind hand. How many of the following five-card hands can be dealt with this 53-card deck?
(a) A flush (all the same suit)
(b) Five of a kind
(c) Four of a kind
(d) Three of a kind
(e) Full house (three of one kind, two of another)
(f) Two pairs

*23. Repeat Problem 22 if the joker *replaces* the 2 of clubs, so there is a 52-card deck again. (Be careful—this problem is trickier than it looks.)

24. Find the number of ways a set with 12 elements can be partitioned into subsets with the following specifications:
    (a) Five elements in one set and seven in the other
    (b) Six elements in one set and six in the other
    (c) Three subsets with four in each subset
    (d) Four subsets: two with four in each and two with two in each

25. In major league baseball before expansion, the National League had eight teams, and each team played each of the other teams in the league 22 times.
    (a) How many games did each team play?
    (b) How many games were played altogether by all the teams in one season?

26. Currently in the National League there are 12 teams divided into two divisions of 6 teams each. Each team plays 18 games against teams in its own division and 12 games against teams in the other division.
    (a) How many games does each team play?
    (b) How many games are played altogether by all the teams in one season?

*27. Prove Theorem 2.12 by means of induction.

*28. Prove Theorem 2.13. (*Hint*: What does $_nC_r$ mean?) This theorem can be proved by using the *binomial theorem*; but that is actually a case of overkill, since it can be proved quite easily by a simple consideration of the meaning of the terms in the sum and the fact that $|P(A)| = 2^{|A|}$.

*29. Prove Theorem 2.14.

*30. Prove Theorem 2.15.

P31. Write a program to list all the permutations of the set $\{a, b, c, d, e\}$ taken 5 at a time.

P32. Write a program to list all the combinations of the set $\{a, b, c, d, e\}$ taken 3 at a time.

*P33. Write a program to generalize the processes of Problems 31 and 32 as much as possible. (This is a very open-ended type of problem—you define and execute the problem as best you see fit.)

P34. Write a program to compute $_nP_r$ and $_nC_r$ for as many values of n and r as possible.

CHAPTER SUMMARY

In this chapter the basic terminology and operations of set theory were introduced. Venn diagrams were used to illustrate both the properties of the operations and the relationships which can exist between sets.

The product of sets was used to introduce the idea of a relation as a set of ordered pairs. The properties of relations were discussed with special focus on the concept of an equivalence relation and the way in which an equivalence relation induces a partition of a set.

The last three sections of the chapter discussed the foundations of the process of counting elements of sets and counting the number of ways in which various processes can take place. The methods introduced in these sections are particularly valuable in the process of analyzing the efficiency of an algorithm.

KEY TERMS

The following key terms were discussed in this chapter:

antisymmetric (2.5)
asymmetric (2.5)
box diagram (2.7)
cardinal number (2.2)
combination (2.8)
complement (2.3)
De Morgan's laws (2.3)
difference of sets (2.3)
disjoint sets (2.2)
domain (2.5)
element (2.2)
empty set (2.2)
equal sets (2.2)
equivalence class (2.5)
equivalence relation (2.5)
factorial (2.8)
function (2.5)
intersection (2.3)

lemma (2.7)
ordered pair (2.5)
partial order (2.5)
partition (2.5)
power set (2.8)
proper subset (2.2)
range (2.5)
reflexive (2.5)
relation (2.5)
set (2.2)
subset (2.2)
symmetric (2.5)
transitive (2.5)
tree diagram (2.7)
union (2.3)
universal set (2.2)
Venn diagram (2.4)

3
Boolean Algebras

3.1 LOGIC AND ELECTRONIC CIRCUITS

Electronic "Gates"

In addition to describing the logic of a computer program, propositional logic is very important in the design of the electronic circuitry which is associated with computer hardware.

Electronic switches or gates are designed to operate in much the same way as logical connectives. For example, an OR gate works as the logical connective $\vee$, by receiving input from two (or more) pins in which there are two possible input states (normally high or low voltage) and producing an output current according to the rule that the output will be "true" provided that at least one of the inputs is "true." The convention is that an input having low voltage (within a specified range) is treated as "false," and an input with high voltage (again, within a specified range) is treated as "true." The true state is usually denoted by a 1, and the false state is usually denoted by a 0.

In the same way, it is possible to construct gates which behave like the $\wedge$ and $\sim$ operators of logic. (See Fig. 3.1.)

The processing of 0s and 1s in this fashion is used to provide the computational functions that are built into computing equipment. (The notations of

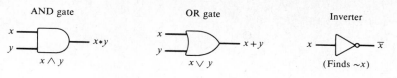

Figure 3.1 Symbols for logic gates.

electrical engineering and of mathematics with regard to logic are quite different. Engineers use multiplicative notation to denote AND, a plus to denote OR, and a bar above an entity to indicate its negation.)

Logic gates are constructed with various numbers of inputs, so that, for example, an AND gate with four inputs is available. The output of a four-input AND gate will be 1 (that is, true) in the case in which all four inputs are 1, and the output is 0 in all other cases. The net effect in terms of mathematical logic is a gate which accepts the inputs for p, q, r, and s and produces as output $p \wedge q \wedge r \wedge s$. In a similar fashion, OR gates with various numbers of inputs are available as well. Our ideas of mathematical logic can be used in the design of digital circuits.

EXAMPLE 3.1

In order to operate a particular printer for a microcomputer, it is necessary that the Data In switch be activated and the Paper Out switch not be activated. The system also operates with a "panic switch," which overrides these considerations. (The panic switch is used to produce a form feed or line feed on the printer manually.) We can set this up as a problem in logic by denoting the following propositions:

p: The paper is out.

s: The data are present.

x: The panic switch is turned on.

When the proposition $(\sim p \wedge s) \vee x$ is true, the printer will be activated. This means that a switch setup which first takes the inputs of p and s and then combines that result with x will do the job. The truth of p must be reversed before we do the combination of p with the other inputs. Our circuit looks like Fig. 3.2. Engineers denote this circuit by $\bar{p}s + x$.

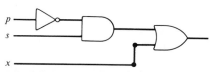

Figure 3.2 Circuit for control of printer.

Simplifying Circuit Designs

The ideas of logic can be used to find different ways of producing the same results with different circuits. The distributive property for logical expressions tells us that $p \vee (q \wedge r)$ is equivalent to $(p \vee q) \wedge (p \vee r)$. This means that the original propositional expression for our solution to Example 3.1 [which is $(\sim p \wedge s) \vee x$] is equivalent to $(\sim p \vee x) \wedge (s \vee x)$, and so the circuit $(\bar{p} + x)(s + x)$ should function in exactly the same way. That circuit looks like the one in Fig. 3.3.

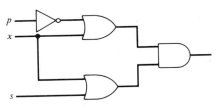

Figure 3.3 $(\bar{p} + x)(s + x)$.

In this particular instance, clearly the first design is preferable to the second, because it is simpler in some sense. In Section 3.6, we discuss the concept of simplification or minimization of switching circuits. The goal is to find what might be called a "simplest" circuit which produces the desired results.

EXAMPLE 3.2

Suppose that we wish to add two one-digit binary inputs to produce a two-digit binary output. The binary system of numeration is the system which is primarily used in computer architecture, because it allows all numbers to be represented by the symbols 0 and 1. Here we can show how gates can be used to produce a numeric result from logical inputs. The addition table for one-digit binary numbers is quite small:

+	0	1
0	0	1
1	1	10

The entry in the lower right corner of the table reflects the fact that $1 + 1 = 2$, which is 10 in binary notation.

This table shows that we will need two wires for the output—one for the 2s digit and one for the 1s. We denote the wires d_0 for the 1s digit and d_1 for the 2s digit. Getting d_1 is pretty easy, because it is the AND operation applied to the two inputs. Getting d_0 is more complicated. If one or fewer of the two digits is a 1, then the result is the usual OR operation. The problem occurs

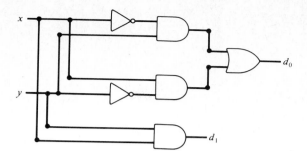

Figure 3.4 A circuit to add two 1-digit numbers.

when both inputs are 1. In that case, even though the OR operation gives us a 1, what we need is a 0 result. This is the condition described by the "exclusive OR" connective, and one might be justified in presuming that it is a standard type of gate. In any case, we can construct one from the gates which we have already discussed, since the exclusive OR will be true when the first input is true and the second false, or the other way around. See Fig. 3.4. This can be constructed as $(x \wedge \sim y) \vee (\sim x \wedge y)$. In the terms of circuit design, $d_0 = x\bar{y} + \bar{x}y$. Thus we find that we get

$$d_0 : x\bar{y} + \bar{x}y \qquad \text{and} \qquad d_1 : xy \qquad \blacksquare$$

By using similar principles, it is possible to construct circuits which add two-digit binary numbers to produce a three-digit binary number.

Using De Morgan's Laws and the Distributive Laws

In the process of simplifying circuits, *De Morgan's laws*

1. $\sim(p \wedge q)$ is equivalent to $\sim p \vee \sim q$.
2. $\sim(p \vee q)$ is equivalent to $\sim p \wedge \sim q$.

and the *distributive laws*

3. $p \wedge (q \vee r)$ is equivalent to $(p \wedge q) \vee (p \wedge r)$.
4. $p \vee (q \wedge r)$ is equivalent to $(p \vee q) \wedge (p \vee r)$.

are extremely useful in simplifying expressions so as to produce circuits which are more efficiently designed.

By making use of the properties listed above, we can show that any propositional form can be written in an equivalent "disjunctive normal form," which means that the form is a disjunction of conjunctions in which each basic proposition or its negation appears.

EXAMPLE 3.3

Write the proposition $p \wedge (q \vee \sim r)$ in disjunctive normal form.

SOLUTION

We solve this problem by writing a sequence of propositions, each equivalent to the one preceding it.

$$p \wedge (q \vee \sim r)$$

$$(p \wedge q) \vee (p \wedge \sim r) \qquad (**)$$

$$[(p \wedge q) \wedge (r \vee \sim r)] \vee [(p \wedge \sim r) \wedge (q \vee \sim q)]$$

$$[(p \wedge q \wedge r) \vee (p \wedge q \wedge \sim r)] \vee [(p \wedge q \wedge \sim r) \vee (p \wedge \sim q \wedge \sim r)]$$

$$(p \wedge q \wedge r) \vee (p \wedge q \wedge \sim r) \vee (p \wedge \sim q \wedge \sim r) \qquad \blacksquare$$

The last proposition in Example 3.3 is in disjunctive normal form because each of the expressions joined by $\vee$ is a conjunction in which each of the basic propositions p, q, r or its negation occurs. In the discussion below, the statement labeled with (**) is extremely important.

The process of putting a proposition in disjunctive normal form is not quite as formidable as a first reading of Example 3.3 might make it seem. We proceed in two stages. The first is to use De Morgan's laws and the distributive properties to convert the proposition to a disjunction of conjunctions (that is, the statement **). If that proposition is already in disjunctive normal form, we can stop. If it isn't, then some of the conjunctions do not include all the basic propositions. For each "missing" variable in a "term," we AND the conjunction with the disjunction of that variable and its negation. (For example, in the first conjunction in ** we took the conjunction with $r \vee \sim r$.) Since the disjunction of a proposition with its negation is always true, this act does not change the truth value of the original proposition. We apply the distributive property of conjunction over disjunction and consolidate like terms. The end result will be in disjunctive normal form.

The disjunctive normal form is convenient for analyzing circuits and comparing two to see whether they are indeed equivalent. The minimization procedures discussed earlier are used on expressions in disjunctive normal form.

The NAND Gate

One final point should be made: In the construction of circuits (or equivalently logical expressions) a single kind of gate (or connective) known as the NAND provides the capability of constructing *any* other gate (or connective).

In particular the AND, OR, and NOT can all be constructed from the NAND. The NAND operator is defined by Table 3.1. It is possible to show by constructing truth tables that

1. $\sim p$ is equivalent to the expression p NAND p.
2. $p \wedge q$ is equivalent to $(p$ NAND $q)$ NAND $(p$ NAND $q)$.
3. $p \vee q$ is equivalent to $(p$ NAND $p)$ NAND $(q$ NAND $q)$.

TABLE 3.1

p	q	P NAND q
T	T	F
T	F	T
F	T	T
F	F	T

In the problems you will have a chance to construct truth tables for 1 and 3. We provide the truth table for 2:

p	q	$(p$ NAND $q)$			NAND	$(p$ NAND $q)$		
T	T	T	F	T	T	T	F	T
T	F	T	T	F	F	T	T	F
F	T	F	T	T	F	F	T	T
F	F	F	T	F	F	F	T	F
			1		3		2	

(As we have done before, the columns are numbered in the order in which they are evaluated, and all columns except the final result are shaded.) The final result of this truth table is identical to that of $\wedge$.

From the results of statements 1 to 3 above, it is apparent that if a NAND gate can be constructed, any of the other three gates can be put together by using only NAND gates.

We have just scratched the surface of the topic of circuit design. Many of the principles of circuit design, as we have seen, are founded in the basic ideas of the propositional logic. In this chapter we continue to refine these ideas by developing a theory known as the *theory of boolean algebras* which will permit us to simplify the ideas discussed above by introducing a mathematical theory for which those techniques are a special case.

Problem Set 1

Design switching circuits for each of the following. (The notation is that of electrical engineering.)

1. $x\bar{y}z$

2. $x + \bar{y}\bar{z}$

3. $x(\overline{y\bar{z}})$

4. $(x + y)(z + w)$

*5. $\overline{xy + xz}$

*6. $[x(y + x\bar{y})] + (x\bar{y} + \bar{x} + z)$

Write a description for each circuit in electrical engineering notation:

7.

8.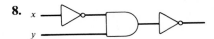

9.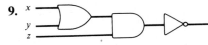

*10.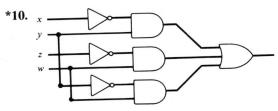

*11. Put the proposition $[p \wedge (\sim q \vee r)] \vee q$ in disjunctive normal form.

*12. Put the proposition $\sim [(p \wedge q) \vee r]$ in disjunctive normal form.

13. Construct a truth table to show that $\sim p$ is equivalent to p NAND p.

14. Construct a truth table to show that $p \vee q$ is equivalent to $(p$ NAND $p)$ NAND $(q$ NAND $q)$.

*15. Show that the operation NOR defined by p NOR $q \leftrightarrow \sim (p \vee q)$ has the same property that NAND has in that the basic three operation of $\vee$, $\wedge$, and $\sim$ can be expressed in terms of NOR.

*16. Describe a circuit which takes in the values x_1, x_0, y_1, and y_0, which represent two two-digit binary numbers, and produces three outputs d_2, d_1, and d_0, which represent the sum of the two numbers (the subscripts represent the power of 2 represented by that digit).

3.2   BOOLEAN ALGEBRAS

In Chapter 1 we introduced the algebra of statements in propositional logic. (By "algebra" we mean the study of the operations which combine particular kinds of objects. In algebra we are interested in the properties of the operations and the way the operations affect those objects.)

In Chapter 2 we introduced the algebra of sets. In the first section of this chapter we introduced an application of logic to the design of circuits in computers.

We first summarize some of the properties which we discovered about propositional logic and set theory.

Logic

1. Commutative laws:

$$p \lor q \leftrightarrow q \lor p$$

$$p \land q \leftrightarrow q \land p$$

2. Associative laws:

$$(p \lor q) \lor r \leftrightarrow p \lor (q \lor r)$$

$$(p \land q) \land r \leftrightarrow p \land (q \land r)$$

3. Distributive laws:

$$p \land (q \lor r) \leftrightarrow (p \land q) \lor (p \land r)$$

$$p \lor (q \land r) \leftrightarrow (p \lor q) \land (p \lor r)$$

4. Identity laws:

$$p \lor F \leftrightarrow p \qquad (F \text{ represents a proposition which is always false})$$

$$p \land T \leftrightarrow p \qquad (T \text{ represents a proposition which is always true})$$

5. Negation laws: For every proposition p there is a negation of p, denoted $\sim p$, such that

$$p \lor \sim p \leftrightarrow T$$

$$p \land \sim p \leftrightarrow F$$

A number of theorems can be proved about logic from these facts, including De Morgan's laws, and other properties of the negations of propositions.

Set Theory

1. Commutative laws:

$$A \cup B = B \cup A$$

$$A \cap B = B \cap A$$

2. Associative laws:

$$(A \cup B) \cup C = A \cup (B \cup C)$$

$$(A \cap B) \cap C = A \cap (B \cap C)$$

3. Distributive laws:

$$A \cap (B \cup C) = (A \cap B) \cup (A \cap C)$$

$$A \cup (B \cap C) = (A \cup B) \cap (A \cup C)$$

4. Identity laws:

$A \cup \varnothing = A$ ($\varnothing$ represents the empty set, or set with no elements)

$A \cap U = A$ (U represents the universal set, or set of all elements under consideration)

5. Complement laws: For every set A there is a complement of A, denoted A', such that

$$A \cup A' = U$$

$$A \cap A' = \varnothing$$

There are also a number of other theorems which we could prove from these facts, including De Morgan's laws, and other properties of the complements of sets.

Boolean Algebra: A Formal Definition

The facts about set theory and logic parallel each other closely, which suggests that the algebra of the two theories is really the same and only the notations are different. That is indeed the case, because propositional logic and set theory are actually instances of a more general theory known as *boolean algebra*. Boolean algebra is named after George Boole. We provide a formal definition of a boolean algebra.

DEFINITION

A **boolean algebra** is an ordered 6-tuple $(S, 0, 1, +, *, ')$, in which S is a set of elements, 0 and 1 are elements of S, $+$ and $*$ are binary operations on S ($+$ and $*$ operate on pairs of elements from S to provide a result in S), and $'$ is a unary operation ($'$ operates on single elements of S to provide an element of S as a result). The boolean algebra must satisfy the following axioms:

1. Commutative laws:

$$x + y = y + x$$

$$x * y = y * x$$

2. Associative laws:

$$(x + y) + z = x + (y + z)$$

$$(x * y) * z = x * (y * z)$$

3. Distributive laws:

$$x * (y + z) = (x * y) + (x * z)$$

$$x + (y * z) = (x + y) * (x + z)$$

4. Identity laws:

$$x + 0 = x$$

$$x * 1 = x$$

5. Complement laws:

$$x + x' = 1$$

$$x * x' = 0$$

(Of course, in each case, the free variables are to be quantified universally.)

It is traditional in the notation for boolean algebras to write $x * y$ as xy. We follow that convention in most cases, except when it may seem prudent to emphasize that we are not dealing with ordinary multiplication. (Boolean algebra notation is quite close to the notation used by electrical engineers.) To avoid excessive use of parentheses, the traditional notation for boolean algebra gives the operation of multiplication a higher priority, so that by $xy + z$ we mean $(xy) + z$, not $x(y + z)$. The other tradition that is often used is to name a boolean algebra simply as its set of elements. When we do this, we usually specify which operations we intend to use, although if that is clear, we may not explicitly state which operations we are using.

The collection of all subsets of a universal set U, with operations $\cup$, $\cap$, and $'$, forms a boolean algebra in which U acts as 1 and $\varnothing$ acts as 0. The collection of all compound propositions which can be formed from n simple propositions forms a boolean algebra with the operations of $\vee$, $\wedge$, and $\sim$. In this algebra the universally true statement T acts as 1, and the universally false statement F acts as 0.

The theory of boolean algebras provides us with a good example of the application of a general theory to a more specific one. Any theorem which we prove about boolean algebras can be translated into a theorem about propositional logic or about set theory.

As an example, we can look at the following theorem which has several parts, each of which has important realizations in the theories of sets and of propositional logic.

THEOREM 3.1

The following properties hold for any boolean algebra $(S, +, *, 0, 1, ')$:

(a) Idempotent laws:

$$x + x = x$$
$$x * x = x$$

(b) Null laws:

$$x + 1 = 1$$
$$x * 0 = 0$$

(c) Absorption laws:

$$x + (x * y) = x$$
$$x * (x + y) = x$$

(d) Involution law:

$$(x')' = x$$

(e) Complements of identities:

$$0' = 1$$
$$1' = 0$$

(f) De Morgan's laws:

$$(x + y)' = x' * y'$$
$$(x * y)' = x' + y'$$

PROOF

We leave several parts of the theorem as exercises, and we ask that you interpret this theorem as it applies to propositional logic and set theory.

(a) $x + x = (x + x)1$	(Identity law)
$= (x + x)(x + x')$	(Complement law)
$= x + xx'$	(Distributive law)
$= x + 0$	(Complement law)
$= x$	(Identity law)
(b) $x + 1 = (x + 1)1$	(Identity law)
$= (x + 1)(x + x')$	(Complement law)
$= x + 1x'$	(Distributive law)
$= x + x'$	(Identity law)
$= 1$	(Complement law)

(c) $x + xy = x1 + xy$ (Identity law)

$\qquad\quad = x(1 + y)$ (Distributive law)

$\qquad\quad = x(1)$ [Part (b) of this theorem]

$\qquad\quad = x$ (Identity law)   ∎

Duality and Boolean Algebras

There is a *duality principle* which applies to boolean algebras. In the definition of a boolean algebra, each axiom has two parts, and the only difference between the two parts is that the roles of the operations of $+$ and $*$ are interchanged and the roles of 0 and 1 are interchanged. Thus, for any theorem about boolean algebras, both a statement *and* its dual will be true. The *dual* of a statement is obtained by replacing all $+$'s with $*$'s, all $*$'s with $+$'s, all 0s with 1s, and all 1s with 0s. A proof of the dual can always be constructed by using the dual of each statement in the proof of the original statement.

Problem Set 2

1. Give the dual for the following expressions:
   (a) $[(a + b) + (a + c)]ac'$
   (b) $ab'c' + ab'c$
   (c) $[(a + b)c]'(a + b)$
   (d) $a(a' + b)$

2. Give the dual for the following expressions:
   (a) $ac + bd$
   (b) $a'c + ab' + c$
   (c) $[(a' + c')b](b + c)$
   (d) $(c + d)(a + b')(bc)$

3. Give the dual for the following equalities:
   (a) $a + ba = a$
   (b) $a + [(b' + a)b]' = 1$
   (c) $ab' + b = a + b$
   (d) $a1 + (0 + a') = 0$

4. Give the dual for the following equalities:
   (a) $(abc)' + a = 1$
   (b) $[(a' + b)(b' + c)](a' + c)' = 0$
   (c) $c'(a' + c') = c'a'$
   (d) $(a' + b)(a + b') = a'b' + ab$

In each of the following, x and y represent arbitrary elements of the boolean algebra B.

5. Prove $xx = x$.

6. Prove $x0 = 0$.

7. Prove $x(x + y) = x$.

8. State the dual of each fact in Problems 5 to 7.

*9. Prove that complements in a boolean algebra are unique. (*Hint*: For any x show that if $x + y = 1$ and $xy = 0$ and also $x + z = 1$ and $xz = 0$, then $y = z$.)

10. Prove $(x')' = x$.

11. Prove $0' = 1$ and $1' = 0$.

*12. Prove $(x + y)' = x'y'$.

*13. Prove $(xy)' = x' + y'$.

14. Verify that if A is any set, then $P(A)$ forms a boolean algebra with the operations of union, intersection, and complement.

15. Verify that the set of all compound propositions which can be formed from the basic propositions p, q, r, and s is a boolean algebra with the operations of disjunction, conjunction, and negation, where 1 is T (the universally true statement) and 0 is F (the universally false statement).

16. Explain why the collection of all circuits with a fixed number of inputs forms a boolean algebra, if the operation of $+$ means to combine two circuits with an OR gate, the operation of $*$ means to combine the circuits with an AND gate, and the operation of $'$ means to invert the output of the circuit.

17. Reinterpret Theorem 3.1 in terms of propositional logic by making the appropriate translation of terminology.

18. Reinterpret Theorem 3.1 in terms of set theory by making the appropriate translation of terminology.

19. Let $S = \{1, 2, 3, 6\}$. Define $x + y$ as the least common multiple of x and y. Define $x * y$ as the greatest common divisor of x and y, and define x' as $6/x$. Show that S with the given operations forms a boolean algebra.

20. Let $S = \{1, 2, 4, 8, 16\}$, with $+$ and $*$ defined as above. Define x' as $16/x$. Show that S with the given operation does not form a boolean algebra.

3.3 SOME BASIC THEOREMS

Boolean Algebras as Posets

Boolean algebras have many properties which not only are algebraic in nature but also are related to orderings. In fact, a boolean algebra can be given an order relation so that it becomes a special kind of poset (*partially ordered set*) known as a *lattice*.

DEFINITION

If $B = (A, +, *, 0, 1, ')$ is a boolean algebra, the **order generated by B** is the relation $< +$ given by $a < + b$ if and only if $a + b = b$.

THEOREM 3.2

The set A together with the ordering $< +$ is a poset.

PROOF

To prove that a relation is a partial ordering, we must prove that the relation is reflexive, antisymmetric, and transitive.

Reflexive By the idempotent law $a + a = a$ for all a, so we have $a < + a$ as required.

Antisymmetric We must show that if $a < + b$ and $b < + a$, then $a = b$. So, we suppose that $a < + b$ and $b < + a$. This means that $a + b = b$ and $b + a = a$. By the commutative property, since $a + b = b + a$, it follows that $a = b$.

Transitive We must show that if $a < + b$ and $b < + c$, then $a < + c$. If $a < + b$, it follows that $a + b = b$. If $b < + c$, then $b + c = c$. By substituting in the second equation, it follows that $(a + b) + c = c$ or by the associative property $a + (b + c) = c$, but since $b + c = c$, this means that $a + c = c$.

As with most facts in boolean algebra, there is a "dual" to the definition of $< +$. It states $ab = a$ iff $a < + b$. The proof of this is left as an exercise. ■

DEFINITIONS

a is a **lower bound** for b and c in a poset with order relation $\leq$ if and only if $a \leq b$ and $a \leq c$.

a is a **greatest lower bound** (glb) of b and c if a is a lower bound for b and c and if d is any lower bound of b and c and so $d \leq a$.

Upper bound and least upper bound (lub) are defined similarly.

THEOREM 3.3

In a boolean algebra with the $< +$ ordering, $a + b$ is the least upper bound of a and b, and ab is the greatest lower bound.

PROOF

We show that $a + b$ is the lub of a and b, and we leave the proof that $a * b$ is the glb as an exercise. It is clear that $a + b$ is an upper bound of a and b because $a + (a + b) = a + b$ and $b + (a + b) = a + b$. To show that $a + b$ is the least upper bound, we suppose that c is an upper bound of a and b. Now $a < + c$ and $b < + c$ means that $a + c = c$ and $b + c = c$. Combining these two yields $(a + c) + (b + c) = c + c$. Making use of the properties of a boolean algebra, we see that $(a + b) + (c + c) = c + c$; but this means that $(a + b) + c = c$, or in other words $a + b < + c$. The proof of the glb property can be handled quite easily as the dual of this if we make use of the fact that $a < + b$ if and only if $ab = a$. ∎

Lattices

EXAMPLE 3.4

Consider $P(\{a, b, c\})$ as a boolean algebra. The elements of the algebra are $\varnothing$, $\{a\}$, $\{b\}$, $\{c\}$, $\{a, b\}$, $\{a, c\}$, $\{b, c\}$, and $\{a, b, c\}$. The relation $< +$ is the familiar one of $\subset$, 0 is $\varnothing$, 1 is $\{a, b, c\}$, $+$ is $\cup$, and $*$ is $\cap$. ∎

Example 3.4 illustrates Theorem 3.3. The elements of the algebra are sets. The least upper bound of a pair of sets is their union. The greatest lower bound of a pair of sets is their intersection. This example, and the statement of Theorem 3.3, leads us to the following definition:

DEFINITION

A **lattice** is a poset in which any pair of elements has both a least upper bound and a greatest lower bound.

It is interesting to note, although we do not elaborate on it here, that it is possible to define a boolean algebra as a special kind of lattice. (The extra properties needed are the properties of the complement operation and the distributive property.) We do not dwell on this fact, and we have included it here and in the problems mainly for the sake of completeness.

Notation

A number of different notations are used for boolean algebras, usually depending on the applications. In applications relating to logic operations, the notations of $\vee$, $\wedge$, and $\sim$ are frequently used; in circuit design, the notations of $+$, $*$, and $^{-}$ are often used.

Power Sets and Atoms

In the algebra of Example 3.4, the elements $\{a\}$, $\{b\}$, and $\{c\}$ play a special role. In terms of order properties of the algebra, they are the smallest nonzero elements. Furthermore, every nonzero element of this algebra can be found uniquely as the union of some of these sets. It is a remarkable fact that every finite boolean algebra has elements like this, and every boolean algebra "acts like" the power set of the collection of elements which have this property.

DEFINITION

An **atom** of a boolean algebra is a nonzero element a which has the property that for each element x of the algebra either $ax = a$ or $ax = 0$.

THEOREM 3.4

If B is a boolean algebra and a is any element of the algebra, then a is an atom if and only if $x < + a$ implies that $x = a$ or $x = 0$.

PROOF
This is left as an exercise. ∎

Although we do not prove it here, the following theorem is an important characterization of finite boolean algebras.

THEOREM 3.5

Let B be a finite boolean algebra, and let A be the set of atoms of B. Then B is algebraically identical to $P(A)$. ∎

A corollary to this theorem is that every finite boolean algebra has 2^n elements, where n is the number of atoms.

Building Boolean Algebras from Simple Pieces

The most basic boolean algebra is the boolean algebra

$$B = (\{0, 1\}, +, *, 0, 1, \,')$$

where the operations are defined by

$$
\begin{aligned}
0 + 0 &= 0 & 0 * 0 &= 0 \\
0 + 1 &= 1 & 0 * 1 &= 0 \\
1 + 0 &= 1 & 1 * 0 &= 0 \\
1 + 1 &= 1 & 1 * 1 &= 1 \\
0' &= 1 & 1' &= 0
\end{aligned}
$$

We can form other boolean algebras from B. For any value of n, let the set S be the set of ordered n-tuples of 0s and 1s. The operations on S are defined componentwise, so that $(x_1, \ldots, x_n) + (y_1, \ldots, y_n) = (x_1 + y_1, \ldots, x_n + y_n)$, and similarly for $*$ and $'$. This set has 2^n elements, and the n-tuples which have a single 1 entry are the atoms of the algebra. There are exactly n of them. This boolean algebra acts just like $P(A)$, where A is a set with n elements; and since every finite boolean algebra acts like $P(A)$ for some choice of A (Theorem 3.5), we have actually demonstrated that every finite boolean algebra can be constructed from the boolean algebra B. This fact is most conveniently used in the way that implementers of Pascal have dealt with the representation of the data type SET.

In representing SETs in Pascal, each word which is used to store a SET is treated as an n-tuple of 0s and 1s, and the operation of union of sets is accomplished by using the OR (boolean addition) operation on the operands on a bit-by-bit basis. The operations of intersection and complement are similarly accomplished with the AND (boolean multiplication) and NOT (boolean complement) operations.

The next section focuses on boolean functions. Boolean functions provide the concept behind the applications of boolean algebra to circuit design.

Problem Set 3

1. Prove that $a < + b$ if and only if $ab = a$.

*2. Explain how to construct the dual to a statement about a boolean algebra if the statement includes reference to the partial order $< +$.

3. Propose a formal definition for upper bound in a poset.

4. Propose a formal definition for least upper bound in a poset.

5. Prove that if c and d are both least upper bounds for a and b in any poset, then $c = d$.

6. Prove that if c and d are both greatest lower bounds for a and b in any poset, then $c = d$.

7. Prove that if a and b are elements of the boolean algebra B, then ab is the greatest lower bound of a and b.

8. In a boolean algebra we can define the relation $> +$ by saying that $a > + b$ if and only if $a + b = a$. Prove that the relation $> +$ provides a partial order.

9. Prove that $a > + b$ if and only if $ab = b$. (Here $> +$ is the relation defined in Problem 8.)

10. What is the relationship between $< +$ and $> +$?

*11. Prove that if $a > + b$ and $b > + a$, then $a = b$.

12. Prove that a boolean algebra is a complemented distributive lattice when the order relation is $< +$.

13. Prove that a complemented distributive lattice is a boolean algebra if the operations are those of computing least upper bound and greatest lower bound.

14. Prove Theorem 3.4. (Remember that this is an "if and only if" statement. You must prove that if a is an atom, the condition holds; and you must also prove that if the condition holds, then a is an atom.)

15. In the boolean algebra consisting of all propositions which can be produced from the basic propositions p, q, r, and s, what are the atoms? Prove your results.

3.4 BOOLEAN FUNCTIONS

DEFINITION

A **boolean function** is a function whose domain is a set of n-tuples of 0s and 1s and whose range is the basic boolean algebra $\{0, 1\}$.

The applications of boolean algebra in the design of switching circuits have their roots in the concept of a boolean function.

Boolean Expressions

DEFINITION

A **boolean expression** on the variables $\{x_1, \ldots, x_n\}$ is a polynomial expression using those variables and the operations of a boolean algebra. The variables are assumed to have as their domain the set $\{0, 1\}$, and the operations on the variables are the boolean operations defined on this set.

Every boolean expression defines a boolean function. The boolean expression on $\{x, y\}$ given by $xy + x'y'$ defines a boolean function if we just substitute values for x and y. If $f(x, y) = xy + xy'$, then f is completely described by

x	y	$f(x, y)$
0	0	0
0	1	0
1	0	1
1	1	1

Since every boolean expression produces a boolean function, it is reasonable to attempt to determine whether every boolean function can be described by a boolean expression. A similar situation does *not* occur with ordinary real-valued functions of a real variable, since it is possible to prove that there are real-valued functions which cannot be described by algebraic expressions. However, as we will soon discover, not only can every boolean function be described by a boolean expression, but also every boolean function can be described by a very special kind of boolean expression.

We first need to introduce some terminology.

DEFINITIONS

A **boolean variable** is any variable whose domain is the set $\{0, 1\}$.

A **literal** is a boolean variable or its complement.

A **minterm** on n variables is a product of n literals in which each variable is represented either by the variable or by its complement.

A **maxterm** on n variables is the sum of n literals.

A boolean expression is a **minterm normal expression** on n variables if it is the sum of minterms on those n variables.

The term "minterm" comes from the fact that a minterm represents the greatest lower bound (*min*imum) of its factors. A maxterm is the least upper bound (*max*imum) of its terms.

Counting Boolean Functions and Expressions

We now state and prove some theorems which make use of our counting properties and Chapter 2 in order to determine the number of boolean functions and boolean expressions which can be produced for n variables.

THEOREM 3.5

There are 2^{2^n} boolean functions on n variables.

PROOF

This is a nice counting argument. There are 2^n n-tuples of 0s and 1s, since we must make one of two choices for each of n positions. For each of the possible n-tuples, we must choose the value of the function; thus we must make one of two choices for each of 2^n n-tuples, or a total of 2^{2^n} different possibilities for boolean functions. ∎

THEOREM 3.6

There are 2^n minterms on n variables.

PROOF

This is left as an exercise. (*Hint*: In a minterm each variable must be represented by either the variable or its complement. Use the methods of Chapter 2 to count the number of choices which can be made.) ∎

THEOREM 3.7

There are 2^{2^n} minterm normal expressions on n variables.

PROOF

A minterm normal expression is a sum of minterms. Each minterm on n variables can be included or not in the expression. (It does no good to include it more than once, since $x + x = x$.) Since we have 2^n minterms and we need to make two choices for each of them, there are altogether 2^{2^n} possible minterm normal expressions. ∎

Characterizing Boolean Functions

Because of the fact that the minterms in a minterm normal expression give us the n-tuples at which the expression is 1—and if two expressions have different values at different n-tuples, then they represent different functions—we can conclude that two different minterm normal expressions represent different functions. We have determined that there are the same number of functions represented by minterm normal expressions as there are boolean functions. From this information we derive Theorem 3.8.

THEOREM 3.8

Every boolean function can be expressed as a minterm normal expression, and every boolean expression is equivalent to a minterm normal expression. ∎

The process for finding the minterm normal expression for a particular function or boolean expression is quite easy. A minterm will have a value of 1 only when all the variables listed without complements are 1 and those which are listed with complements have a value of 0. To build the minterm normal expression, we need only find the n-tuples for which the function is not zero and add the minterms corresponding to those ordered n-tuples.

EXAMPLE 3.5

Find a minterm normal expression for the function described by the table below:

x	y	z	$f(x, y, z)$
0	0	0	1
0	0	1	0
0	1	0	0
0	1	1	0
1	0	0	1
1	0	1	0
1	1	0	0
1	1	1	1

The nonzero values for the function occur for the triples $(0, 0, 0)$, $(1, 0, 0)$, and $(1, 1, 1)$. These correspond to the minterms $x'y'z'$, $xy'z'$, and xyz, respectively. Thus the minterm normal expression for this function is $x'y'z' + xy'z' + xyz$. ■

An immediate corollary of Theorem 3.8 is that any boolean function can be written as a boolean expression.

We say that two boolean expressions are *equivalent* if they represent the same boolean function. Two boolean expressions are equivalent if and only if they have the same minterm normal expressions.

There are two ways of finding a boolean expression in minterm normal form that is equivalent to a given boolean expression. The first is the method illustrated above, of simply evaluating the expression for all possible ordered n-tuples and then producing the appropriate minterms for each n-tuple which produces a nonzero value of the function. The second technique is to use algebraic methods and the laws of boolean algebras to produce the equivalent minterm normal expression.

This latter method mimics the method we discussed in Section 3.1 to put a logical expression in disjunctive normal form. We first apply the properties of a boolean algebra—particularly De Morgan's laws and the distributive property of $*$ over $+$—to produce an expression which is a sum of terms. If a variable or its complement is not present in a particular term, then we multiply that term by 1 in the form of the sum of the variable and its complement. We again use the distributive and idempotent laws to produce a new expression which is the sum of products. If all "missing" variables have been taken care of, the resulting expression will be in minterm normal form.

EXAMPLE 3.6

Use both the algebraic and the computational method to produce a minterm normal expression equivalent to $(x'y)'(x + z)$.

SOLUTION

Algebraic Method   We can use algebraic methods in the following way:

$$(x'y)'(x + z) = (x'y)'x + (x'y)'z \qquad \text{(Distributive law)}$$

$$= [(x')' + y']x + [(x')' + y']z \qquad \text{(De Morgan's law)}$$

$$= (x + y')x + (x + y')z \qquad \text{(Involution law)}$$

$$= xx + y'x + xz + y'z \qquad \text{(Distributive law)}$$

$$= x + y'x + xz + y'z \qquad \text{(Idempotent law)}$$

$$= x + xy' + xz + y'z \qquad \text{(Commutative law)}$$

**
$$= x + y'z \qquad \text{(Absorption law)}$$

$$= x(y + y')(z + z') + (x + x')y'z \qquad \text{(Complement and identity laws)}$$

$$= (xy + xy')(z + z') + xy'z + x'y'z \qquad \text{(Distributive law)}$$

$$= xyz + xyz' + xy'z + xy'z' + xy'z + x'y'z \qquad \text{(Distributive law)}$$

$$= xyz + xyz' + xy'z + xy'z' + x'y'z \qquad \text{(Idempotent law)}$$

(In a manner identical to that in Section 3.1, we reached **, which is a sum of terms, then introduced the "missing" factors, and simplified again.)

Computational Method   Using the computational method, we evaluate the expression at each of the eight points in the domain. A process identical to the one used in the evaluation of a truth table can be used to do this:

x	y	z	(x	'	*	y)	'	*	(x	+	z)
0	0	0	0	1	0	0	1	0	0	0	0
0	0	1	0	1	0	0	1	1	0	1	1
0	1	0	0	1	1	1	0	0	0	0	0
0	1	1	0	1	1	1	0	0	0	1	1
1	0	0	1	0	0	0	1	1	1	1	0
1	0	1	1	0	0	0	1	1	1	1	1
1	1	0	1	0	0	1	1	1	1	1	0
1	1	1	1	0	0	1	1	1	1	1	1

$$\qquad\qquad 1\ 2 \qquad 3\ 5 \qquad 4$$

The order of evaluation of the columns is indicated by the numbers beneath the columns, and the last evaluated column (5) contains the actual values of the function. Looking at column 5, we see that the expression is 1 for the following ordered triples: (0, 0, 1), (1, 0, 0), (1, 0, 1), (1, 1, 0), and (1, 1, 1). Those triples correspond to the minterms $x'y'z$, $xy'z'$, $xy'z$, xyz', and xyz. The minterm normal expression is $x'y'z + xy'z' + xy'z + xyz' + xyz$. This is (except for the order of the terms) the same minterm normal expression found by the algebraic method.

The fact that any boolean function can be written as a boolean expression means that every boolean function can be written in terms of the operations of $*$, $+$, and $'$. Any set of boolean operations which is sufficient to describe all boolean functions is said to be *functionally complete*. It is clear that a set of boolean operations is functionally complete if and only if the operations of $*$, $+$, and $'$ can be expressed in terms of the operations in the set.

In the next section we look at some applications of the theorems related to boolean functions.

Problem Set 4

1. Prove Theorem 3.6.

2. Prove that two boolean expressions are equivalent if and only if they can be shown to have the same minterm normal expansions.

Find the value of each of the following boolean functions for each possible n-tuple.

3. $f(x, y, z) = [(x + y)z]'(x + y)$

4. $g(x, y, z) = x'z + x'y + z'$

5. $h(x, y, z) = (x + y)(x' + y)1$

6. $j(x, y, z) = (xy + z)(y + z) + z$

7. $k(x, y, z) = xz + y' + y'z + xy'z$

8. $[(xy + xy'z) + xz')]z$

9–14. Find minterm normal expressions for the functions in Problems 3 to 8.

15–20. Use algebraic manipulations to find the minterm normal expressions for the functions in Problems 3 to 8.

21. Prove that the set consisting of the single operation NAND is functionally complete by producing expressions for x', $x * y$, and $x + y$ using only the operation NAND. Here x NAND y is defined to be $(xy)'$.

22. Prove that the set consisting of the single operation NOR is functionally complete by producing expressions for x', $x * y$, and $x + y$ using only the operation NOR. Here x NOR y is defined to be $(x + y)'$.

*23. Prove that the set of all boolean functions on n variables with the natural operations is a boolean algebra.

*24. Find the atoms of the boolean algebra described in Problem 23.

The previous two sections included quite a few results which were presented as theoretical results. The fact is that every one of them has immediate applications in the boolean algebras which we have studied. This section provides us with a good opportunity to see how theoretical mathematics can be used to apply results to specific instances of a particular theory.

To make these applications, we can make the following observations:

1. Propositional logic can be viewed as a kind of infinite boolean algebra whose atoms are simple statements and whose operations are disjunction, conjunction, and negation. The universally true statement T acts as 1, and the universally false statement F acts as 0. The problem of determining the truth of a compound statement can be regarded as that of finding the value of a boolean function whose value depends on the truth or falsity of the underlying basic statements. (We interpret 1 as true and 0 as false.)

2. The set of all subsets of a given set A is a boolean algebra. And we can transform it into propositional logic, by noting that for each x in the set A and for each subset S of a, the statement $x \in S$ is a proposition. And $S = T$ if and only if, for all elements of A, the propositions $x \in S$ and $x \in T$ are equivalent.

3. Every switching circuit with n inputs is really a boolean function with n variables.

With these facts in mind, we can look at the theorems of the previous sections in terms of particular applications of boolean algebra.

THEOREM

Every boolean function can be written as a boolean expression.

1. *Logic*: The operations of $\vee$, $\wedge$, and $\sim$ are sufficient to describe all compound statements.

2. *Circuits*: Every logic circuit can be constructed by the use of AND gates, OR gates, and inverters. ∎

THEOREM

Every boolean function can be written as an equivalent minterm normal expression.

1. *Logic*: Every proposition can be written in the form of a disjunction of conjunctions of basic propositions or their negations (disjunctive normal form).

2. *Sets*: Every set can be written as the union of intersections of sets or their complements. ∎

THEOREM

Two boolean functions are equivalent if and only if they have the same minterm normal expressions.

1. *Logic*: Two propositions are equivalent (have the same truth tables) if and only if they have the same disjunctive normal form.

2. *Circuits*: Two circuits provide the same outputs if and only if they can be expressed in the same disjunctive normal form. ∎

THEOREM

The NAND operation is functionally complete.

1. Any compound proposition can be expressed in terms of the operation NAND.

2. The set operation # defined by $A \; \# \; B = \{x \mid \sim(x \in A) \vee \sim(x \in B)\}$ is sufficient to describe all boolean set operations.

3. *Circuits*: All circuits can be constructed from NAND gates. ∎

Almost all the theorems about propositional logic and set theory from Chapters 1 and 2 can be seen to be specific instances of theorems about boolean algebras.

Problem Set 5

Write the disjunctive normal form for the propositions whose truth tables are listed below.

1.

p	q	r	$P(p, q, r)$
T	T	T	T
T	T	F	F
T	F	T	T
T	F	F	F
F	T	T	T
F	T	F	F
F	F	T	F
F	F	F	F

2.

p	q	r	$Q(p, q, r)$
T	T	T	F
T	T	F	F
T	F	T	F
T	F	F	F
F	T	T	T
F	T	F	T
F	F	T	F
F	F	F	F

3.

p	q	r	$R(p, q, r)$
T	T	T	T
T	T	F	T
T	F	T	F
T	F	F	F
F	T	T	T
F	T	F	T
F	F	T	F
F	F	F	F

4.

q	r	$S(q, r)$
T	T	T
T	F	F
F	T	F
F	F	T

Write the sets indicated in the Venn diagrams below as the union of intersections of the sets A, B, and C.

5.

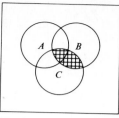

6.

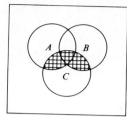

7.

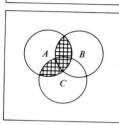

8.

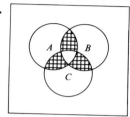

Show that the following circuits are equivalent:

9.

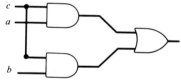

10.

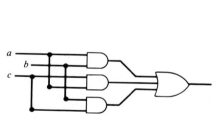

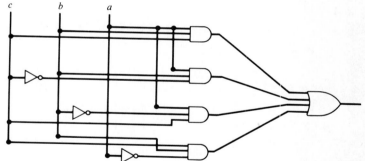

Express each of the following propositions by use of the NAND operator only. (Some authors use | to indicate the NAND operator.)

11. $p \lor q \land \sim r$

12. $(\sim p) \land (q \lor r)$

Express each of the following sets by using only the operation # :

13. $(P \cap Q) \cap R'$

14. $P' \cup (Q \cap R)$

Use only NAND gates to construct the following circuits. NAND gates are denoted as follows:

15. $(x + y) + z$

16. $x(y + z)$

Express the following propositions by use of the NOR operator only:

17. $p \vee q \vee \sim r$

18. $(\sim p) \wedge (q \vee r)$

Use only NOR gates to construct the following circuits. NOR gates are denoted as follows:

19. $(x + y) + z$

20. $x(y + z)$

3.6 MINIMIZATION

Simplifying Boolean Expressions

In Section 3.1 we introduced the problem of using switching circuits to perform logical functions. In Example 3.1 we found a circuit to control the action of a printer (see Fig. 3.5, left), and in Example 3.2 we found a circuit to perform the operation of addition of one-digit binary numbers (Fig. 3.5, right).

   In Section 3.4, we found that the minterm normal expression for $(x'y)'(x + z)$ was $xyz + xyz' + xy'z + xy'z' + x'y'z$. We display a circuit for that expression in Fig. 3.6.

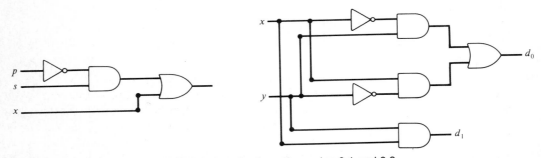

Figure 3.5 Circuits from Examples 3.1 and 3.2.

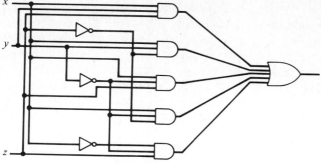

Figure 3.6 Circuit for $(x'y)'(x + z)$.

In each case we would like to know whether we have produced the most efficient way of designing the circuit. In the design of a switching circuit, regardless of whether the circuit is needed for a single custom-made job, or if it is being used as a part of an integrated circuit which will be mass-produced, an important issue is the cost of producing the circuit. It is considered desirable to design circuits which are in the form of a sum of products. But even within that general guideline, we would like to produce circuits as cheaply as possible. It is generally accepted that the most economical designs take as few sums as possible, and the terms of the sums are products of as few literals as possible. Creating a circuit which satisfies these goals is referred to as *minimization*. To minimize a circuit, we need to find a minimized boolean expression for it according to the same rules of minimization.

In the three problems described above, it is fairly obvious that the first two are minimal solutions and the third is not. The boolean expression $x + y'z$ is equivalent to the expression in the third example, but it has only two terms consisting of one and two literals. You should verify that $x + y'z$ is equivalent to the original expression.

In many cases it is not obvious how to minimize a boolean expression. A couple of different approaches can be used. One approach involves the use of a refinement of the Venn diagram known as a *Karnaugh map*, and the other approach uses an algorithm to do the job. The best known algorithm is the *Quine-McClusky algorithm*, and it is typically used to minimize expressions with a large number of variables. The typical approach is to do the job by hand by using Karnaugh maps if the number of variables is small (usually five or less) and to apply the Quine-McClusky algorithm as implemented in a software package if the number of variables is more than five. We do not discuss the Quine-McClusky algorithm; instead we concentrate on Karnaugh maps. There are, however, a number of similarities between the approach that we use with the maps and the operation of the Quine-McClusky algorithm.

Karnaugh Maps

A Karnaugh map represents all the possible minterms available for a boolean expression with the given number of variables. In many respects, Karnaugh maps represent an effort to generalize the idea of a Venn diagram, and in particular it is possible to use a Karnaugh map to represent boolean expressions with up to six variables, as compared to the three or four sets which can be represented on a Venn diagram. Karnaugh maps are rectangles, and each possible minterm is represented by a box, or cell. If there are n variables, then the Karnaugh map will have 2^n cells. In using the Karnaugh map, checks are placed in a cell if the minterm to which the cell corresponds is included in a minterm normal expression for the boolean expression (or function) in question. A Karnaugh map for two variables is quite simple. See Fig. 3.7. The upper half of the map represents minterms which include the factor x, while the lower half represents those with x'.

Figure 3.7 Karnaugh map for two variables.

In a Karnaugh map with two variables, the second variable y is handled in the same way with the left and right columns of cells. A check is placed in the upper left if the minterm xy is included in the minterm normal expression of the function under study. If both boxes on the top row are checked, we have the sum $xy + xy'$, which can be replaced by the single term x. Algebraically $xy + xy'$ is $x(y + y') = x1 = x$. It will always be the case that if *all* minterms corresponding to a term which involves fewer literals (as occurred above with $xy + xy'$) are checked, then that sum of minterms may be replaced with the simpler term. This situation occurs if the expression includes a sum of minterms represented by rectangular blocks of contiguous cells.

To represent the expression $(xy + x'y)'$ with a Karnaugh map, we determine a minterm normal expression for the expression. In this case it is $x'y + x'y'$. We draw the map and check the appropriate boxes. See Fig. 3.8.

Figure 3.8 Karnaugh map for $(xy + xy')'$.

To minimize this expression, we note that the entire bottom row (and nothing else) is checked. The bottom row corresponds to the boolean expression x' (to see this, note that the bottom row of a Karnaugh map with one variable is x'), so x' is the minimized expression. In Karnaugh maps with two variables, contiguous blocks of two cells can be replaced with a single literal.

Three Variables

With three variables, a Karnaugh map looks a bit more complicated. See Fig. 3.9. In this case, x is again represented by the top row and y by the left half, but now z is represented by the first and last columns (z' is the *middle* two columns). This map has some very important properties which hold for all Karnaugh maps. Moving from one row or column to an adjacent one results in a change of one of the literals from the variable to its complement or from the complement to the variable, in each of the cells; but that change occurs only in one variable, and the other factors remain the same for the cells in the same relative positions. The first column and last column are regarded as adjacent to each other.

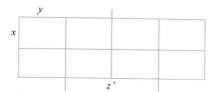

Figure 3.9 Karnaugh map for three variables.

Any block of four boxes represents a single literal. The top row of four boxes contains all minterms involving x, and if they are summed, the result is x. In a similar fashion, the first and last columns (remember they are regarded as adjacent) represent the literal z. The middle two columns represent z'. Any block of two boxes represents the product of two literals. For example, the first column represents xz. The first two entries in the bottom row represent the product $x'y$. Let us make a Karnaugh map for the expression $xyz + xy'z + xyz' + xy'z' + x'y'z$:

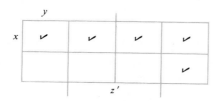

All of row 1 is checked. This portion of the map can be expressed as the single literal x. All that remains is the box in the lower right. We could write our expression as $x + x'y'z$, but it is better to note that the entire last column is checked. This last column represents two cells, so it is the product of two literals ($y'z$), and we can write the expression as $x + y'z$. (We put z on the edges of the map so as to make the upper left corner xyz. Some authors make z the *middle* two columns. That choice is only a matter of taste and does not affect how the process actually works.)

Four Variables

A similar strategy can be followed with four variables, except that we now must have 16 boxes, and we make the top and bottom rows adjacent and allow them to represent a variable. A typical arrangement for four variables is shown in Fig. 3.10. Now, eight adjacent boxes represent a single literal; four, the product of two literals; and two, the product of three literals. The idea is to attempt to find a "covering" of the marked cells with as few blocks as possible. (This represents the minimization of the number of terms.) We also want to make the blocks used to cover the checked cells as large as possible. (This represents the minimization of the number of factors in each term.)

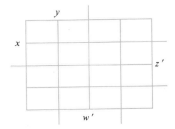

Figure 3.10 Karnaugh map for four variables.

We take, for example, the expression whose minterm normal expression is $wxyz + wxyz' + wx'yz + wx'yz' + w'x'y'z' + w'xy'z' + w'x'yz' + w'xyz' + wx'y'z$ (Fig. 3.11, left). We proceed to first mark all blocks of size 8, and finding none, we mark blocks of size 4. There are two, the first column and the center block of 4 (Fig. 3.11, right). Next we mark blocks of 2 that are *not* completely contained in a block of larger size, and we note that the box in the lower right is adjacent to the box in the lower left. Finally we mark any single boxes that have not been marked, and in this case we find none.

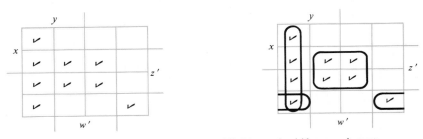

Figure 3.11 *Left*, Karnaugh map. *Right*, marked Karnaugh map.

Having marked the map in this fashion, we now try to select enough of the blocks we marked to cover all the checked cells. In this example we need all the blocks. In other cases it may be possible to select a proper subset of marked blocks, in which case we would try to choose as few blocks as possible, and use blocks of as large a size as possible. In this case our selected blocks correspond to yw, $z'w'$, and $x'zw$ which means that our expression is given by $yw + z'w' + x'zw$.

We can prescribe a recipe for the process (not quite an algorithm, because we don't resolve how to make some of the choices that need to be made) in the following way:

RECIPE MINIMIZE

Mark the map, placing a check in each cell that represents a minterm in
minterm normal expression;
$n := 2^k$ {k = number of distinct variables in expression}
while n > 1 **do**
n := n/2;
Mark all rectangular blocks of size *n* that are not totally contained in a block
already marked **od** ;
Select enough of the marked blocks to cover all checked cells, such that
(1) The number of blocks is as small as possible
(2) Among those choices with the same number of blocks, the size of the
blocks is as large as possible.
end.

This is not quite an algorithm, because we make no effort to describe how to accomplish 1 and 2. Sometimes it is obvious how to achieve 1 and 2, and sometimes it is not. Except for that issue, this "recipe" is quite similar to the Quine-McClusky algorithm. A little experience often helps in making the appropriate choices.

EXAMPLE 3.7

Construct a minimal switching circuit for the boolean expression
$xyzw + xyz'w + xyzw' + xyz'w' + x'yz'w' + x'yzw' + xy'z'w' + x'y'z'w'$.
First, we mark the minterms on the map:

We look for blocks of size 8. There are none. We look for blocks of size 4. We find three:

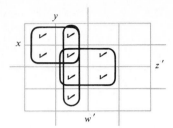

All the checked cells are marked at this point, so we do not have to look for blocks of size 2 or single blocks.

In this case, all the marked blocks are needed, and the final expression is $xy + w'y + w'z'$. We can then construct the circuit as $xy + \bar{w}y + \bar{w}\bar{z}$ (Fig. 3.12).

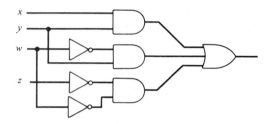

Figure 3.12 The completed circuit.   ■

If we need to minimize an expression not in minterm normal form, then first we reduce it to minterm normal form and then we proceed as before.

More Than Four Variables

Karnaugh maps can be constructed with five variables by placing two maps for four variables side by side. The first map will be the map in which we take the first variable and all combinations of the last four variables, and the second map takes the complement of the first variable and all combinations of the last four. In this situation, each cell is considered to be adjacent to the cell in the same position on the other map. For six variables, the same idea is repeated once more, now producing four maps each with 16 cells and the same adjacency rules. For many people, visualizing patterns with five- and six-variable Karnaugh maps is very difficult. The main thing needed to deal with these kinds of maps seems to be a good deal of experience and "feel" for such maps. We restrict the problems and examples to four variables.

EXAMPLE 3.8

155

3.6 MINIMIZATION

Construct a minimal circuit for the boolean expression
$xyz + x'yz + x'yz' + xy'z' + x'y'z' + xy'z$.

First, we construct and mark the Karnaugh map:

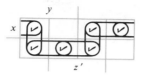

When we mark the blocks in this example, we see that there are actually two
ways of selecting blocks to cover the checked cells, one choice corresponding
to the expression $x'y + xz + y'z'$ (see Fig. 3.13) and the other corresponding to
the expression $x'z' + yz + xy'$. In this case, either choice is acceptable because
both expressions are the sums of three terms, each with two literals.

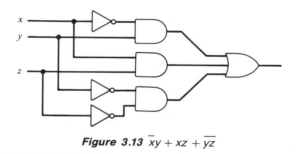

Figure 3.13 $\overline{x}y + xz + \overline{yz}$

EXAMPLE 3.9

Find a boolean expression to minimize $wx'z' + wy' + w'y + w'z + w'x$.

The map in Fig. 3.14 is much more complicated than those of previous
examples, because in this case each cell is in two or more marked blocks. The
trick is to select as few blocks as possible. After a bit of staring, perhaps we
can see that four blocks are enough if we take zw', xw', $y'w$, and $xz'w'$. After a
bit more head scratching, we can convince ourselves that there is no way to

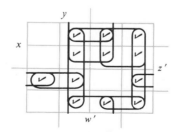

Figure 3.14 Marked Karnaugh map.

cover all the checked cells with just three blocks, because the blocks we have to work with are not enough to cover the 12 marked blocks. (To use three blocks would require three blocks of size 4, and any choice of blocks of that size will leave out some checked cells.) This is the best solution. In the problems you will have the opportunity to try some more of these kinds of problems. ∎

Problem Set 6

Find boolean expressions for the following Karnaugh maps. (There is no need to minimize.)

1.

2.

3.

4.

Find Karnaugh maps for the following expressions:

5. $x + y$

6. $x + xy'$

7. $xy'z' + xy + z$

8. $x + x'y' + xyz' + w$

Use Karnaugh maps to minimize the following boolean expressions:

9. $xy'z' + xy'z$

10. $xyz + xyz' + xy'z$

11. $xyz + xyz' + x'yz' + x'y'z'$

12. $x'yz' + x'y'z' + xy'z' + xyz + xy'z$

13. $(x + y)' + z + x(yz + y'z')$

14. $yz' + xz'w + xyzw + x'y'z + x'y'w + x'w' + xyzw' + xy'z'w'$

15. $xyzw + x'yzw' + xy'zw + xy'zw + w'z'$

16. $xyzw + xyz'w + xy'z'w + x'yzw + x'y'z'w' + x'y'zw + x'w'z' + xy'w'$

17. $yzw + y'zw + x'yz'w + xyzw' + xy'zw' + x'y'z'w'$

18. $xyz' + (xyz') + xy'(w + z) + (z + w)'$

Construct minimal circuits equivalent to the following:

19. $y\bar{z}(x + \bar{w}) + y\bar{z}(x + \bar{w}) + \bar{x}yz$

20. $\overline{w + \bar{x}y + \bar{z}(w + \bar{x}y)}$

CHAPTER SUMMARY

Chapter 3 provides a logical extension to the ideas from Chapters 1 and 2 by abstracting the properties common to propositional logic and set theory to discuss a new kind of mathematical object known as a boolean algebra. A boolean algebra is a mathematical system with two operations which satisfies most of the same properties as propositional logic and its basic operations of conjunction and disjunctions, and as does set theory with its basic operations of union and intersection.

We proved several of the classic theorems of boolean algebra, and we focused on applications of these theorems to logic and set theory, as well as introducing the relationship between boolean algebra and digital switching circuits. In the context of circuit design, we considered the problem of simplification of circuits, and we utilized the method of the Karnaugh map to provide a method of simplifying a boolean expression (or circuit) with respect to certain criteria.

KEY TERMS

In Chapter 3 we introduced the following terms:

AND gate (3.1)
atom (3.3)
boolean algebra (3.2)
boolean expression (3.4)
boolean function (3.4)
complemented lattice (3.3)
distributive lattice (3.4)
duality (3.2)
equivalent expressions (3.4)
functionally complete (3.4)
greatest lower bound (3.3)
Karnaugh map (3.6)

lattice (3.3)
least upper bound (3.3)
lower bound (3.3)
maxterm (3.4)
minimization (3.6)
minterm (3.4)
NAND gate (3.1)
order generated by B (3.3)
OR gate (3.1)
poset (3.3)
upper bound (3.3)

4
Matrices

4.1 WHAT IS A MATRIX?

To commence our discussion of matrices, we begin with an example.

EXAMPLE 4.1

The U.S. Sprocket Corporation produces three models of sprockets—Model I, Model II, and the Deluxe. In making sprockets three kinds of metal are used—copper, steel, and aluminum. A Model I requires 2 grams of each metal, while a Model II requires 1 gram of copper, 1 gram of steel, and 2 grams of aluminum. A Deluxe sprocket takes 2 grams of copper, no steel, and 2 grams of aluminum.

A number of questions might be asked about the material requirements.

1. If copper costs $0.05 per gram, steel costs $0.03, and aluminum costs $0.10, how much does the raw material for each model cost?

2. If an order is placed for 500 Model I's, 600 Model II's, and 150 Deluxes, how much of each raw material is needed?

3. What is the cost of filling the order described in 2?

4. Suppose U.S. Sprocket also manufactures gearboxes in Models A and B. Model A needs two Model I sprockets, one Model II, and three Deluxe, while Model B uses five Model II sprockets and six Deluxe. How much of each metal is needed to produce the sprockets for the two different types of gearboxes?

The reader who is well versed in a programming language may recognize that the information described above could be conveniently represented in a two-dimensional array or doubly subscripted variable. The reader who has not encountered these ideas is probably a bit in the dark at this stage but is probably no more confused than the others about how to conveniently answer the questions posed above. Here we will find a convenient computational way to deal with these questions, even if the number of products and raw materials becomes large. This is the idea of a *matrix* and the fundamental computational techniques for dealing with matrices. (Note the singular is *matrix*, while the plural is *matrices*. There is no such word as "matrice," even if students *have* been heard to use it on occasion.)

In this chapter we introduce the idea of a matrix as a mathematical tool for dealing with cross-correlated information, and we discuss some of the computational algorithms used to deal with matrices.

DEFINITION

A **matrix** is a rectangular array (or table) of numbers.

We refer to matrices by the number of rows and columns in the table. For example, a 2 × 3 (read, "2 by 3") matrix is one that has two rows and three columns. The first number listed is always the number of rows, and the second number is the number of columns. Thus, an $n \times m$ matrix is one with n rows and m columns.

DEFINITIONS

The **dimension** or **shape** of a matrix is the number of rows and columns in the matrix.

A **square** matrix is a matrix with the same number of rows as columns.

A **row vector** is a matrix with only one row.

A **column vector** is a matrix with one column.

In Example 4.1, if we label the rows of a table by copper, steel, and aluminum, and the columns by Model I, Model II, and Deluxe, we have this arrangement:

	Model I	Model II	Deluxe
Copper			
Steel			
Aluminum			

To complete the table, all we do is put into the appropriate spot the correct information. In the intersection of the row labeled "Copper" and column labeled "Model I," we put the amount of copper needed to make a Model I, and so on. Our completed table or matrix looks like this:

	Model I	Model II	Deluxe
Copper	2	1	2
Steel	2	1	0
Aluminum	2	2	2

We have described the information from Example 4.1 in a 3×3 matrix. When we are dealing with mathematical ideas, the rows and columns of a matrix are usually just given numbers instead of labels, and the usual notation is to use *subscripts* to refer to particular positions in the matrix. If we call the matrix above A, then $A_{2,3}$ refers to the number found in row 2, column 3 of the matrix (in this case, a 0). The first subscript refers to the row number, and the second refers to the column number. So $A_{i,j}$ refers[1] to the entry in row i, column j of the matrix A. Sometimes we use the notation $[a_{i,j}]$ to denote a matrix and use a formula of the form $a_{i,j} = $ (expression) to describe the entries of the matrix.

Matrices and Programming Languages

Programming languages such as FORTRAN, BASIC, and Pascal provide the capability for dealing with matrices by the use of what are usually called subscripted variables or arrays. In these languages, we can declare a variable to be dimensioned as an $m \times n$ array, and then we refer to entries in that array as $A(I, J)$, or $A[I, J]$. This is identical to the mathematical concept of a matrix, except that the "subscripts" are written inside the parentheses or brackets instead of as subscripts. In fact, in these languages, people often say that arrays are subscripted variables. A subscripted variable with two subscripts is really nothing more than a matrix.

4.2 MATRIX ARITHMETIC

Matrix Addition

We introduce matrix arithmetic in a couple of stages. The first two operations that we define (in this section) will seem natural, while the third operation

---

[1] We use a comma to separate the i and j in $A_{i,j}$, even though many authors do not, because it helps avoid confusion. When A_{23} appears, one might not be sure whether A_{23} is the 23d entry in a one-dimensional array or row 2, column 3 of a matrix.

(defined in the next two sections) needs a bit more motivation. The first operation is matrix addition.

If we have matrices such as

$$\begin{bmatrix} 1 & 2 \\ 3 & 4 \end{bmatrix} \quad \text{and} \quad \begin{bmatrix} 1 & -1 \\ 2 & 3 \end{bmatrix}$$

and want to add them, there seems to be a pretty obvious way to proceed. We simply create a new 2×2 matrix whose values at each position are found by adding the entries in the same position in the two matrices. It seems reasonable to get

$$\begin{bmatrix} 1 & 2 \\ 3 & 4 \end{bmatrix} + \begin{bmatrix} 1 & -1 \\ 2 & 3 \end{bmatrix} = \begin{bmatrix} 1+1 & 2-1 \\ 3+2 & 4+3 \end{bmatrix} = \begin{bmatrix} 2 & 1 \\ 5 & 7 \end{bmatrix}$$

The operation of addition is natural. We recognize that in order to add two matrices, it is necessary that those matrices be the same shape. Attempting to add

$$\begin{bmatrix} 1 & 2 \\ 3 & 4 \end{bmatrix} \quad \text{and} \quad \begin{bmatrix} 1 & 2 & 3 \\ 4 & 5 & 6 \\ 7 & 8 & 9 \end{bmatrix}$$

doesn't seem to make sense, so we don't allow it.

The formal mathematical description of matrix addition will *seem* to be much more complicated than what we just described, but we include it here for the sake of completeness.

DEFINITION

If A and B are both matrices of the same dimension, the **sum of the matrices** $(A + B)$ is a matrix C of the same dimension as A and B such that for all i and j, $C_{i,j} = A_{i,j} + B_{i,j}$.

An algorithm for the addition of two matrices is easy to describe. (We assume that A, B, and C are all $m \times n$ arrays.)

ALGORITHM SUM

```
begin
for i:=1 to M do
    for j:=1 to N do
        C[i,j]:=A[i,j] + B[i,j]
        od
od
end.
```

EXAMPLE 4.2

Harriet's Super Store carries two different brands of toothpaste: Bright Smile and Crust. Each of those two brands is available in three different sizes: King (the smallest), Giant (medium), and Family Size (large). Harriet's inventory in toothpaste can be described as a matrix, by using the rows to represent the brand of toothpaste and the columns to represent the size. In each entry of the matrix, we place the number on hand of that item. Harriet's inventory for toothpaste is then

$$
\begin{array}{c}
\phantom{\text{Bright Smile}} \quad \text{King} \quad \text{Giant} \quad \text{Family} \\
\begin{array}{c} \text{Bright Smile} \\ \text{Crust} \end{array}
\begin{bmatrix} 3 & 5 & 6 \\ 2 & 4 & 7 \end{bmatrix}
\end{array}
$$

We can, by making the same use of rows and columns, also use a matrix to represent the number of items arriving in a new shipment. The matrix

$$
\begin{array}{c}
\phantom{\text{Bright Smile}} \quad \text{King} \quad \text{Giant} \quad \text{Family} \\
\begin{array}{c} \text{Bright Smile} \\ \text{Crust} \end{array}
\begin{bmatrix} 10 & 15 & 20 \\ 15 & 6 & 10 \end{bmatrix}
\end{array}
$$

could represent the number of items arriving at the store. The sum of these matrices represents the new inventory. The sum is the matrix

$$
\begin{array}{c}
\phantom{\text{Bright Smile}} \quad \text{King} \quad \text{Giant} \quad \text{Family} \\
\begin{array}{c} \text{Bright Smile} \\ \text{Crust} \end{array}
\begin{bmatrix} 13 & 20 & 26 \\ 17 & 10 & 17 \end{bmatrix}
\end{array}
$$

Row 1, column 1 represents the fact that after the new shipment is shelved, the number of king-size tubes of Bright Smile is 13 (3 + 10). Similar statements can be made for the other entries of the matrix. ∎

Note: You have every right to ask whether we could use the rows for sizes and the columns for brands in Example 4.2. If we did that and left everything else the same, we would have this initial inventory matrix:

$$
\begin{array}{c}
\phantom{\text{King}} \quad \text{Bright Smile} \quad \text{Crust} \\
\begin{array}{c} \text{King} \\ \text{Giant} \\ \text{Family} \end{array}
\begin{bmatrix} 3 & 2 \\ 5 & 4 \\ 6 & 7 \end{bmatrix}
\end{array}
$$

As long as we are consistent throughout the problem, this way works just as well as the other. The matrix looks a little different, but all we really have done is trade the roles that the rows and columns play. This leads us to a definition.

DEFINITION

The **transpose** of a matrix is the matrix obtained by turning the rows of the matrix into columns.

We denote the transpose of the matrix A by A^T, and the following relations hold:

1. If A is $m \times n$, then A^T is $n \times m$.

2. For all i and j, $A^T_{i,j} = A_{j,i}$.

A few matrices have the property that when you take their transpose, you end up with the same matrix. Matrices which have that property are said to be *symmetric* matrices.

DEFINITION

A matrix A is **symmetric** if and only if $A = A^T$.

The following matrix is an example of a *symmetric matrix*:

$$\begin{bmatrix} 1 & 2 & 5 \\ 2 & -1 & 7 \\ 5 & 7 & -3 \end{bmatrix}$$

A symmetric matrix must be a square matrix, but (of course) not all square matrices are symmetric.

Scalar Multiplication

Another operation of importance for dealing with matrices is *scalar multiplication*. In this process we multiply a matrix by a number (or scalar, as numbers are often referred to in matrix algebra). If matrix A is multiplied by the scalar x, we write the product as xA. We perform the operation by multiplying all the entries in A by the value x. If A is the matrix

$$A = \begin{bmatrix} 1 & 4 & 6 \\ 2 & -1 & 0 \end{bmatrix}$$

then

$$3A = \begin{bmatrix} 3 & 12 & 18 \\ 6 & -3 & 0 \end{bmatrix}$$

In Example 4.2, if Harriet wanted to double her order the next time she purchased toothpaste, this process could be described by taking the order matrix and multiplying it by 2. It is a reasonably simple task (and thus left to the reader) to write an algorithm to do scalar multiplication of a matrix.

Some computer languages provide built-in capability for matrix manipulation. Some versions of BASIC provide as part of the language the commands to add, perform scalar multiplication, take the transposes of matrices, as well as do some other matrix manipulations. This is a convenience, although all that really happens is that the people who write the compiler (or interpreter) make it convenient for you to do the manipulations by not forcing you to write the algorithms to do those tasks.

Problem Set 2

Suppose that

$$U = [2 \quad -7 \quad 1] \qquad V = [3 \quad 0 \quad -4] \qquad W = [1 \quad -5 \quad 8]$$

$$X = \begin{bmatrix} 1 & 1 & 3 \\ 0 & 5 & -3 \end{bmatrix} \qquad Y = \begin{bmatrix} 1 & -2 & 3 \\ 6 & 1 & 1 \end{bmatrix} \qquad Z = \begin{bmatrix} 1 & 5 \\ -2 & 3 \\ 1 & 0 \end{bmatrix}$$

Perform the following computations:

1. $U + V$
2. $V + W$
3. $2U$
4. $-W$
5. $U - V$
6. $W - U$
7. $3U - 4V$
8. $5U + 3V$
9. $2U + 3V - 5W$
10. $3U - V + 8W$
11. $X + Y$
12. $2Z$
13. $X + Z$
14. $3X + 2Y$
15. $2Y - 3X$
16. $2V + Z$
17. U^T
18. $X^T + Y^T$
19. $(2X)^T$
20. $X + Y^T$
21. $X^T + Z$
22. $2Y^T - 3Z$

23. Prove that matrix addition is commutative, or $A + B = B + A$ for all matrices A and B.

24. Prove that matrix addition is associative, or $(A + B) + C = A + (B + C)$ for all matrices A, B, and C.

25. Write a formal definition of the operation of scalar multiplication.

26. Prove the associative property of scalar multiplication, that is, $a(bA) = (ab)A$.

27. Prove that scalar multiplication distributes over matrix addition, that is, $a(A + B) = aA + aB$ for all real numbers a and all matrices A and B.

28. Prove that for any pair of matrices for which $A + B$ is defined $(A + B)^T = A^T + B^T$.

*29. Prove that if A and B are symmetric matrices, so is $A + B$.

30. Prove that for any matrix A and any scalar a, $aA^T = (aA)^T$.

31. Write an algorithm to multiply a matrix by a scalar.

P32. Write a Pascal procedure to store the transpose of matrix A in array B.

P33. Write Pascal procedures to add two $m \times n$ matrices and to multiply an $m \times n$ matrix by a scalar.

P34. Write a Pascal procedure to determine whether a given $n \times n$ matrix is symmetric.

P35. Use your procedures from Problems 32 and 33 to write a program to print out the results of Problems 1 to 5.

P36. Use your procedures from Problems 32 and 33 to write a program to print out the results of Problems 18 to 22.

4.3   MATRIX MULTIPLICATION

In this section, we begin by looking back at the first three questions from Example 4.1. To refresh our memories about this example, we include the details again.

The U.S. Sprocket Corporation produces three models of sprockets—Model I, Model II, and the Deluxe. In making sprockets three kinds of metal are used: copper, steel, and aluminum. A Model I requires 2 grams of each metal, while a Model II requires 1 gram of copper, 1 gram of steel, and 2 grams of aluminum. A Deluxe sprocket takes 2 grams of copper, no steel, and 2 grams of aluminum.

In this section we focus on the following questions:

1. If copper costs $0.05 per gram, steel costs $0.03, and aluminum costs $0.10, how much does the raw material for each model cost?

2. If an order is placed for 500 Model I's, 600 Model II's, and 150 Deluxes, how much of each raw material is needed?

3. What is the cost of filling the order described in 2?

In answering these questions, we will begin to develop a new technique for manipulating matrices. As we recall from Section 4.1, we can form a *materials matrix* describing the amount of material required for each item:

$$\begin{bmatrix} 2 & 1 & 2 \\ 2 & 1 & 0 \\ 2 & 2 & 2 \end{bmatrix}$$

In this case the rows represent the materials, and the columns represent the types of items being produced.

To determine the amount of money spent in producing each item, we must multiply for each model (column) the amount of each raw material needed by its cost, and then we add those results. Thus for a Model I, we obtain the cost by the calculation

$$2 \times 0.05 + 2 \times 0.03 + 2 \times 0.10 = 0.36$$

This represents the amount of copper times the cost of copper plus the amount of steel times the cost of steel plus the amount of aluminum times the cost of aluminum. Similarly, the cost of a Model II is obtained by

$$1 \times 0.05 + 1 \times 0.03 + 2 \times 0.10 = 0.28$$

Finally, the cost of a Deluxe is

$$2 \times 0.05 + 0 \times 0.03 + 2 \times 0.10 = 0.30$$

We summarize these results in a row vector, with columns ordered to match the arrangement of the columns in the original matrix:

$$[0.36 \quad 0.28 \quad 0.30]$$

What did we do? In each column of the matrix, we multiplied the top entry by 0.05, the middle by 0.03, the bottom by 0.10, and then we added the results. It would be reasonable to express this process as somehow having taken the cost per unit of material and multiplied it by the amount of material per item to obtain the cost per item. To express this, we indicate the cost of material as a *row* vector (in the order copper, steel, aluminum)

$$[0.05 \quad 0.03 \quad 0.10]$$

(*Note*: An argument could be made that this should be a column vector, but a row vector works better in this application.) This leads us to write the following:

$$[0.05 \quad 0.03 \quad 0.10] \begin{bmatrix} 2 & 1 & 2 \\ 2 & 1 & 0 \\ 2 & 2 & 2 \end{bmatrix} = [0.36 \quad 0.28 \quad 0.30]$$

This does seem to work, and we describe the process of multiplying a row vector (on the left) by a matrix (on the right) in the following way:

For each column of the matrix, take the corresponding entries of the row vector and multiply; add these results and place the total in the same column of a row vector as the column of the matrix that was operated on. This is what we did in answering question 1. We note that to accomplish this, the row vector on the left must have the same number of columns as the matrix has rows (in our example, that means we had to have exactly one cost for each kind of material, a not unreasonable expectation), and the result will have the same number of columns as the matrix.

Writing the mathematics formally, we have the following definition:

DEFINITION

Product of a row vector with a matrix: If $[a_i]$ is a row vector with n columns and $[b_{i,j}]$ is a matrix with n rows and m columns, their product is

$$[a_i][b_{i,j}] = [c_j]$$

where

$$c_j = \sum_{k=1}^{n} a_k b_{k,j}$$

and $[c_j]$ has m columns.

Question 2 deals with determining how much material is needed. It would seem reasonable to expect to be able to produce a column vector whose rows corresponded to the rows of our original matrix. In doing the obvious calculation for copper, in the order Model I, Model II, Deluxe, we get

$$2 \times 500 + 1 \times 600 + 2 \times 150 = 1900$$

Doing the same thing for steel, we get

$$2 \times 500 + 1 \times 600 + 0 \times 150 = 1600$$

and for aluminum we have

$$2 \times 500 + 2 \times 600 + 2 \times 150 = 2500$$

It would again seem reasonable to express this operation as having taken

Material per item × number of items = total materials

If we express the number of items ordered as a *column* vector, we could write

$$\begin{bmatrix} 2 & 1 & 2 \\ 2 & 1 & 0 \\ 2 & 2 & 2 \end{bmatrix} \begin{bmatrix} 500 \\ 600 \\ 150 \end{bmatrix} = \begin{bmatrix} 1900 \\ 1600 \\ 2500 \end{bmatrix}$$

In this case, we have multiplied a matrix (on the left) by a column vector (on the right), and we obtained a column vector. The result in row 1 of our answer is found by multiplying the entries in the column vector by the corresponding entries in row 1 of the matrix and then adding the results. This process is similar to the preceding one, in that we take rows from the left, multiply by columns on the right, and place the result in a corresponding place in the end result. In order for this process to work, there must be the same number of rows in the column vector as there are columns in the matrix, and our answer will have the same number of rows as the matrix. We leave it as an exercise to write a formal mathematical description of the process that we just completed.

Finally, question 3 asks us to compute the total cost of all this. If we were to phrase this as a multiplication problem, we might expect that

Cost of material × material per item × number of items = cost

would be a reasonable solution. This would lead us to attempt to compute

$$[0.05 \quad 0.03 \quad 0.10] \begin{bmatrix} 2 & 1 & 2 \\ 2 & 1 & 0 \\ 2 & 2 & 2 \end{bmatrix} \begin{bmatrix} 500 \\ 600 \\ 150 \end{bmatrix}$$

as a reasonable solution. There are two slight problems with this approach. The first is that we have no way of knowing which two quantities to multiply first, and the second is that once we do the first multiplication, we might end up with some kind of multiplication that we can't handle.

Let us call the row vector A, the matrix B, and the column vector C. As before, AB gives us the row vector $[0.36 \quad 0.28 \quad 0.30]$, so we are left with the problem of computing

$$[0.36 \quad 0.28 \quad 0.30] \begin{bmatrix} 500 \\ 600 \\ 150 \end{bmatrix}$$

It seems reasonable, in light of what we have done, to view this problem as one of multiplying a row vector by a matrix. In this case, since the matrix has one column, the answer should be a row vector with one column, which sounds suspiciously like a number. (If we view the problem as one of multiplying a matrix by a column vector, we get a column vector with one row, which is the same thing as a row vector with one column.) Taking the row times the column and adding, we get

$$0.36 \times 500 + 0.28 \times 600 + 0.30 \times 150 = 393.00$$

Note, in this computation, that each of the products involved in this sum represents the cost of a model times the number of that model ordered. If we compute BC first, we then must compute

$$[0.05 \quad 0.03 \quad 0.10] \begin{bmatrix} 1900 \\ 1600 \\ 2500 \end{bmatrix}$$

This yields

$$0.05 \times 1900 + 0.03 \times 1600 + 0.10 \times 2500 = 393.00$$

The result is the same, and we note that each product in this sum represents the cost of material times the amount of material required.

At this point we have discussed two related ideas in matrix multiplication: (1) how to multiply a row vector on the left by a matrix on the right and (2) how to multiply a column vector on the right by a matrix on the left.

In the next section we take up the general problem of multiplying matrices. As we might expect by now, the process involves taking rows from the left and columns from the right, and it requires that the number of columns on the left match the number of rows on the right.

Problem Set 3

$$\text{Let } A = \begin{bmatrix} 1 & -1 & 2 \\ 0 & 4 & 3 \end{bmatrix} \qquad B = \begin{bmatrix} 3 & 0 & 4 \\ 1 & -2 & 1 \end{bmatrix}$$

$$C = \begin{bmatrix} 2 & 1 & 0 & -3 \\ 4 & 2 & 1 & 5 \\ 0 & 0 & 3 & -1 \end{bmatrix} \qquad D = \begin{bmatrix} 2 \\ -1 \\ 3 \end{bmatrix}$$

Find each of the following (if possible):

1. AD 2. BD 3. CD 4. $D^T A^T$

5. $D^T C$ 6. $(A + B)D$ 7. $AD + BD$ 8. $D^T D$

9. A computer has two different kinds of programs in its batch queue. Program A requires 60 seconds of input time, 10 seconds of central processing unit (CPU) time, and 300 seconds of output time. Program B requires 10 seconds of input time, 600 seconds of CPU time, and 10 seconds of output time.
   (a) If there are three people who need to run program A and 1 person who needs to run program B, what total input, CPU, and output times are required by the current queue?
   (b) If input time costs $1 per second, CPU time costs $10 per second, and output time costs $1 per second, how much does each job in the queue cost?
   (c) What is the total cost of the programs in the queue?

10. Jovita's Rent-a-Car has offices in Seattle, Washington, and in Denver, Colorado. In Seattle, subcompacts rent for $21 a day, compacts for $29 per day, and full-size cars for $35 per day. In Denver the figures are $21, $27, and $40, respectively.
    (a) A salesperson must spend 4 days in Seattle and 5 days in Denver on a trip. If the salesperson rents the same kind of car in each location, what will the total cost be?

(b) Suppose that three salespeople make the trip, with two renting compact cars and 1 renting a full-size car. What will the total cost be?

In Problems 11 to 14, assume that the matrices in question are such that all indicated products are actually defined.

*11. Prove that if R_1 and R_2 are row vectors, then $(R_1 + R_2)M = R_1M + R_2M$.

*12. Prove that if R is a row vector, a is a scalar, and M a matrix, then $(aR)M = a(RM) = R(aM)$.

*13. Prove that if R is a row vector and M is a matrix, then $(RM)^T = M^TR^T$. Also prove that if C is a column vector then $(MC)^T = C^TM^T$.

*14. Use Problems 11 to 13 and the statements proved in the problems at the end of Section 4.2 to prove that if C_1 and C_2 are column vectors and M is a matrix, then
(a) $M(C_1 + C_2) = MC_1 + MC_2$
(b) $a(MC_1) = (aM)C_1$

15. Write an algorithm to describe the process of multiplying a row vector by a matrix.

16. Write an algorithm to describe the process of multiplying a column vector by a matrix.

P17. Write a procedure in Pascal which will multiply a row vector by a matrix.

P18. Write a procedure in Pascal which will multiply a matrix by a column vector.

P19. Use the procedures produced in Problems 17 and 18 to produce a program which will compute the results of Problems 1 to 3.

P20. Write a Pascal program to compute the results of Problems 4 to 8.

4.4 MORE MATRIX MULTIPLICATION

We now look at question 4 of Example 4.1. Suppose U.S. Sprocket also manufactures gearboxes in Models A and B. Model A needs two Model I sprockets, one Model II, and three Deluxe while Model B uses five Model II and six Deluxe. How much of each metal is needed to produce the sprockets for the two different types of gearboxes?

There are two matrices involved in answering this question. The first one is the materials matrix which we used extensively in Section 4.3:

$$\begin{array}{c} \\ \text{Copper} \\ \text{Steel} \\ \text{Aluminum} \end{array} \begin{array}{ccc} \text{I} & \text{II} & \text{Deluxe} \\ \left[\begin{array}{ccc} 2 & 1 & 2 \\ 2 & 1 & 0 \\ 2 & 2 & 2 \end{array}\right] \end{array}$$

The second is the new matrix

$$\begin{array}{c} \\ \text{Model I} \\ \text{Model II} \\ \text{Deluxe} \end{array} \begin{array}{cc} \text{A} & \text{B} \\ \left[\begin{array}{cc} 2 & 0 \\ 1 & 5 \\ 3 & 6 \end{array}\right] \end{array}$$

In this matrix, the columns represent the models of gearboxes, while the rows represent the amount of each kind of sprocket used in the manufacture of a given type of gearbox. The columns are presented in the order Model A, Model B, while the rows are in the order Model I, Model II, Deluxe, which is the same order as used in the *columns* in the other matrix.

If we want to compute the amount of copper needed to manufacture the sprockets for a Model A gearbox, we must multiply the amount of copper used in each type of sprocket by the number of each type of sprocket needed to make a Model A and then sum the results. This is accomplished by multiplying row 1 of our "old" matrix by column 1 of our "new" matrix in exactly the same way as we multiplied rows and columns in Section 4.3. A few things might be noted. We "labeled" the first row of our old matrix "Copper." We "labeled" the first column of our new matrix "A." Our result deals with copper in Model A. The columns of the old matrix and the rows of the new matrix were labeled with the same labels *in the same order.*

We further note that the process of finding the amount of steel due to using sprockets in model A consists of multiplying the steel row of our old matrix by the Model A column of our new matrix. In fact we can construct a *raw-materials matrix* to contain all the possible results as follows:

$$\begin{array}{c} \\ \text{Copper} \\ \text{Steel} \\ \text{Aluminum} \end{array} \begin{array}{cc} \text{Model A} & \text{Model B} \\ \left[\begin{array}{cc} 11 & 17 \\ 5 & 5 \\ 12 & 22 \end{array}\right] \end{array}$$

In each case the result in a particular row and column of the result can be found by taking that row from the old matrix and multiplying it by that column of the new one. Similarly, aluminum in Model B was found by taking the aluminum from the old matrix and multiplying by the Model B column from the new matrix. This works out in general because the rows of the new matrix were designed to align with the columns of the old matrix. This scheme suggests a way of multiplying matrices.

We first *informally* describe multiplication of matrices. To multiply matrix A by matrix B (the result is denoted AB):

1. Matrix A must have the same number of columns as B has rows.

2. To find row i, column j of the product, multiply row i of A by column j of B and sum the results.

3. Matrix AB has the same number of rows as A and the same number of columns as B.

If the matrices are labeled, the labels on the columns of A should correspond to the labels on the rows of B.

The following is the formal definition of matrix multiplication:

> **DEFINITION**
>
> If $A = (A_{i,j})$ is an $n \times m$ matrix and $B = (B_{j,k})$ is an $m \times r$ matrix, then matrix AB is defined as matrix $C = (C_{i,k})$, where $C_{i,k} = \sum_{j=1}^{n} A_{i,j} B_{j,k}$.

There are several mathematical properties that matrix multiplication satisfies which are similar to the properties satisfied by ordinary multiplication of real numbers.

1. Associative property:

$$(AB)C = A(BC)$$

2. Distributive properties:

$$A(B + C) = AB + AC$$
$$(B + C)A = BA + CA$$

3. Identity property: The $n \times n$ matrix $I = [a_{i,j}]$ in which $a_{i,j} = 1$ if $i = j$ and $a_{i,j} = 0$ if $i = j$ has the property $IA = A$ for all $n \times m$ matrices A and $BI = B$ for all $m \times n$ matrices B.

This type of matrix multiplication does *not* satisfy the commutative property; that is, AB and BA will *not* always be the same matrix. In many cases we find that AB and BA even have different dimensions or that one of the two is not even defined.

A formal description of the matrix multiplication algorithm follows:

ALGORITHM PRODUCT

{Finds the product of an n × k matrix and a k × m matrix.}
begin
for 1 := 1 **to** n **do**
   for j:= 1 **to** m **do**
      C[i,j] := 0;
      for r:= 1 **to** k **do**
         C[i,j] := C[i,j] + A[i, r]*B[r,j]
      od
   od
od
end.

Many versions of BASIC do provide one with the ability to multiply matrices with a single line of code, although implementing the above algorithm is a relatively simple task.

It might seem strange that the process of multiplying matrices that we have described is of much greater importance than the "obvious" matrix multiplication scheme that one would generalize from matrix addition. But many processes which are described by matrices result in the need to combine matrices in this fashion. Matrix multiplication is one of the most important mathematical operations used in the study of a number of the topics of discrete mathematics.

Problem Set 4

$$\text{Let } A = \begin{bmatrix} 1 & 0 & 1 \\ 1 & 2 & 1 \\ 0 & -1 & 1 \end{bmatrix} \qquad B = \begin{bmatrix} 0 & 1 & 1 \\ 2 & 1 & 0 \\ 1 & 0 & 2 \end{bmatrix}$$

$$C = \begin{bmatrix} 3 & -1 & -1 & 0 \\ 2 & 1 & 1 & 1 \\ 0 & 2 & 1 & 0 \end{bmatrix} \qquad D = \begin{bmatrix} 1 & -1 & 3 \\ -2 & 0 & 0 \\ 1 & 4 & 0 \\ 2 & 1 & 1 \end{bmatrix}$$

$$E = \begin{bmatrix} 1.5 & -0.5 & -1 \\ -0.5 & 0.5 & 0 \\ -0.5 & 0.5 & 1 \end{bmatrix} \qquad F = \begin{bmatrix} 1 & 0 & 0 \\ 0 & 1 & 0 \\ 0 & 0 & 1 \end{bmatrix}$$

Compute the following:

1. AB	**2.** BA	**3.** AF	**4.** FA
5. AC	**6.** BC	**7.** CD	**8.** DC
9. $(AB)E$	**10.** $A(BE)$	**11.** $A(B+E)$	**12.** $B(A+E)$
13. $AB+AE$	**14.** $BA+BE$	**15.** $(AB)^T$	**16.** $B^T A^T$

*17. Prove that for all matrices $(AB)^T = B^T A^T$.

*18. If I is the $n \times n$ matrix such that entries of the form $a_{i,i} = 1$ but all other entries are 0, prove that for any $n \times n$ matrix A, $AI = IA = A$. Matrix I is called the $n \times n$ *identity*.

*19. Matrix B is the inverse of A if $AB = BA = I$ (see above); then we say that B is the inverse of A, and we write $B = A^{-1}$. Show that $(A^{-1})^{-1} = A$.

*20. Use the results of Problems 17, 18, and 19 to show that $(A^{-1})^T = (A^T)^{-1}$.

P21. Write a Pascal procedure to multiply two matrices based on the algorithm PRODUCT. Use your procedure to write a program which produces the results of Problems 1 to 10.

P22. Use your Pascal procedure for multiplication along with the procedures you developed in the earlier sections of this chapter to write a program to produce the results of Problems 11 to 16.

CHAPTER SUMMARY

In this chapter we introduce the concept of a matrix and the mathematical operations used in dealing with them. The first section introduces the basic terminology and notation of matrices. In the following sections we introduce the operations of matrix addition, scalar multiplication, and matrix multiplication. We also present the algorithms which are used to perform the operations, and we discuss the properties possessed by these operations. It is most surprising that the operation which to most students seems the least natural is the one which is probably of greatest value in applications of matrices to other branches of mathematics and in applications of mathematics to other areas of study. Matrix multiplication will be of considerable value to us in our study of graphs and directed graphs.

Having completed this chapter, the reader should now be familiar with each of the following:

column vector (4.1)

dimension of a matrix (4.1)

matrix (4.1)

matrix addition (4.2)

matrix multiplication (4.3, 4.4)

row vector (4.1)

scalar multiplication (4.2)

square matrix (4.1)

symmetric matrix (4.2)

transpose of a matrix (4.2)

5
Graph Theory

5.1 WHAT ARE GRAPHS?

The idea of a graph is one of the most useful concepts in discrete mathematics and one of the more mathematically interesting. In fact, almost all of us have seen and used graphs at one time or another. For example, consider Fig. 5.1, the route map for X Airlines.

EXAMPLE 5.1

A number of questions are natural to ask about the map in Fig. 5.1.

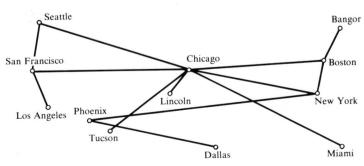

Figure 5.1 Airline route map.

1. Can a person get to Miami from Dallas?

2. How many times must you change planes to get from Boston to Los Angeles?

3. If the Baggage Handlers Union goes on strike in Chicago and closes down the airport, is it still possible to get from Boston to Los Angeles?

4. If X Airlines decided to stop all flights from Boston to New York, can you still get from Bangor to Phoenix?

5. Could you travel all X Airline's routes without having to retrace your steps by following the same route twice? ■

There are literally dozens of other questions that could be asked about this map, and most turn out to be examples of the kinds of questions that mathematics asks about graphs. The answers to questions like the ones above can often be found in the general theory of graphs.

Let's take another example.

EXAMPLE 5.2

In this case Fig. 5.2 represents a computer network in which the circles represent microcomputers. The squares represent minicomputers which are used to switch signals from one circuit to another as well as to process data. The diamond represents the mainframe computer at the heart of the network, and the lines represent the circuits connecting the network. A person who wishes to access the network dials up one of the minis from her or his micro and is switched via available lines to make a connection with the computer desired. The system can be used to send messages between micros, or the user can call on one of the larger computers to do information processing. Some natural questions arise:

1. Can the micro in Seattle send a message to the micro in Miami?

2. How much switching is required to get a message sent from Boston to Los Angeles?

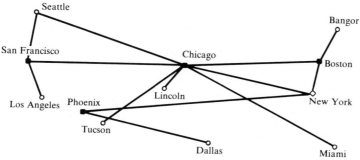

Figure 5.2 Computer network.

3. If the mini in Chicago goes down, can a message still be sent from Boston to Los Angeles?

4. If the circuits between Boston and New York fail, can a message still be sent from Bangor to Phoenix?

5. Is it possible to propagate a chain of signals that reaches every circuit in the network exactly once? ∎

Obviously these are not the only questions that one would want to ask about these pictures, or perhaps not even the most important, but they do point out the scope of the ideas involved in graph theory. It is interesting to note that the same mathematical object can be used to study applications in fields as seemingly unrelated as transportation and communications. The fact that such diverse applications can be studied through the use of the same theory shows one of the useful features of mathematical abstraction.

It is exciting to discover that one theory provides us with information about so many different applications. A key to making the whole scheme work is to describe the mathematical system carefully enough so as to make it understandable in terms of previous theory and yet useful in the areas in which we wish to apply it. In the case of graph theory, the terminology and notation of set theory serve our purposes well.

5.2 BASIC IDEAS AND DEFINITIONS

Vertices and Edges

The idea of a graph is straightforward. What we need to do is to formalize the idea that we have some points (called *vertices*) which are joined directly to other points by means of line segments (called *edges*). Many of the interesting topics in graph theory involve the process of tracing sequences of edges (called *paths*) through a graph. It is important to be precise in our definitions so that there is no ambiguity which can occur in discussions of the topic. Our definition of a graph is not all inclusive, since it ignores some important possible situations (such as one-way edges, a loop on a vertex, or more than one edge connecting the same pair of vertices); but it does get us started considering the basic ideas of graphs.

> **DEFINITIONS**
>
> A **graph** is an ordered pair (V, E), in which V is a set called the **vertex set** and E is a set consisting of subsets of V with two elements. The set E is called the **edge set**, and we say that *a and b are joined by an edge* if $\{a, b\}$ belongs to E.

We sometimes refer to the elements of V as *nodes*, especially in applications of graph theory which refer to communications networks.

In thinking about a graph or representing one with a diagram, we consider the set V to contain points or vertices (like the cities or computers of our examples in Section 5.1), and then we draw arcs between vertices which are joined by an edge.

EXAMPLE 5.3

Consider the graph $G = (V, E)$ where $V = \{a, b, c, d\}$ and $E = \{\{a, b\}, \{b, c\}, \{c, d\}\}$. This graph can be represented in any of the three ways shown in Fig. 5.3.

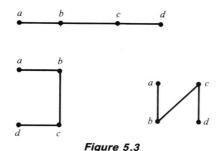

Figure 5.3

Although the diagrams look different, we should note that they all have the same properties; in fact, if the bottom two were straightened out, the result in each case would be the diagram on top. In each diagram it is possible to go from a to d by means of a route which takes us through b and c; and if any of the segments are broken, it will no longer be possible to go from a to d. ■

We say that each of the diagrams in Fig. 5.3 is a pictorial representation of G, because each gives a visual representation of the mathematical properties which are associated with G. Note that from a pictorial representation of a graph it is easy to recover the formal description of the graph. As a consequence, we usually work with the pictorial representations of graphs, unless it is necessary to deal with technical or theoretical aspects of graph theory.

Paths

DEFINITION

A **path** is a sequence of vertices of a graph in which each successive pair of vertices is an edge of the graph.

In graph G of Example 5.3, *abcd* is a path, because $\{a, b\}$, $\{b, c\}$, and $\{c, d\}$ are all edges of G. A path is said to be a path *from* the first listed vertex *to* the last listed vertex. It is also possible to define a path as a sequence of *edges* in which each successive pair of edges has a vertex in common. This alternate definition of a path is also quite useful, and we often switch between the two points of view, as is convenient. The path *abcd* described above becomes from that point of view the path $\{a, b\}\{b, c\}\{c, d\}$. The following definitions provide some important terminology about paths relating to the vertices (or edges) which make them up:

DEFINITIONS

A **simple path** is one in which no edge is repeated in the path.

An **elementary path** is one in which no vertex is repeated.

A **circuit** is a path which starts and ends at the same vertex.

A **simple circuit** is a simple path which is a circuit.

A **cycle** is a circuit in which the initial vertex occurs twice and no other vertex is repeated.

A warning needs to be issued here. The terminology and notation in graph theory are *very diverse*. The terms we have used here are ones which we have seen in other textbooks, but the terms are by no means standard. Because graph theory is such a relatively new area of mathematical study, the terminology has not had time to become standardized. Terms such as "trail," "route," "journey," and so on are often found in graph theory books and mathematical papers, and the meanings of these terms may (or may not) coincide with the meanings that we have given for other terms. You really need to be very diligent when reading about graph theory to make sure that you know what the author means by a term. The following terminology has become much more standardized.

DEFINITIONS

A graph is **connected** if given any two distinct vertices there is a path from one to the other.

A graph which is not connected is said to be **disconnected**.

In everyday language we would say that a graph is connected if it "hangs together" in one piece. In the case of an airline route map, connected would mean that any two cities in the system can be connected by a sequence of flights involving only that airline. In the case of a communications network, connected means that any two nodes in the system can communicate with each other. If a graph is disconnected, the graph will fall into "pieces" which we call *components*.

THEOREM 5.1

Let $G = (V, E)$ be a graph. Define the relation R on V by xRy if and only if $x = y$ or there is a path from x to y. R is an equivalence relation.

PROOF

This is left as an exercise. ∎

THEOREM 5.2

Let $G = (V, E)$ and R be defined as in Theorem 5.1. G is connected if and only if there is only a single equivalence class for the relation R.

PROOF

To prove a statement of the form $p \leftrightarrow q$, we must prove both $p \to q$ and $q \to p$. We start by assuming that G is connected.

For any distinct pair of vertices a and b there is a path from a to b, and thus aRb is true. Since for all a, aRa is also true, we have $\forall x \; \forall y (xRy)$, which means that V is the only equivalence class for R.

If there is only one equivalence class for R and x and y are distinct vertices, it follows that xRy, since x and y are in the same equivalence class. Since x and y are not equal, there must be a path from x to y. Since x and y were arbitrary, there is a path connecting any two distinct vertices of G, so G is connected. ∎

Subgraphs

> **DEFINITIONS**
>
> A **subgraph** of a graph (V, E) is a graph (V', E') in which V' is a subset of V and E' consists of edges in E which join only vertices of V'.
>
> If S is a subset of V, the subgraph **induced by** S is the graph with vertex set S and an edge set consisting of *all* the edges of G which join vertices of S.

If G is the graph described in Example 5.3 ($V = \{a, b, c, d\}$ and $E = \{\{a, b\}, \{b, c\}, \{c, d\}\}$), the graph (V', E') where $V' = \{a, b, c\}$ and $E' = \{\{a, b\}, \{b, c\}\}$ is a subgraph of G (in fact, it is the subgraph induced by the set $\{a, b, c\}$).

Informally, a *subgraph* is a graph which has some of the vertices of the original graph and some of the edges of the original graph. Here G' is a subgraph of G:

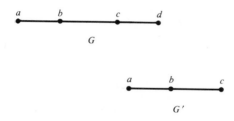

Components

One area of interest in considering a graph is the determination of what happens when the graph is not connected. The concept of a component is designed to help us discuss that situation and is rather easy to grasp. A component is supposed to be a subgraph of the original graph which is connected and as large as possible. In fact, the definition is often used that a component of a graph is a maximal connected subgraph, and you may well wish to think of a component of a graph in those terms. A graph which is not connected will fall into a number of components. The following discussion provides an alternative definition of a component, and it is interesting because it causes us to go back to the ideas of equivalence classes for an equivalence relation, as discussed in Chapter 2.

THEOREM 5.3

If C is an equivalence class of the relation R described in Theorem 5.1, then the subgraph induced by C is connected.

PROOF

The proof of Theorem 5.3 is very similar to the proof of the second half of Theorem 5.2. We let a and b be distinct vertices in C. Now aRb, and so there is a path $ax_1x_2, \ldots, x_nb$ from a to b. In fact, for all i, aRx_i (why?), so each x_i is also in C. Thus the path is a path from a to b in C. ∎

DEFINITION

A **component** of a graph is a subgraph induced by an equivalence class of the relation R described in Theorem 5.1.

COROLLARY

A graph is connected if and only if it has only one component.

PROOF
This is left as an exercise. ∎

COROLLARY

A component of a graph is a connected subgraph.

PROOF
This is left as an exercise. ∎

Theorem 5.3 and its two corollaries show us that the definition of a component as discussed above really provides us with a *maximal* connected subgraph. That is a connected subgraph of the original graph to which *no more of* the original graph can be added without making the subgraph disconnected.

EXAMPLE 5.4

Consider the graph G drawn here:

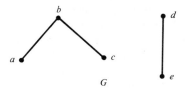

The subgraph induced by the vertices a, b, c is a component of G. We should note that it is connected, but attempting to add any more of G to it would make it disconnected. The subgraph with vertices d and e is also a component of G. The components are the "biggest" connected pieces of the graph. ∎

Degrees of Vertices

DEFINITIONS

A vertex is **incident to an edge** if the vertex is an element of that edge. (Pictorially speaking, a vertex is incident to an edge if it is one of the endpoints of the edge.)

The **degree** of a vertex is the number of edges to which it is incident. (In some cases it is convenient to allow graphs to have loops, or edges, of the form $\{a, a\} = \{a\}$. A loop is considered to be incident to its single vertex twice since that vertex serves as both endpoints. Our graphs will usually *not* have loops.)

THEOREM 5.4

In any graph, the sum of the degrees of the vertices is twice the number of edges.

PROOF

All we have to note is that each edge must be counted twice when we calculate the sum of the degrees of the vertices—once for each of its endpoints. Actually, what we just said is probably sufficient, but to illustrate a common proof technique in graph theory, we make the proof more formal. To do so, we proceed by induction. Let $P(n)$ be the statement that the sum of the degrees of the vertices for a graph with n edges is $2n$.

Basis Step If a graph has no edges, then the degree of each vertex is zero and the sum is zero, which is exactly the statement of $P(0)$.

Induction Step We assume $P(k)$ and we show that $P(k + 1)$ follows. To do so, we take a graph G with $k + 1$ edges, and we need to show that the sum of the degrees of its vertices is $2k + 2$. To do this, we consider a graph G' which is identical to G, except G' is missing one of the edges that belongs to G. Suppose that the missing edge is $\{a, b\}$. By the assumption that $P(k)$ is true (the induction hypothesis), since G' has k edges, the sum of the degrees of the vertices of G' is $2k$. Now, in order to form G from G', all that is required is to add the edge $\{a, b\}$ back into G'. This will increase the degree of the vertex a by 1 and increase the degree of b by 1, thus increasing the total by 2 and making the sum of the degrees of all vertices $2k + 2$. This completes the induction step. ■

The above proof was verbose, but it illustrates a very useful technique for induction proofs in graph theory, namely removing something to get back to the induction hypothesis and then fitting it back in to make the statement work in the next higher case. Since the variable n in the statement of $P(n)$ was the number of edges, a proof like this is sometimes called *induction on the number of edges*. Similar techniques are applied for induction on the number of vertices. A good number of theorems in graph theory are proved in this manner.

We now state a corollary of this theorem which will be of value later.

COROLLARY

In any graph the number of vertices which have odd degree is an even number.

PROOF

The sum of the degrees of the vertices is an even number, and for this to be so the number of odd terms in that sum cannot be an odd number. ■

Problem Set 2

For the graphs represented in Problems 1 to 4, do the following:
(a) Write a formal description of the graph (as an ordered pair of sets).
(b) Find the degree of each vertex.
(c) Count the number of edges.
(d) Verify Theorem 5.1.

1.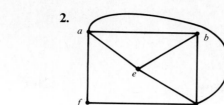

2.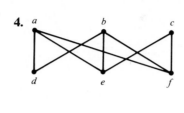

3.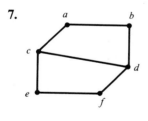

4.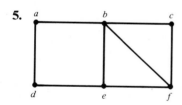

For each graph in Problems 5 to 8, do the following (if possible—if it is not possible, say so):

(a) Find an elementary path from *a* to *f*.

(b) Find a simple path from *a* to *f* which is not elementary.

(c) Find a path from *a* to *f* which is not simple.

5.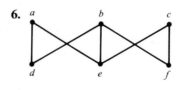

6.

7.

8.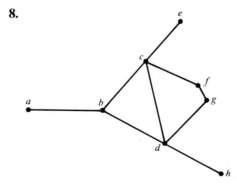

For each graph in Problems 9 to 12, do the following:

(a) Find a circuit which is not a cycle.

(b) Find a circuit which is not simple.

(c) Find a simple circuit.

9.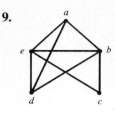

10.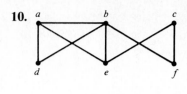

11.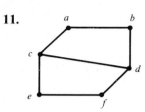

12.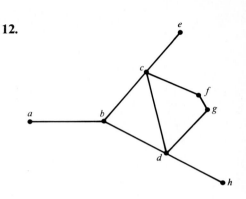

13. Using the graph from Problem 5, find the subgraph induced by the vertex set $\{a, b, c, f\}$.

14. Using the graph from Problem 8, find the subgraph induced by the vertex set $\{a, c, d, f\}$.

15. Using the graph in Problem 7, find each of the subgraphs induced by the subsets of the vertex set obtained by deleting a single vertex from the vertex set.

16. Repeat the instructions for Problem 15, using the graph in Problem 6.

17. Consider the following figures.

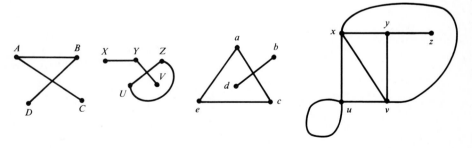

(a) Which are representations of graphs (as we have defined them)?
(b) Which are representation of connected graphs?
(c) For those graphs which are not connected, indicate the components.

*18. Prove that if there is a path from a to b in a graph, then there is an elementary path from a to b.

19. Prove Theorem 5.1.

20. Justify the statement in the proof of Theorem 5.3 which said that each x_i was contained in C.

21. Prove the first corollary to Theorem 5.3. (A graph is connected if and only if it has a single component.)

22. Prove the second corollary to Theorem 5.3. (A component of a graph is a connected subgraph.)

5.3   PLANAR GRAPHS

Graphs and Integrated Circuits

A further application illustrates the wide variety of uses which can be made of graphs in computing. Modern computer circuitry is largely constructed on integrated-circuit chips and printed-circuit boards. A chip is produced by creating several layers of miniaturized circuits in a silicon wafer by means of a masking process and photolithography. A crucial consideration in the design of a chip is that the miniature wires which are produced in each layer should not cross each other except at intended points. A person designing a chip is faced with the problem of determining whether a particular arrangement of electronic circuits can "fit" into one level or whether more than one layer is necessary to implement that particular design. Obviously, the number of layers in the chip design should be made as small as possible in order to minimize the number of steps and the costs involved in the manufacturing process.

If we regard the intended crossing points as vertices and the "wires" as edges, the problem can be viewed as a problem in graph theory.

EXAMPLE 5.5

Suppose that the design of a chip requires transistors to be placed at six locations called a, b, c, d, e, and f and organized into two groups a, b, c, and d, e, f with connections to be made in all possible combinations between the two groups, but no connections are to be made within the groups. A natural question to ask is, Does it require more than one layer? We are looking at a graph like the one below, and we are asking the question, Can the graph be drawn in a plane without having the edges cross between vertices?

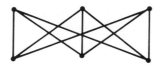

We might ask a similar question about an arrangement which required us to make all possible connections between a set of five transistors, as in the following diagram:

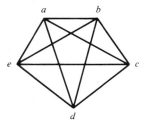

> **DEFINITIONS**
>
> A **planar graph** is a graph which *can* be drawn in the plane with no edges crossing between vertices.
>
> A **nonplanar graph** is one in which such a drawing cannot be made.

Beginners sometimes forget one important point—the fact that *one* picture of a graph has crossings in it does *not* necessarily mean that some other picture of that same graph might not avoid the crossings. One nonplanar representation of a graph *does not* force us to conclude that the graph is nonplanar.

In this section we deal only with graphs that are connected. This restriction creates no major difficulties. In particular, we should note that if all the components of a graph are planar, then the graph itself must be planar. For example, the graph

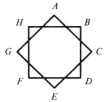   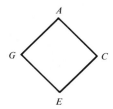

has two components, the first containing vertices *A*, *C*, *E*, and *G* and the second containing *B*, *D*, *F*, and *H*. Each component is planar, and putting them side by side in the plane yields a planar representation of the original graph. We thus really only need to answer questions about the planarity of connected graphs.

We will be able to state a theorem which completely characterizes planar graphs. The theorem does this by describing a property which all nonplanar graphs have and which no planar graphs have. Then we can examine a graph for this property. If it has it, the graph is nonplanar—if it doesn't, it is planar.

Looking at some connected planar graphs, we see that some interesting features begin to emerge. If we visualize a pair of scissors cutting along the edges of the planar graph in Fig. 5.4, we see that the plane would then fall into a number of pieces. Each of these pieces is called a *region*, or *face*, of the graph. The boundary of each region is a cycle; and if we remove an edge from one of the cycles, we then decrease the number of faces by 1. (*Note*: We should not forget that there is a "big" region which lies to the "outside" of the graph.)

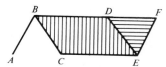

Figure 5.4 Planar graph.

To be able to prove theorems about the properties of planar graphs, we first prove a theorem about the effect of removing edges from cycles on connectivity.

THEOREM 5.5

Let $G = (V, E)$ be a connected graph, and suppose that the path $abx_1, \ldots, x_n a$ is a cycle. If $G' = (V, E')$, where $E' = E - \{\{a, b\}\}$ (that is, E' is the same as E except that the edge $\{a, b\}$ has been removed), then G' is connected.

PROOF

Let x and y be vertices of G. To show that G' is connected, we must show that there is a path from x to y that uses only the edges of E'. Since G is connected, there is a path from x to y using edges from E. If $\{a, b\}$ is not included in the path, then this path is the one we need. If $\{a, b\}$ is included, we can replace each occurrence of ab by the sequence $ax_n, \ldots, x_1 b$. If ba occurs, it can be replaced by the sequence traversed in reverse order. This new path does not contain the edge $\{a, b\}$, because $\{a, b\}$ was removed from the cycle. ∎

In Fig. 5.4, we see that each cycle is the boundary of at least one face, and each face may also include some edges contained completely within its boundaries. Those edges which are in boundary cycles "touch" two faces, and those which are not in a boundary cycle "touch" only one face. Furthermore, the boundary cycles must include at least three edges. All this leads us to the following observation:

THEOREM 5.6

Suppose that G is a connected planar graph. If G has e edges and a planar representation of G has r faces, then $r \leq \frac{2}{3}e$.

PROOF

Let S represent the sum produced by the following process. For each face, count the number of edges which touch that face. Add those numbers. Since each edge touches at most two faces, each edge will be counted in S at most twice, and so $S \leq 2e$. However, the boundary of each face must include at least three edges, so S must count at least three for each face and thus S must be at least 3 times the number of faces. In other words, $3r \leq S$. Putting those two facts together, we get $3r \leq 2e$, which is equivalent to the desired result. ∎

Trees

Next, we introduce the idea of a special kind of connected graph which is known as a tree.

> **DEFINITION**
>
> A **tree** is a connected graph which has no cycles.

Trees are very useful graphs for representing information and describing processes in computer science. Here we state and prove two useful results about trees.

THEOREM 5.7

A tree always has one fewer edge than vertices.

PROOF

To prove this statement, we proceed by induction on the number of edges, again using the scheme of deleting an edge, observing the induction hypothesis to hold in the smaller case, and then fitting the deleted edge back in. The problem is that in deleting an edge from a tree, we will disconnect it and may end up with a component that does not have exactly the number of vertices required for the induction hypothesis. This is a good example of the kind of situation in which strong induction is particularly useful.

Basis Step Clearly a tree with no edges has one vertex, as is required for $P(0)$.

Induction Step We assume $P(n)$ for all $n \leq k$, and show that $P(k + 1)$ must follow. Let $G = (V, E)$ be a tree with $k + 1$ edges. We must show that G has one fewer edge than vertices. Let $\{a, b\}$ be any edge of G, and form a new graph $G' = (V, E')$ where E' is $E - \{\{a, b\}\}$. By removing one edge, we have disconnected the graph G'. (See Problem 21.) Each of the components of G' has k or fewer edges. For G' there are two components, and each component is a tree. For each component, the number of edges is 1 less than the number of vertices. In the two components taken together, the total number of edges is 2 fewer than the number of vertices. Graph G has one more edge than G', and so G has only one fewer edge than vertices. ∎

THEOREM 5.8

Every tree has at least two vertices of degree 1.

PROOF (Not by Induction)

If $G = (V, E)$ is a tree with v vertices and e edges, then $e = v - 1$ by Theorem 5.7. By Theorem 5.4, the sum of the degrees of the vertices is twice the number of edges, so that number must be $2(v - 1) = 2v - 2$. If all vertices had degree 2 or more, the sum of the degrees of the vertices would have to be at least $2v$, hence at least one of the vertices of G has degree 1. If only one of the vertices had degree 1, then the sum of the degrees would be at least $2(v - 1) + 1 = 2v - 1$, which is still too large. ∎

Euler's Formula

In 1752 Euler proved a theorem about connected planar graphs which can be used to show that a graph is not planar.

THEOREM 5.9   Euler's Formula

For a connected planar graph with v vertices, e edges, and r faces, the equation $v - e + r = 2$ holds. (This theorem actually holds for graphs with loops at vertices and with multiple edges as well.)

Before we prove the theorem, an example will help to clarify the situation. In the graph below, $v = 6$, $e = 7$, and $r = 3$; thus $v - e + r = 2$.

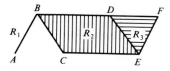

Let us now proceed to the actual proof.

PROOF

This is proved neatly by induction on the number of edges. We let $P(n)$ be the statement that Euler's formula holds for all connected planar graphs with n edges.

Basis Step   If a graph has no edges and is connected, there is only one possibility:

$$\bullet$$
$$A$$

Clearly, in this graph we have one region, one vertex, and no edges, so the formula holds.

Induction Step   We assume $P(k)$ and take $G = (V, E)$ to be a connected graph with $k + 1$ edges. We take one of the edges of G and delete it, and we make use of $P(k)$. Our real challenge is to choose an edge in such a way as not to disconnect the graph. First we attempt to find a vertex of degree 1. If such a vertex (let us call it a, and its incident edge $\{a, b\}$) exists, then a new graph $G' = (V', E')$ can be formed by letting V' be $V - \{a\}$ and E' be $E - \{\{a, b\}\}$. This graph has 1 less vertex and 1 less edge than G, but since no cycles were broken, it has the same number of faces as G. In particular, it has k edges, so $P(k)$ applies (the required proof that G' is connected is left as an exercise); thus Euler's formula holds and gives us $v - 1 - (e - 1) + r = 2$. Simplifying, we see that this yields $v - e + r = 2$.

If G has no vertices of degree 1, then G is not a tree and thus must include at least one cycle. In that case we form the graph $G'' = (V, E'')$ where E'' is E with one edge from a cycle removed. By Theorem 5.5 the graph obtained in this way is still connected. Breaking a cycle in this graph will reduce the number of regions by 1. This graph has k edges. Hence by the induction hypothesis $P(k)$ we find that once again Euler's formula holds. For G'' there are v vertices, $e - 1$ edges, and $r - 1$ faces; so by Euler's formula $v - (e - 1) + r - 1 = 2$. Simplifying, this becomes $v - e + r = 2$.   ∎

Complete Graphs and Complete Bipartite Graphs

To consider the key theorem of planarity of graphs, we need to consider two standard types of graphs. These types of graphs are known as complete graphs and complete bipartite graphs.

DEFINITIONS

A **complete graph** on n vertices, denoted K_n, consists of n vertices all of which are connected to each other. Figure 5.5 is a representation of K_4.

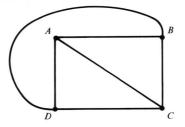

Figure 5.5 K_4.

A **bipartite graph** is one in which the vertices are divided into two sets in such a manner that there are no edges connecting vertices in the same set. The graph in Fig. 5.6 is bipartite.

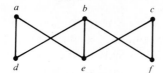

Figure 5.6 Bipartite graph.

A complete bipartite graph on m and n vertices, denoted $K_{m,n}$, is a bipartite graph in which one of the sets has m vertices and the other n, and all possible edges exist between pairs of vertices in the opposite sets. Figure 5.7 is a representation of $K_{2,4}$. The graphs represented in Example 5.5 were $K_{3,3}$ and K_5. The question asked in that example can be rephrased: Are $K_{3,3}$ and K_5 planar? The following theorem will lead to a solution of part of that problem.

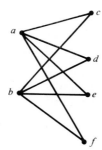

Figure 5.7 $K_{2,4}$.

THEOREM 5.10

In a connected planar graph with e edges and v vertices, $3v - e \geq 6$.

PROOF
From Euler's formula $v - e + r = 2$, but from Theorem 5.6 we note that $r \leq \frac{2}{3}e$, so that

$$2 = v - e + r \leq v - e + \tfrac{2}{3}e = v - \frac{e}{3}$$

Multiplying through by 3, we see that $6 \leq 3v - e$. ∎

It took a lot of work, but at last we have a test which can tell us in some cases that a graph is nonplanar. If we apply the formula above to K_5, we have $v = 5$ and $e = 10$. And $3v - e$ gives $15 - 10 = 5$, and for planar graphs $3v - e$ must be at least 6! Thus K_5 is not planar, and the second part is solved. Indeed, clearly *any* graph which includes K_5 must be nonplanar.

Turning our attention to $K_{3,3}$, we are not quite as lucky. We find that $v = 6$, $e = 9$, and $3v - e = 18 - 9 = 9$. This is a *nonresult* because the theorem is phrased "If G is planar, then .... " In this case the conclusion ($6 \leq 3v - e$) is true, but that does not tell us anything about the truth of the hypothesis. (We have a situation in which $p \rightarrow q$ and q are true, but it is a *fallacy* to conclude that p is true.) To attack $K_{3,3}$, we observe that every cycle in that graph must have length of at least 4 (why?). We leave the completion of the problem as an exercise, but we do state that $K_{3,3}$ is nonplanar.

We formally state the results discussed above as theorems:

THEOREM 5.11

K_n is planar if and only if $n < 5$. ■

THEOREM 5.12

$K_{m,n}$ is planar if and only if $m < 3$ or $n < 3$. ■

Kuratowski's Theorem

To complete our discussion of planar graphs, we note a result of Kuratowski. To understand the statement of Kuratowski's theorem, we introduce a process which we call *modification* by vertices of degree 2. If $\{a, b\}$ is an edge in graph G, then we *add* a vertex of degree 2 if we add a new vertex x to the vertex set and we replace the edge $\{a, b\}$ in the edge set with the pair of edges $\{a, x\}$ and $\{x, b\}$. This has the effect of adding the vertex x to the graph between a and b:

The process of *removing* a vertex of degree 2 is exactly the reverse. If x is a vertex of degree 2 incident to the edges $\{a, x\}$ and $\{b, x\}$, then the new graph has the vertex x removed from the vertex set and the two edges incident to x are replaced with the edge $\{a, b\}$. The process then appears as follows:

Intuitively it is clear that neither of the above operations has any effect on the planarity of the graph.

DEFINITION

Two graphs are **homeomorphic** if one can be obtained from the other by means of a finite sequence of modifications by vertices of degree 2 (and possibly also relabeling the vertices of one of the graphs).

THEOREM 5.13   Kuratowski's Theorem

A graph is nonplanar if and only if it contains a subgraph homeomorphic to K_5 or $K_{3,3}$.   ∎

The remarks preceding the statement of Kuratowski's theorem essentially prove the "if" part, but the "only if" part is beyond the scope of this book and thus is not proved here.

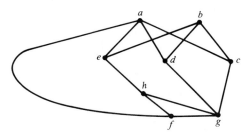

Figure 5.8

The graph G illustrated in Fig. 5.8 can be shown quite easily to be nonplanar by using Kuratowski's theorem. We note that G contains the subgraph shown in Fig. 5.9. If we remove the vertex h, since it is a vertex of degree 2, the subgraph in Fig. 5.9 can be seen to be $K_{3,3}$. And so G contains a subgraph homeomorphic to $K_{3,3}$, which means that G is nonplanar.

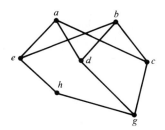

Figure 5.9

Problem Set 3

1. For each of the following representations of graphs, count the number of vertices, edges, and faces. For each graph, verify Euler's formula.

 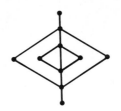

2. Repeat Problem 1 with the following graphs.

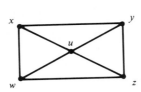 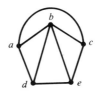 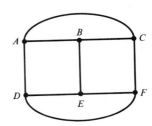

3. Verify Theorem 5.6 for each graph in Problem 1.

4. Verify Theorem 5.6 for each graph in Problem 2.

5. Draw two trees with four vertices.

6. Draw two trees with five vertices.

7. Two graphs are *isomorphic* if they are "the same graph except for the names of the vertices." This means that the graphs have identical representations except that the vertices may have different labels. For example, graphs G_1 and G_2 are isomorphic, because simply trading the labels on

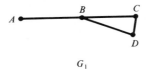 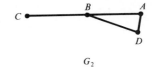

vertices A and C in G_1 gives us a picture identical to the one in G_2. (It is possible to make a much more formal definition in terms of one-to-one functions, but our brief description should suffice.) Show that graphs G_3 and G_4 *cannot* be isomorphic. (*Hint*: Consider the degrees of the vertices.)

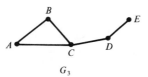 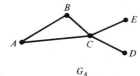

8. Show that the following pair of graphs cannot be isomorphic:

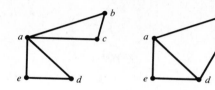

9. Show that the following graphs are isomorphic by identifying the vertices in the two graphs that should be "matched":

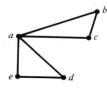
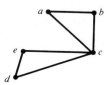

10. Show that the following graphs are isomorphic in the same manner as you did the pair of graphs in Problem 9:

Show that the following pairs of graphs are homeomorphic by making the appropriate modifications by vertices of degree 2:

11.

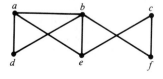
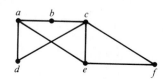

12.

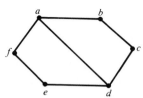
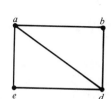

13.

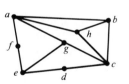
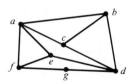

14.

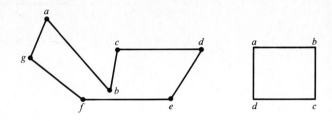

Draw planar representations of each of the following graphs, or explain why it is not possible to do so:

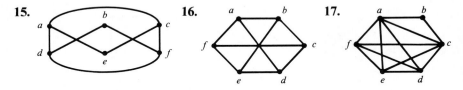

15.   **16.**   **17.**

18. Draw a representation of K_6.

19. Draw a representation of $K_{3, 5}$.

20. Prove that every tree satisfies Euler's formula by using Theorem 5.7.

*21. Prove that if an edge is removed from a tree, then the resulting graph is disconnected with two components. (*Hint:* To show the graph is disconnected, try an indirect proof by assuming that removing an edge does not disconnect the graph.)

22. Prove that a circuit in a bipartite graph must have an even number of edges.

23. Prove that a cycle in a bipartite graph must have at least four edges.

*24. Prove that for a planar bipartite graph $r \le e/2$. (Use Problem 23.)

25. Use Problem 24 to prove Theorem 5.12.

5.4   CONNECTIVITY

Most of the problems stated in Examples 5.1 and 5.2 deal with the question of how well a graph is connected in the sense of "Can you get there from here?" or "If something is removed from the graph, is it still possible to get there from here?" In this section we look at some of the methods which are available for studying such problems. In particular, we introduce some algorithms which can supply the answers to those questions, although these algorithms have to be refined later in order to provide more efficient ways of dealing with graphs. The main reason for introducing the algorithms at this point is to convince ourselves that it is possible to write computer programs which will solve problems about graphs. We first note a useful fact:

If G is a graph in which there is a path from a to b, then there is an elementary path from a to b.

PROOF

Let $ax_1, \ldots, x_k b$ be a path from a to b in graph G. If no vertex is repeated in the path, then we are done. If some vertex, say x_i, occurs more than once, the path can be shortened in order to eliminate the replication. Let x_j represent the last occurrence of this vertex in the path. Note that $ax_1, \ldots, x_i x_{j+1}, \ldots, x_n$ will be a path from a to b, and the vertex x_i occurs only once. We repeat this process for each vertex which occurs more than once, and the end result is an elementary path.

■

COROLLARY

In a graph with n vertices, if there is a path between vertices a and b, then there is a path with $n - 1$ or fewer edges between those vertices.

PROOF

This is left as an exercise.

■

Algorithms PATH and CONNECT

One consequence of this corollary is that it makes it quite easy to design an algorithm to test whether two vertices are connected by a path. We assume that it is possible to design an algorithm to test whether a sequence of edges in a graph is a path; and we assume that given the set of edges for a graph, we can design an algorithm to construct any sequence of edges. We will be able to prove the algorithm (PATH) that we describe works in the sense that in some finite time it will determine whether there is a path joining the two vertices.

ALGORITHM PATH

```
{Returns true if a path exists and false if no path is found.}
{Requires the user to supply vertices a and b.}
begin
result := false;
repeat
Construct a sequence of n − 1 edges in G;
if the sequence is a path and a is a vertex in the path then
        if b is a vertex in the path
                    then result := true
        fi
fi
until all sequences of n − 1 edges in G have been constructed;
PATH := result
end.
```

There are many algorithms for making this same test which work much more efficiently. The key fact is that this one will work in a finite amount of time for any graph, because there are only a finite number of sequences that need to be tested for any graph. This is a first pass at testing for the existence of a path between two vertices—later we will do much better. This particular algorithm is an example of what is called an *exhaustive algorithm* because it works by exhausting all possible situations which can occur. Exhaustive algorithms tend to be very inefficient.

By making use of the algorithm PATH, we can construct another algorithm which will test to see whether a graph is connected. The scheme is pretty straightforward. We check all pairs of vertices to make sure of the existence of a path between them—if one pair fails, we return the value false; otherwise, the result will be true.

ALGORITHM CONNECT

{Tests a graph for connectivity, uses the algorithm PATH.}
begin
success: = true;
repeat
    success: = path(a,b) and success
for all possible pairs (a,b) of vertices;
CONNECT: = success
end.

Obviously CONNECT is even less efficient than PATH, and the amount of time required to implement it is probably very large. The reader may also note that we have dodged an important issue in the process, namely, how graphs are to be represented in a computer. Right now that is not important, since we can regard these algorithms as instructions to be carried out by a human being who can look at the graph while carrying out the instructions. Ultimately computer representation of graphs will have to be tackled.

Suitable use of the algorithm PATH can provide us with the answers to questions 1, 3, and 4 of Examples 5.1 and 5.2.

Distance and Diameter

DEFINITIONS

The **distance** between vertices in a graph is the smallest number of edges in any path connecting the vertices.

The **diameter** of a graph is the greatest distance found between any pair of vertices in the graph.

We leave it to the reader to construct algorithms to determine both the distance between a pair of vertices in a graph and the diameter of a given graph.

In the language of graph theory, question 2 in Examples 5.1 and 5.2 is asking to find the distance between a pair of vertices in the graph.

Spanning Trees and Cut Sets

At this point, we want to consider some generalizations of the fourth question in Examples 5.1 and 5.2. In particular, we focus on the issue of how much removal of edges a graph can stand before it becomes disconnected. To explore this question, we introduce the idea of a cut set.

DEFINITION

If $G = (V, E)$ is a connected graph, then a **cut set** for G is a set of edges S which has the property that

1. $(V, E - S)$ is not connected.
2. For any proper subset T of S, $(V, E - T)$ is connected.

A cut set is a smallest set which, when removed, disconnects a graph.

The implications for applications of graph theory are apparent. Any edge which is part of a cut set with a small number of edges represents a threat to the connectivity of the graph. If that edge is removed from the graph, it becomes that much easier to disconnect the graph. In a communications network, that would have the effect of making it easier to cut off communications between some of the points in the network. In a transportation network, it would become more difficult to move between all points in the network.

We begin our study of cut sets by considering trees. The removal of an edge from a tree will disconnect it, which means that each edge is a cut set by itself. This suggests that it would be interesting to see how trees are "built into" connected graphs.

DEFINITIONS

A **spanning graph** for $G = (V, E)$ is a subgraph $G' = (V', E')$ such that $V' = V$.

A **spanning tree** for G is a spanning graph for G which is a tree.

In Fig. 5.10, G' is a spanning tree for G.

There is an interesting relationship among the seemingly distinct ideas of spanning trees, cut sets, and cycles. The following two theorems illustrate that fact.

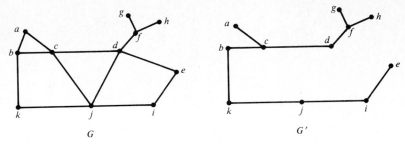

Figure 5.10

THEOREM 5.15

If $T = (V, E')$ is a spanning tree for $G = (V, E)$ and Q is a cut set for G, then Q contains at least one element of E'.

PROOF
If Q had no edges in common with E', then E' would be a subset of $E - Q$ and thus T would be disconnected, which is a contradiction. ∎

THEOREM 5.16

If $T = (V, E')$ is a spanning tree for $G = (V, E)$ and $ax_1, \ldots, x_k a$ is a cycle in G, then at least one of the edges in the cycle must be in $E - E'$.

PROOF
This is left as an exercise. ∎

Constructing Spanning Trees and Cut Sets with Algorithms

Next, we describe an algorithm to construct spanning trees. We state this in the form of a theorem, and we use the proof of the theorem to supply us with the algorithm.

THEOREM 5.17

Every connected graph has at least one spanning tree.

PROOF
Consider a connected graph. If that graph has no cycles, then it is a tree and thus is its own spanning tree. If the graph has cycles, we can pick any edge from one of its cycles and remove if from the graph. The resulting graph is a connected (by Theorem 5.5) subgraph of the original graph. Repeat this process until the subgraph obtained has no cycles. This graph will be a spanning tree for the original graph. ∎

If we wished to be more careful about this proof, we might phrase the proof as an induction on the number of edges in the graph. The reader is invited to do this in order to get a better feel of how induction proofs on graphs are constructed.

Our proof of this theorem is actually an algorithm. We write it up as follows:

ALGORITHM SPAN

{Takes a connected graph and finds a spanning tree.}
begin
while cycles exist in G **do**
    remove an edge from a cycle of G
od
end.

There is a lot of choice built into this scheme, and this does point out the fact that a connected graph will usually have several spanning trees.

We will go through an example of the use of this algorithm by finding a spanning tree for the graph of Fig. 5.11. This graph has cycles, so we pick an edge from one of its cycles and remove it, in this case $\{a, b\}$. The resulting graph is Fig. 5.12. The new graph still has cycles, so we remove another edge from a cycle, this time $\{c, j\}$, and obtain Fig. 5.13. Next we remove $\{d, j\}$ (see Fig. 5.14). Finally we remove $\{d, e\}$, and the result is a tree (Fig. 5.15).

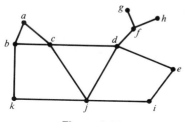

Figure 5.11

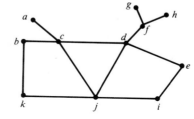

Figure 5.12

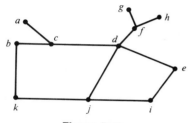

Figure 5.13

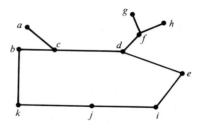

Figure 5.14

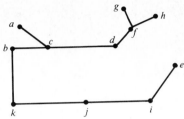

Figure 5.15 Spanning tree.

Before we go on to other business, we can observe that this algorithm actually provides us with a new proof of Euler's formula. In a planar graph, every time we break a cycle, we reduce the number of faces by 1. We end up with a tree, which has a single face, so the number of faces in the original graph must be the number of steps in the process plus 1. Recall that for any tree the number of edges is 1 less than the number of vertices, so our process must end with $v - 1$ edges. If we start with e edges, the number of steps required must be $e - (v - 1)$. Thus we have the equation $r = e - (v - 1) + 1$. This simplifies to the more familiar $v - e + r = 2$.

Next, we look at a scheme that is guaranteed to produce cut sets. Suppose that G is connected and that A and B are two disjoint subsets of V which together include all vertices of G. If A and B are each vertices of *connected* subgraphs of G, then we can determine a cut set S from these two sets of vertices by including in S *all* edges of G which join a vertex of A to a vertex of B. Using the graph from Fig. 5.11, we can let $A = \{a, b, c, k, j, i, e\}$ and let $B = \{d, f, g, h\}$. Those two sets do meet the criteria listed above and lead us to the cut set $S = \{\{c, d\}, \{j, d\}, \{d, e\}\}$. Thus $\{c, d\}$ is the only edge in S which belongs to the spanning tree we created earlier (Fig. 5.16).

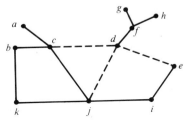

Figure 5.16 Graph with cut set.

For each edge in a spanning tree we can create sets like A and B described above just by deleting an edge from the spanning tree. We then let A be the vertices in one component of the part of the spanning tree which is left and B be the vertices in the other component, and we form S as above.

DEFINITIONS

A **fundamental cut set** corresponding to the edge E is a cut set formed in the manner described above by the removal of the edge E from a spanning tree.

A **system of fundamental cut sets** is the collection of all the fundamental cut sets which can be constructed from one spanning tree.

The cut set found in Fig. 5.16 was the cut set corresponding to the edge $\{c, d\}$.

The discussion in the previous paragraphs describes an algorithm for constructing cut sets for connected graphs. It is a good exercise to try to write a description of the process of constructing fundamental cut sets in the form of an algorithm.

Fundamental Cycles

If we take a spanning tree for a graph G and add one more edge to the tree, the resulting graph will not be a tree and so will have at least one cycle. A cycle in G which can be found in this way from a spanning tree is called a *fundamental cycle*, and the added edge is called the *chord* of the cycle.

The relationship among fundamental cycles, fundamental cut sets, and spanning trees is illustrated very nicely by Theorem 5.18.

THEOREM 5.18

Let T be a spanning tree for G and $C = abx_1, \ldots, x_n a$ be a fundamental cycle with chord $\{a, b\}$. Let F be a fundamental cut set determined by T. Now $\{a, b\}$ is an element of F if and only if the edge used to determine f was one of the edges in the cycle C.

PROOF

If we remove the edge $\{x_i, x_{i+1}\}$ from T, there is no path from b to a in the remaining graph. So a and b will be in different sets, thus guaranteeing that $\{a, b\}$ will be one of the edges in S. Conversely, if an edge not listed in C is deleted, we still can find a path from a to b in the remaining graph, so they will remain in the same component, and $\{a, b\}$ does not get into the resulting cut set. ∎

Problem Set 4

1. Consider the algorithm PATH, and suppose that G has 10 edges and 5 vertices.
   (a) How many possible sequences of four edges can be formed in G? (A sequence of edges may or may not be a path.)
   (b) How many tests are needed to determine whether a sequence of edges is a path?
   (c) Use the results of (a) and (b) to determine the maximum number of operations needed to execute PATH on this graph. (*Hint*: Use the multiplication principle of counting.)

*2. Repeat Problem 1 in the case of a graph with n vertices and e edges.

3. Consider the algorithm CONNECT as applied to a graph with 10 edges and 5 vertices.
   (a) How many pairs of vertices must be tested by the algorithm?
   (b) What is the maximum number of operations needed to complete the algorithm on this graph? (*Hint*: Use the multiplication property again.)

*4. Repeat Problem 3 in the case of a graph with n vertices and e edges.

The maximum number of operations required to complete an algorithm is one measure of the efficiency of the algorithm. The term often applied to this is the *complexity* of the algorithm. In Section 6.5 we develop some other algorithms for testing for existence of paths and connectivity, and their complexity can be compared with that of these algorithms. Our real concern is how well the algorithm does for large amounts of data, and hence we usually are concerned with the part of the complexity function which "dominates" as n gets large. If we had two algorithms, one with a complexity of n^5 and the other with a complexity of $0.2(n!)$, we would regard the first algorithm as more efficient because as n becomes large, the "cost" of the second algorithm far outpaces that of the first (even though for small values of n, the second has smaller values).

5. Consider the graph in Fig. 5.17. Find the distances between vertices (a) a and k, (b) g and d, (c) b and i, and (d) a and d. (e) Find the diameter of the graph.

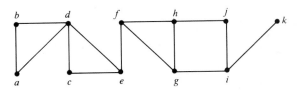

Figure 5.17

6. Consider the graph in Fig. 5.18. Find the distances between vertices (a) a and j, (b) g and d, (c) b and i, and (d) b and c. (e) Find the diameter of the graph.

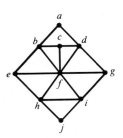

Figure 5.18

7. Find a spanning tree for the graph in Fig. 5.17.

8. Find a spanning tree for the graph in Fig. 5.18.

9. Using the spanning tree produced in Problem 7, find a system of fundamental cut sets for the graph.

10. Using the spanning tree produced in Problem 8, find a system of fundamental cut sets for the graph.

11. Using the graph in Fig. 5.17, produce a fundamental cycle and use it to illustrate Theorem 5.18.

12. Using the graph in Fig. 5.18, produce a fundamental cycle and use it to illustrate Theorem 5.18.

13. Construct an algorithm to find the distance between two vertices.

14. Use your algorithm from Problem 13 to help construct an algorithm to find the diameter of a graph.

*15. A *weighted graph* is one in which numbers (such as a number measuring the distance between the vertices connected by the edge) are assigned to each edge. The total weight of a path is the sum of the numbers associated with its edges. The distance between two vertices in a weighted graph is the minimum of the total weights of paths connecting the two vertices. Construct an algorithm for finding the distance between vertices in a weighted graph.

*16. Use your algorithm in Problem 13 to help construct an algorithm for finding the diameter of a weighted graph.

*17. Prove Theorem 5.16.

*18. Use induction to prove Theorem 5.17.

*19. Express the algorithm for constructing cut sets by using the algorithm language employed in this text.

5.5 COMPUTER REPRESENTATION OF GRAPHS

In Section 5.4, we alluded to the fact that one problem with the algorithms that we presented was that we did not provide a means for representing the graphs in a computer. There actually are a number of different ways of representing graphs in the computer, and the means chosen for a particular application often depends on the purpose. Complicated graphs, such as ones which include labels associated with the vertices and/or edges, can require quite complex data structures to represent the information contained in them. In Example 5.1, for example, for each edge we would need to store not only the fact that the edge existed but also the length of the trip between the two vertices. In Example 5.2, we would want to include for each vertex the type of computer located at that vertex.

In real-life applications, the information needed to be stored for each edge and vertex may be quite extensive. It is also quite likely that the information may change over time; thus the representation of the graph should be a form that can be modified quite easily. In regard to Example 5.1, we note that airlines in recent years have found their route maps changing drastically from time to time as the airlines tried to cope with the realities of our deregulated airline structure. Fortunately, for our purposes we will not need to use very sophisticated techniques to represent graphs—we are mainly concerned with the theoretical aspects of graphs. And it turns out that a matrix will do the job for us, although in courses in data structures it is often necessary to produce much more complex organizations of information in order to make the full power of graph theory available for a particular organization of data. Consider graph G in Fig. 5.19.

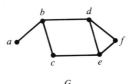

G

Figure 5.19

Incidence Matrix

There are two matrices which provide a lot of information about the graph. The first is called the *incidence matrix*. In the incidence matrix, a row is set aside for each vertex, and a column is set aside for each edge. The incidence matrix is then a 0–1 matrix, in which a 1 is placed as the entry in row i, column j only if the vertex i is incident to the edge j. The *incidence matrix for G* is

$$
\begin{array}{c}
\quad\;\; 1 \;\; 2 \;\; 3 \;\; 4 \;\; 5 \;\; 6 \;\; 7 \\
\begin{array}{c} a \\ b \\ c \\ d \\ e \\ f \end{array}
\left[
\begin{array}{ccccccc}
1 & 0 & 0 & 0 & 0 & 0 & 0 \\
1 & 1 & 1 & 0 & 0 & 0 & 0 \\
0 & 0 & 1 & 1 & 0 & 0 & 0 \\
0 & 1 & 0 & 0 & 1 & 1 & 0 \\
0 & 0 & 0 & 1 & 1 & 0 & 1 \\
0 & 0 & 0 & 0 & 0 & 1 & 1
\end{array}
\right]
\end{array}
$$

The incidence matrix is a convenient way of representing information about edges, and it can give information directly about the degrees of vertices. By summing a row of the matrix, we have counted the number of edges to which the vertex represented by that row is incident, and thus we have found the degree of that vertex. If we have the incidence matrix for a graph, we can reconstruct the graph itself quite easily. But the incidence matrix does not lend itself very conveniently to the investigation of the properties of graphs that we considered in Section 5.4 because it does not provide a convenient means for determining which pairs of vertices are connected by paths. This matrix does not provide us with the information that the algorithms of Section 5.4 require. We are led to consider another matrix, called the *adjacency matrix*.

Adjacency Matrix

The adjacency matrix is also a 0-1 matrix, but the adjacency matrix has a row *and* a column for each vertex. A 1 is placed in row i, column j provided that there is an edge connecting vertex i with vertex j. The *adjacency matrix for G* is

$$
\begin{array}{c c}
 & \begin{array}{c c c c c c} a & b & c & d & e & f \end{array} \\
\begin{array}{c} a \\ b \\ c \\ d \\ e \\ f \end{array} &
\left[
\begin{array}{c c c c c c}
0 & 1 & 0 & 0 & 0 & 0 \\
1 & 0 & 1 & 1 & 0 & 0 \\
0 & 1 & 0 & 0 & 1 & 0 \\
0 & 1 & 0 & 0 & 1 & 1 \\
0 & 0 & 1 & 1 & 0 & 1 \\
0 & 0 & 0 & 1 & 1 & 0
\end{array}
\right]
\end{array}
$$

This matrix appears to contain redundant information. (The symmetry of the matrix suggests that we could ignore everything below the diagonal and still retain the same information.) The graph is symmetric because of the fact that an edge from A to B is both $\{A, B\}$ and $\{B, A\}$, so we must have 1s in both row B, column A and row A, column B. The adjacency matrix generalizes very easily to the situation in which one-way edges or multiple edges are allowed. Most importantly, this matrix turns out to be the one which will provide us with the information needed to actually improve upon the algorithms of the previous section. Our first theorem about the incidence matrix points out some useful facts.

THEOREM 5.19

If M is the adjacency matrix for graph G, then the i, j entry of the matrix M^2 contains the number of paths of length 2 which connect vertex i to vertex j in G.

PROOF

Recall how the i, j entry of matrix M^2 is computed. We take row i of M, form the product of that row with column j of M, and sum the results. A 1 will occur in this sum for every value k such that $M_{i, k}$ and $M_{k, j}$ are both 1. This occurs when there is an edge between i and k and an edge between k and j. But that means there is a path of length 2 passing through vertex k which connects i and j. So for each k such that a path of length 2 passes through k, a 1 is added to the sum which computes $M_{i, j}^2$. The net result is to count all paths of length 2 from i to j. ∎

It is not hard to see that $M^3 = M^2 * M$ will contain similar information relating to paths of length 3, and this in fact generalizes to include paths of any length. This gives us exactly the handle we need to get back to algorithm PATH. To check for the existence of a path joining vertices i and j in a graph with n vertices, we must look at the first $n - 1$ powers of the adjacency matrix. If any of those powers has a nonzero entry in row i, column j, then we must conclude that a path exists between those two vertices. If no nonzero entry occurs in the i, j slot for any of the first $n - 1$ powers of M, then we conclude that no path exists between the two vertices. For the purposes of describing algorithms, it is easier to just take all those powers of M and add them. If the i, j position of the sum is nonzero, then we know that there was a path of some length joining the two vertices; but if it is a zero, we conclude that there is no path. We have actually proved Theorem 5.20.

THEOREM 5.20

Let G be a graph with n vertices and adjacency matrix M. Let S be the matrix formed by the sum of the first $n - 1$ powers of M. There is a path from i to j in G if and only if S has a nonzero entry in position i, j. ∎

If the matrix S mentioned above had *all* nonzero entries, then the graph G would be connected. Thus we have two algorithms.

ALGORITHM PATH2

{The user must supply as input the initial vertex A and terminal vertex B.}
begin
compute the first n − 1 powers of the adjacency matrix;
sum the first n − 1 powers of the adjacency matrix and store at S;
if S[A,B] < > 0 **then** PATH2: = true
                else PATH2: = false

fi
end.

ALGORITHM CONNECT2

begin
compute the first n − 1 powers of the adjacency matrix;
sum the first n − 1 powers of the adjacency matrix and store at S;
CONNECT: = true;
for i: = 1 **to** n **do**
for j: = 1 **to** n **do**
if S[i,j] = 0 **then** CONNECT: = false
fi od od;
CONNECT2: = CONNECT
End.

These algorithms will work very nicely for determining the answers to questions relating to the connectivity of graphs or existence of paths. The main problem which we find with these algorithms is really one of computational efficiency. Looking at the number of steps involved in performing the multiplication of a matrix by itself, we can see that the number of multiplications for each entry's calculation is n (where n is the number of vertices). Since the product matrix will have n^2 entries, this means that n^3 multiplications must be done in order to compute M^2. In fact, we still need to compute the other powers of M as well, and each of them requires n^3 multiplications. So the net result is to require something like n^4 multiplications to carry out the entire algorithm. The process of addressing the question of the number of computations required to perform an algorithm is one facet of algorithm analysis.

Our two new algorithms do perform better than the algorithms from Section 5.4; but when we deal with the topic of directed graphs, we will find better algorithms. The current algorithms are, however, adequate for our present purposes.

Problem Set 5

1. Suppose that G is the graph in Fig. 5.20. Find the incidence matrix for G.

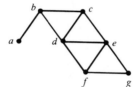

Figure 5.20

2. Suppose that H is the graph in Fig. 5.21. Find the incidence matrix for H.

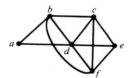

Figure 5.21

3. Suppose that G has the incidence matrix

	A	B	C	D	E	F
1	1	1	1	0	0	0
2	0	0	0	1	1	0
3	0	0	1	0	0	1
4	0	1	0	0	1	0
5	1	0	0	1	0	1

(a) Determine the degree of each vertex.

(b) Draw a representation of the graph.

4. Suppose that G has the incidence matrix

$$
\begin{array}{c}
\begin{array}{ccccccccc}
a & b & c & d & e & f & g & h & i
\end{array} \\
\begin{array}{c}
1 \\ 2 \\ 3 \\ 4 \\ 5 \\ 6
\end{array}
\left[
\begin{array}{ccccccccc}
1 & 1 & 1 & 0 & 0 & 0 & 0 & 0 & 0 \\
0 & 0 & 0 & 1 & 1 & 1 & 0 & 0 & 0 \\
0 & 0 & 0 & 0 & 0 & 0 & 1 & 1 & 1 \\
1 & 0 & 0 & 1 & 0 & 0 & 1 & 0 & 0 \\
0 & 1 & 0 & 0 & 1 & 0 & 0 & 1 & 0 \\
0 & 0 & 1 & 0 & 0 & 1 & 0 & 0 & 1
\end{array}
\right]
\end{array}
$$

(a) Determine the degree of each vertex.

(b) Draw a representation of the graph.

5. Construct the adjacency matrix for the graph in Fig. 5.20.

6. Construct the adjacency matrix for the graph in Fig. 5.21.

7. Construct the adjacency matrix for the graph in Problem 3.

8. Construct the adjacency matrix for the graph in Problem 4.

9. Let G be the graph with adjacency matrix given by

$$
\begin{array}{c}
\begin{array}{ccccccccc}
1 & 2 & 3 & 4 & 5 & 6 & 7 & 8 & 9
\end{array} \\
\begin{array}{c}
1 \\ 2 \\ 3 \\ 4 \\ 5 \\ 6 \\ 7 \\ 8 \\ 9
\end{array}
\left[
\begin{array}{ccccccccc}
0 & 0 & 0 & 1 & 1 & 0 & 0 & 0 & 0 \\
0 & 0 & 0 & 0 & 1 & 1 & 0 & 0 & 0 \\
0 & 0 & 0 & 0 & 0 & 1 & 0 & 0 & 0 \\
1 & 0 & 0 & 0 & 1 & 0 & 1 & 1 & 0 \\
1 & 1 & 0 & 1 & 0 & 1 & 0 & 1 & 1 \\
0 & 1 & 1 & 0 & 1 & 0 & 0 & 0 & 0 \\
0 & 0 & 0 & 1 & 0 & 0 & 0 & 0 & 0 \\
0 & 0 & 0 & 1 & 1 & 0 & 0 & 0 & 0 \\
0 & 0 & 0 & 0 & 1 & 0 & 0 & 0 & 0
\end{array}
\right]
\end{array}
$$

(a) Determine the degree of each vertex.

(b) Draw a representation of G.

(c) Find the incidence matrix of G.

10. Let G be the graph with adjacency matrix given by

$$
\begin{array}{c}
\begin{array}{cccccc}
a & b & c & d & e & f
\end{array} \\
\begin{array}{c}
a \\ b \\ c \\ d \\ e \\ f
\end{array}
\left[
\begin{array}{cccccc}
0 & 1 & 1 & 0 & 1 & 0 \\
1 & 0 & 1 & 1 & 0 & 0 \\
1 & 1 & 0 & 0 & 0 & 1 \\
0 & 1 & 0 & 0 & 1 & 1 \\
1 & 0 & 0 & 1 & 0 & 0 \\
0 & 0 & 1 & 1 & 0 & 0
\end{array}
\right]
\end{array}
$$

(a) Determine the degree of each vertex.
(b) Draw a representation of G.
(c) Find the incidence matrix of G.

11. Suppose the graph H has the adjacency matrix given by

$$
\begin{array}{c}
 \\
1 \\
2 \\
3 \\
4 \\
5
\end{array}
\begin{array}{ccccc}
1 & 2 & 3 & 4 & 5 \\
\left[\begin{array}{ccccc}
0 & 1 & 0 & 0 & 0 \\
1 & 0 & 1 & 0 & 0 \\
0 & 1 & 0 & 0 & 0 \\
0 & 0 & 0 & 0 & 1 \\
0 & 0 & 0 & 1 & 0
\end{array}\right]
\end{array}
$$

Use PATH2 to determine whether there is a path from vertex 1 to vertex 5.

12. Suppose the graph K has the adjacency matrix given by

$$
\begin{array}{c}
 \\
1 \\
2 \\
3 \\
4 \\
5
\end{array}
\begin{array}{ccccc}
1 & 2 & 3 & 4 & 5 \\
\left[\begin{array}{ccccc}
0 & 1 & 1 & 0 & 0 \\
1 & 0 & 0 & 1 & 0 \\
1 & 0 & 0 & 0 & 1 \\
0 & 1 & 0 & 0 & 1 \\
0 & 0 & 1 & 1 & 0
\end{array}\right]
\end{array}
$$

Use PATH2 to determine whether there is a path from vertex 1 to vertex 5.

13. Use algorithm CONNECT2 to determine whether the graph with the adjacency matrix given below is connected.

$$
\begin{array}{c}
 \\
1 \\
2 \\
3 \\
4 \\
5
\end{array}
\begin{array}{ccccc}
1 & 2 & 3 & 4 & 5 \\
\left[\begin{array}{ccccc}
0 & 1 & 1 & 0 & 0 \\
1 & 0 & 0 & 1 & 0 \\
1 & 0 & 0 & 0 & 1 \\
0 & 1 & 0 & 0 & 1 \\
0 & 0 & 1 & 1 & 0
\end{array}\right]
\end{array}
$$

14. Use algorithm CONNECT2 to determine whether the graph with the adjacency matrix given below is connected.

$$
\begin{array}{c}
 \\
1 \\
2 \\
3 \\
4 \\
5
\end{array}
\begin{array}{ccccc}
1 & 2 & 3 & 4 & 5 \\
\left[\begin{array}{ccccc}
0 & 1 & 0 & 0 & 0 \\
1 & 0 & 1 & 0 & 0 \\
0 & 1 & 0 & 0 & 0 \\
0 & 0 & 0 & 0 & 1 \\
0 & 0 & 0 & 1 & 0
\end{array}\right]
\end{array}
$$

*P15. Write a program to implement algorithm PATH2. (*Hint:* You will need to use the matrix multiplication procedure from Chapter 4.) Apply your program to the graphs described in Problems 11 and 12.

*P16. Write a program to implement algorithm CONNECT2. Execute algorithm CONNECT2. Apply your program to the graphs described in Problems 13 and 14.

5.6   SOME HISTORICAL PROBLEMS

The *Original* Graph Theory Problem

In the early 1700s, the Prussian village of Koenigsberg was laid out on the two sides of the river Pregel and on two islands in the middle of the river. The parts of the village were connected by bridges as pictured in Fig. 5.22.

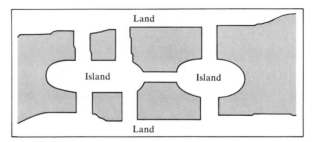

Figure 5.22

According to tradition, the citizens of the village were in the habit of taking walks through the village on Sundays, and a popular pastime was to attempt to find out if there was some route through the village which would enable a person to cross every bridge exactly once. A letter describing the problem was written to the Swiss mathematician Leonard Euler. Euler's solution to this problem, and the techniques that he used, actually marked the start of the study of the topic which we now call *graph theory*. Euler's technique for solving the problem was essentially to take the map of Koenigsberg and transform it into what we would now call a graph, with the shores and islands becoming the vertices of the graph and the vertices being joined by edges in the same manner as they were joined by bridges. Euler's graph looked like Fig. 5.23.

Although in Euler's graph we have a couple of instances of multiple edges (a pair of vertices joined by more than one edge), the theory as we have developed it so far will still work under those circumstances. The only thing which we really haven't done is to make a definition which would formalize the idea

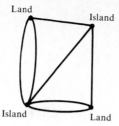

Land   Island

Island   Land

Figure 5.23

of multiple edges. We leave it to the reader to construct a viable definition of a graph which has the possibility of multiple edges. (Such a graph is usually called a *multigraph*.)

To approach the Koenigsberg problem, we need two more definitions:

DEFINITIONS

An **eulerian path** through a graph is a path which contains each edge of the graph exactly one time.

An **eulerian circuit** is an eulerian path which is a circuit.

Theorem 5.21 contains the essence of Euler's work on the Koenigsberg problem.

THEOREM 5.21   Euler's Theorem

A connected graph has an eulerian path if and only if the number of vertices in the graph of odd degree is either 0 or 2. The eulerian path is a circuit if and only if that number is 0; otherwise, the eulerian path is one which leads from one of the vertices of odd degree to the other.

PROOF

Recall that the number of vertices of odd degree must be an even number, so no graph will have exactly one vertex of odd degree. First we prove the "if" part. We show that if a graph has zero or two vertices of odd degree, then it does admit an eulerian path, which is a circuit when the number of such vertices is 0 and is a path connecting the odd vertices when the number of odd vertices is 2.

The proof is by induction on the number of edges in the graph, with $P(n)$ being the assertion that the theorem is as stated above. We can prove $\forall n P(n)$ by means of strong induction.

Basis Step   For a connected graph with only one edge, there are only two possibilities. First, the graph consists of a single vertex with a "loop" (Fig. 5.24a). In that case the degree of the vertex is 2, and following that edge is indeed following a circuit. Second, two vertices are joined by a single edge (Fig. 5.24b). In that case we have two vertices of odd degree, and following the

Figure 5.24 Graphs satisfying the basis condition.

single edge does provide an eulerian path from one odd vertex to the other. (We observe that our proof is valid for graphs which have loops and multiple edges.)

Induction Step We assume that $P(n)$ is true for $n \le k$ and show that it follows that $P(k + 1)$ is true. Take $G = (V, E)$ and assume that G is connected, with $k + 1$ edges, and has two or fewer vertices of odd degree. Our trick here will be to reduce the number of edges to k by consolidating two of them, noting that the induction hypothesis is satisfied for the new graph, and then reconstructing the original graph. The only real problem with this would occur if our graph were to become disconnected in the process.

Since we have already proved $P(1)$, we can assume that G has at least two edges. Thus G has at least one vertex of positive even degree. Suppose that a is such a vertex. We can guarantee two edges are incident to a. (The verification of this is left as an exercise. You need to be careful about the case in which there is a loop at a.) The two edges at a we call $\{a, b\}$ and $\{a, c\}$ (Fig. 5.25, left). Now we form a new graph $G' = (V', E')$, where $V' = V$ and E' is the same as E *except* that $\{a, b\}$ and $\{a, c\}$ have been removed and replaced by a new edge $\{b, c\}$ (Fig. 5.25, right).

Graph G' has k edges and the same number of odd vertices as G. There are now two possibilities: Either G' is connected, or it is not. If it is connected, we can apply the induction hypothesis to G' and find an eulerian path through G'. That path can be turned into an eulerian path through G by replacing the part of the path that uses $\{b, c\}$ by the sequence of vertices *bac*, which uses the edges $\{a, b\}$ and $\{a, c\}$.

If G' is not connected, we have to work a little harder. (The need for the use of strong induction will become apparent here.) Then G' will have two components, one containing a and the other containing b and c. (And b and c must be in the same component, because we have an edge in G' connecting them.) We call these two components G'_a and G'_b. Each of these components is a connected graph with k or fewer edges. Graph G' has exactly the same

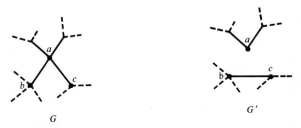

Figure 5.25 The edges $\{a, b\}$ and $\{a, c\}$ replaced by $\{b, c\}$.

number of odd vertices as G (verify); so between the two components there are no more than two vertices of odd degree, and the induction hypothesis applies to each of G'_a and G'_b. If G had no odd-degree vertices, then neither of the two components of G' will have odd vertices; and if G' has two odd vertices, then one of the two components will have two odd vertices and the other will have no odd vertices.

Thus there are three cases: two odd vertices in G'_a, two odd vertices in G'_b, or no odd vertices in either. We take care of the first and leave the other two cases as exercises. If there are two odd vertices in G'_a (and because the induction hypothesis does apply), we know there is an eulerian path $o_1 x_1 \cdots x_m a x_{m+1} \cdots x_k o_2$ which connects the two odd vertices o_1 and o_2 of G'_a. Also by the induction hypothesis, in G'_b there is an eulerian circuit $w_1 \cdots w_p b c w_{p+1} \cdots w_1$. We can remove bc from the circuit and glue it to the path in the other component by means of the following path:

$$o_1 x_1 \cdots x_m a c w_{p+1} \cdots w_1 \cdots w_p b a x_{m+1} \cdots o_2$$

This path includes all the edges of G ($\{a, b\}$ and $\{a, c\}$ are included, and $\{b, c\}$ is not included) and connects the proper vertices. For the "only if" part of the proof, note that if we have an eulerian path $a x_1 \cdots x_m b$, each of the x's will occur on two edges each time it is listed in the path. Thus each must have even degree unless it is either a or b. If $a = b$, then all the vertices have even degree; and if not, then only a and b have odd degree. ∎

If we look at Euler's graph of Koenigsberg, we see that all four vertices have odd degree, and thus no eulerian path is possible.

We also note that this theorem provides us with the answer to question 5 in Examples 5.1 and 5.2.

Hamiltonian Paths

A related, but surprisingly more difficult, problem was raised by the mathematician Hamilton in the following form: Is there a path through a given graph which passes through each vertex in the graph exactly one time? A path with that property is called a *hamiltonian path*. It is somewhat surprising to find that the question of whether a graph has a hamiltonian path is harder to answer than the same question about the existence of an eulerian path. There is no nice, neat set of criteria which will guarantee the existence or non-existence of a hamiltonian path through a graph. There are obviously some sufficient conditions for the existence of a hamiltonian path, for example, any complete graph has a hamiltonian path, as does a complete bipartite graph in which both sets of vertices have the same number of elements. In general, it is very difficult to come up with an efficient algorithmic scheme for making such a test. An exhaustive algorithm will work, but it is most assuredly *not* an efficient algorithm.

The Four-Color Problem

The final classical problem of graph theory that we will take up is the famous four-color map problem. As originally stated to De Morgan (as in De Morgan's laws) in the mid-1800s by one of his students, the four-color map problem asks the following question:

> If we are given a map drawn in the plane, with all the countries represented in the map being connected, what is the minimum number of colors required to color the map if countries which share a common border must be colored with different colors?

The four-color conjecture posed to De Morgan suggested that the number required was four. This led to a considerable amount of activity by a number of mathematicians. De Morgan reportedly conveyed the problem to Hamilton in a letter, but little really happened until the great English mathematician Cayley reported on the problem to the London Mathematical Society in 1878, indicating that he could not solve the problem. Within a year, A. B. Kempe, a lawyer, announced that he had proved the sufficiency of four colors. Unfortunately, Kempe's proof had some holes in it. In fact, what was actually proved by Kempe's argument was that five colors would be sufficient for any planar map. One of the basic means of handling coloring problems is still referred to as a *Kempe chain argument* in honor of Kempe, who was the first to attempt such techniques in the study of graphs.

The four-color problem "floated" around for a long time with no resolution in sight. Interestingly enough, the problem was solved for maps drawn on some nonplanar surfaces such as the torus (or doughnut), but the problem as originally stated for planar maps went unsolved for a long time. Finally, in 1976, a solution was announced by two mathematicians Haken and Appel from the University of Illinois. The controversial point about their solution was the fact that it involved the analysis of nearly 1500 special cases by use of a computer program. The complexity of the computer program—and the possibility that the algorithm might not do as advertised—made Haken and Appel's solution suspect to some mathematicians.

Historically, a number of approaches were taken to this problem, but most people approached it by turning the cartography problem into one of graph theory. The transfer is made by constructing a graph from the map

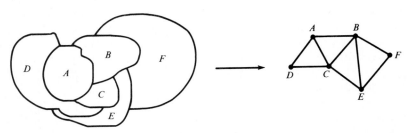

Figure 5.26

under consideration. A vertex is created for each country, and edges are drawn connecting vertices whenever the countries represented by those vertices share a common border. (See Fig. 5.26.) The problem for a graph then becomes one of coloring the vertices of the graph in such a way that two vertices which are connected by an edge (i.e., adjacent vertices) will have different colors. It is easy to see that such a graph will be a planar graph if the original map was at all reasonable (there are no disconnected countries, etc.).

> **DEFINITION**
>
> The **chromatic number** of a graph is the minimum number of colors required to color it according to the rules described above.

The four-color conjecture can then be stated as follows: The chromatic number of a planar graph is always less than or equal to 4. It is clearly not reasonable to prove that here, but we do prove that the chromatic number of a planar graph is always less than or equal to 5. This is, in fact, what Kempe showed, and in part of our argument we make use of a Kempe chain construction.

The first step is to prove a theorem about the degrees of the vertices in planar graphs.

THEOREM 5.22

Every planar graph has at least one vertex of degree 5 or less.

PROOF

If all the vertices of G were of degree 6 or more, then the sum of the degrees of the vertices would be greater than or equal to $6v$. The sum of the degrees of the vertices is, of course, equal to twice the number of edges, so we conclude that $2e \geq 6v$, or $v \leq e/3$. For any planar graph $r \leq 2e/3$, so from $2 = v - e + r$ we get $2 \leq e/3 - e + 2e/3 = 0$. Since we know that $2 \leq 0$ is false, we conclude that there must be some vertex in G with degree 5 or less. ∎

We are now in position to prove the theorem that we wanted.

THEOREM 5.23

Every planar graph has chromatic number less than or equal to 5.

PROOF

We proceed as usual by induction on the number of vertices, so $P(n)$ becomes the statement that every planar graph with n vertices can be colored with five or fewer colors.

Basis Step Obviously a graph with one vertex can be colored with fewer than five colors.

Induction Step   As usual, the plan is to remove something from the graph and then fit it back in, in order to prove our statement. To this end, we take a planar graph G with $k + 1$ vertices and take vertex a to be the vertex of least degree. We let G' be the subgraph that contains all the vertices of G except a and all edges of G except those incident to a. Subgraph G' has k vertices and thus can be colored with five or fewer colors. If the degree of a is less than 5 or if the vertices adjacent to a are colored with fewer than five colors, there is no problem in fitting a back in to the graph. In that case we choose an unused color and assign it to a. If, however, a has degree 5 and each vertex adjacent to a has a different color, then we have to do a little work to color a. The situation is as illustrated in Fig. 5.27.

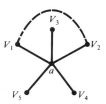

Figure 5.27

We consider the subgraph G'' consisting of all vertices of G' which are colored the same colors as V_1 or V_2 and the edges joining them. (We call the colors red and puce for the sake of giving the colors names.) If V_1 and V_2 are in different components of G', we can simply interchange the colors red and puce in the component containing V_1, which means that we now have two puce vertices in our graph, and a can now be colored red. If V_1 and V_2 are in the same component of G'', then we have a path connecting the two vertices consisting of alternating red and puce vertices. This path can be assumed to be an elementary path. (This path is often called a *Kempe chain*.) This path, together with $V_1 a$ and $a V_2$, forms a cycle in G which encloses at least one vertex of G, say V_3, and has at least one vertex, say V_4, in its exterior. (We let the colors of those vertices be green and indigo, respectively.) If we look at the subgraph that contains all vertices colored green and indigo, we observe that V_3 and V_4 must be in different components. (If this were not true, a path connecting them would cross our original Kempe chain. Since we are dealing with a planar representation of a graph, such a crossing can occur only at a vertex; but the vertices of the Kempe chain are red and puce, not green and indigo.) In this case we can swap green with indigo in all vertices in one of the components and thus color a with the remaining color. This completes the coloring of G. ∎

To determine the chromatic number of a graph by means of an algorithm is a very difficult problem. Some very inefficient algorithms can be written, but the number of operations required for every known algorithm is essentially an exponential function of the number of vertices. It is not known whether algorithms can be found to solve this problem which are more efficient than that.

In fact, the problem of finding the chromatic number of a graph is one of several problems which are called *NP-complete*. It is a fact that if one polynomial time algorithm can be found for any *NP*-complete problem, then a polynomial time algorithm can be found for all of them.

Since no efficient algorithm is known, if one needs to find the chromatic number of a graph, it is necessary to accept either the lengthy algorithm or a faster algorithm which may not provide the exact results.

The second approach is referred to as finding a *heuristic algorithm*. Heuristic algorithms are necessary when hard problems need to be solved with large amounts of data. The following is a heuristic algorithm which provides an approximation to the chromatic number of a graph.

The *Welch-Powell algorithm* finds an approximation to the chromatic number of a graph in the following way:

1. Sort the vertices into decreasing order of degree.

2. Assign a color to the first vertex in the list, and in turn try to assign this color to each vertex in the list. Repeat this process until a color is assigned to each vertex.

The problem of coloring graphs is one with many practical applications. In many universities final examinations are scheduled for each course individually, so that multiple-section courses can all have final examinations at the same time. If a school has 300 different courses during a semester, final examination times must be assigned to each of them. The courses can be viewed as vertices in a graph, and the assignment of an examination period to a course amounts to coloring the vertex. If we view two courses as being joined by an edge if there is a student who is taking both courses, then the problem of assigning a period to an examination so that no student is required to take two examinations at the same time is precisely the problem of coloring this graph according to the usual rules of graph coloring. Finding the chromatic number of the graph in this case is the problem of determining the minimum number of examination periods needed. The use of an exact algorithm is not feasible. For 300 courses, the algorithm, even running on a new fast computer, would most likely not run to completion until after finals' week even if the data were made available at the beginning of the semester. Most universities which schedule finals in this fashion use a scheduling algorithm which is very similar to the Welch-Powell algorithm.

There are many theorems which relate to colorings of graphs in general, and we look at a few in the problems.

Problem Set 6

Determine whether each of the following graphs has an eulerian path. If it does, find it. If it does not, explain why not.

1.
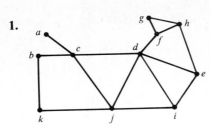

2.
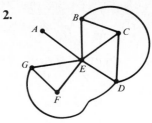

3.
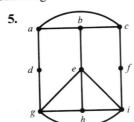

4.
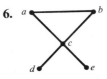

Find hamiltonian paths or explain why there cannot be one in each of the following:

5.
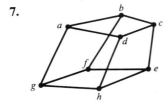

6.
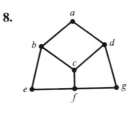

7.
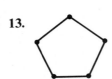

8.

Determine the chromatic number for the following graphs:

9.

10.

11.

12.

13.

14.

*15. Devise a mathematical definition for a multigraph.

16. Prove that every bipartite graph has chromatic number 2.

17. Prove that if the chromatic number of a graph is 2, then the graph is bipartite.

*18. A graph is called an *n*-cycle if it consists of a single cycle with *n* vertices. Prove that the chromatic number of an *n*-cycle is 2 if *n* is even and is 3 if *n* is odd.

19. Prove that the chromatic number of a tree is 2.

20. Write a description of the Welch–Powell algorithm, using the algorithm language we use in this text.

21. Describe another scheduling problem which can be modeled by coloring a graph.

22. Use the Welch-Powell algorithm to estimate the chromatic number of the graph:

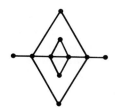

23. Use the Welch-Powell algorithm to estimate the chromatic number of the graph:

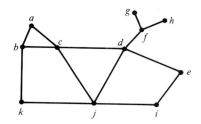

24. Explain the reason for the comment made in the proof of Theorem 5.23 that the path could be assumed to be elementary.

P25. Write a program to take the adjacency matrix of a graph and produce a coloring of the graph by means of the Welch-Powell algorithm. Use the program to find colorings for the graphs in Problems 22 and 23.

CHAPTER SUMMARY

Chapter 5 is the first of two chapters devoted to graph theory. This chapter has dealt with the theory of *undirected* graphs, while the next will consider directed graphs.

A graph consists of a set, usually considered to be a set of points, called the set of vertices, and a set which is considered to be a set of lines or arcs connecting vertices, although it is more formally described as a set of pairs of vertices.

One of the most important concepts of graph theory is that of a path. A path is a sequence of vertices or edges which provides a "route" through the graph connecting two vertices. By means of paths, in graph theory we can discuss the concepts of distance, diameter, and connectivity of many structures which can be modeled by a graph. In Section 5.1 we introduced a couple of examples of situations which can be described by graphs, and we made use of these examples from time to time in the remainder of the chapter. Special considerations and applications of graph theory make the questions of graph planarity and the chromatic number of a graph especially interesting. The famous four-color map problem is particularly interesting because it is an example of a problem in mathematics which is extremely easy to state, but also extremely difficult to solve.

The techniques of proof which we discussed in Chapter 1 were especially important in proving several of the theorems in this chapter. In many instances the technique of mathematical induction was particularly important.

KEY TERMS

Having completed this chapter, you should now be familiar with the following ideas and terminology:

adjacency matrix (5.5)
bipartite graph (5.3)
chromatic number (5.6)
circuit (5.2)
complete bipartite graph (5.3)
complete graph (5.3)
component of a graph (5.2)
connected graph (5.2)
cut set (5.4)
cycle (5.2)
degree of a vertex (5.2)
diameter of a graph (5.2)
disconnected graph (5.2)
distance (5.2)
edge (5.2)
edge set (5.2)

elementary path (5.2)
eulerian circuit (5.6)
eulerian path (5.6)
Euler's formula (5.3)
face (5.3)
four-color problem (5.6)
fundamental cut set (5.4)
fundamental cycle (5.4)
graph (5.2)
homeomorphic graphs (5.3)
incidence matrix (5.5)
induced subgraph (5.2)
Kempe chain (5.6)
Kuratowski's theorem (5.3)
multigraph (5.6)
node (5.2)

nonplanar graph (5.3)
path (5.2)
planar graph (5.3)
simple circuit (5.2)
simple path (5.2)
spanning subgraph (5.4)
spanning tree (5.4)

subgraph (5.2)
system of fundamental cut sets (5.4)
tree (5.3)
vertex (5.2)
vertex set (5.2)
Welch-Powell algorithm (5.6)

6
Directed Graphs and Trees

6.1 INTRODUCTION

In Section 5.1, we presented a problem which could be solved quite easily once the concept of a graph was developed. In the same spirit, we present a problem which can be described quite conveniently with the language of directed graphs.

EXAMPLE 6.1

The map at the top of page 227 describes the network of one-way streets in the downtown area of a city. The streets run north-south and are labeled with numbers, while the avenues run east-west and are labeled with letters.

We can ask a number of questions about traffic flow in the central area of our city:

(a) Is it possible for a vehicle to travel legally from 1st and A to 1st and D, using only downtown streets?

(b) If pedestrian traffic is not possible on the closed section of 1st Street, can a pedestrian (who need not obey one-way signs) get from 1st and A to 1st and D?

(c) What is the fastest way for a vehicle to (legally) travel from 1st and A to 4th and D?

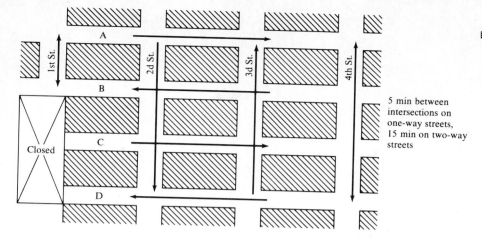

5 min between
intersections on
one-way streets,
15 min on two-way
streets

(d) What routes are possible for a bus "loop" which will travel through downtown streets, starting and ending at 1st and A?

(e) Is it possible for a vehicle to travel from any intersection to any other while remaining on downtown streets? Is it possible for pedestrians? ■

In this chapter, two ideas—directed graphs and weighted graphs—will enable us to develop the terminology for dealing with the questions in Example 6.1a–e and will allow us to find some algorithms for solving the problems. In many cases, our methods will also apply to the type of graph discussed in Chapter 5.

In Example 6.1, a graph is not quite a satisfactory tool for answering these questions. The network of downtown streets can be described by a graph such as the one in Fig. 6.1, but the problem is that it contains no information at all about the one-way nature of some of the streets, nor does it provide us with information about the travel times along each of the streets.

In the sections which follow, we expand our concept of graph to include objects which go beyond Fig. 6.1, and we discover that much of what we do turns out to be an extension of the ideas which we developed in considering ordinary graphs. In Fig. 6.2 we represent all the information necessary to solve the problems in Example 6.1 in a graph which we will call a *weighted directed graph*.

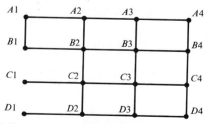

Figure 6.1 Graph representing the streets in Example 6.1.

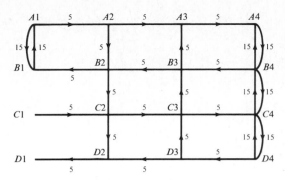

Figure 6.2 Weighted, directed graph.

First we focus on the question of direction for a graph, and later we introduce the idea of a weighted graph. The concept of a weighted graph can also be applied to an ordinary graph.

The problems which follow ask you to begin thinking, based on your experiences with graphs from Chapter 5, about how we might proceed to consider directed graphs.

Problem Set 1

1. Propose a definition for a directed graph. (*Hint*: What objects have we studied which could be used to place direction on an edge?)

2. What should a definition be for a path in a directed graph?

3. Answer the question in Example 6.1a, either by finding a path that works or by producing a convincing argument that no such path is possible.

4. Answer Example 6.1b, either by finding a path that works or by producing a convincing argument that no such path is possible.

*5. Propose a definition for connected directed graphs.

*6. Attempt to determine whether the graph in Fig. 6.2 satisfies your definition in Problem 5.

*7. Solve the problem posed in Example 6.1c.

Problems 8 to 16 refer to the following situation. Sandra's Department Store is configured in the following way: There are 10 floors, with express elevators running between floors 1 and 10 and floors 1 and 5. (The express elevators make *no* intermediate stops.) There are escalators connecting each pair of floors in both directions, but the down escalator between floors 5 and 6 is broken.

8. Draw a directed graph to represent this information. You should include in your graph the fact that some distance must be traveled between the escalator landings and the elevator entries on floors 1, 5, and 10.

9. With this configuration, is it still possible to get from any floor to any other floor? Explain your answer.

10. The escalators take 30 seconds per floor, while the 10th-floor elevator takes 4 minutes and the 5th-floor elevator 2 minutes. There is a 1-minute walk from the elevators to the escalator landings. Represent this information with a directed graph.

11. Find the *fastest* route from the escalator landing on floor 6 to the escalator landing on floor 5 and the *fastest* route from the escalator landing on floor 10 to the escalator landing on floor 5 in view of the information given in Problem 10.

12. Represent the possible routes of travel with a directed graph if the *up* escalator from floor 3 to floor 4 is also broken.

13. Repeat Problem 9 in view of the information in Problem 12.

14. Repeat Problem 11 in view of the information in Problem 12.

15. Because of fire regulations, there are stairways connecting all pairs of floors, which are located a 1-minute walk from the escalators and 2 minutes from the elevators. It takes 2 minutes to go up a floor on the stairs and 1 minute to go down a floor. Represent this new information in a directed graph.

16. Repeat Problem 11 given the new information in Problem 15.

6.2 PATHS, CYCLES, ETC.

Basic Terminology

As was the case in Chapter 5, the second section of this chapter introduces the basic terminology that we use in dealing with directed graphs. Also, instead of including a separate section dealing with computer representation of graphs, we note that an extension of the idea of the adjacency matrix will serve our purposes very nicely in constructing algorithms for solving questions related to directed graphs. The obvious difference between graphs and directed graphs is that for directed graphs, we wish to allow the edges to have a first and last entry; i.e., edges in directed graphs should run from one vertex to another. Much of our earlier terminology carries over easily, and we tie everything back to Chapter 5 whenever that is possible. As before, understanding of the terminology is an important first step to understanding how to use the concepts involved. From now on, we refer to graphs as *undirected* when we need to emphasize that we are not dealing with a directed graph.

DEFINITIONS

A **directed graph** or **digraph** is an ordered pair (V, E) in which V is the **vertex set** and E is a subset, called the **edge set**, of $V \times V$.

We say that an edge (a, b) is an edge **from a to b** and that there is an edge from a to b if $(a, b) \in E$.

The vertex a is called the **initial vertex** of the edge, and b is called the **terminal vertex**.

All edges in a directed graph are regarded as having a direction, and we cannot assume that there is an edge from b to a just because there is an edge from a to b. The only change which we have made, compared to our definition of an undirected graph, is to make edges ordered pairs of vertices instead of two-element sets of vertices. This does not make much difference in terms of our dealing with concepts such as paths, cycles, etc. There will be a small hitch when we deal with connectivity, but even that is easily resolved.

EXAMPLE 6.2

Consider the digraph $G = (V, E)$, where $V = \{a, b, c, d\}$ and $E = \{(a, b), (b, c), (b, b), (c, d)\}$. Draw a representation of G.

SOLUTION

In representing a directed graph with a diagram, we use the same basic technique as we did with undirected graphs, except that we use an arrowhead on each edge to indicate the direction of the edge. Thus we have four points to label as vertices, and we connect the points with arcs between a and b, b and c, and c and d. We also draw a loop at vertex b.

Arrows are drawn on each edge, pointing away from the initial vertex and toward the terminal vertex. ∎

From Example 6.1, clearly the concept of a path should play the same kind of pivotal role in dealing with directed graphs as it did in the theory of undirected graphs. In the diagram in Example 6.2 we can see that the routing from a to b to c to d would appear to be the kind of object that we would like to designate as a path.

DEFINITION

A **directed path** is a sequence of vertices $a_1 a_2 \cdots a_n$ in which (a_i, a_{i+1}) is an edge.

We often refer to a path informally as a sequence of edges, and the transition between the two points of view should be quite obvious. We actually use whichever of the two references is more convenient for the problem we are considering.

DEFINITION

An **undirected path** is a sequence of vertices in which either (a_i, a_{i+1}) or (a_{i+1}, a_i) is an edge.

Undirected paths are those sequences of vertices which would be paths if the graph in question were an undirected graph instead of a directed graph.

On occasion we need to consider the undirected graph which results from ignoring the direction of the edges in the graph. It is a good exercise for the reader to write a formal definition of this graph, which we call the *associated undirected graph*. In Example 6.2, the associated undirected graph is the graph (V, E) where $V = \{a, b, c, d\}$, and $E = \{\{a, b\}, \{b, c\}, \{c, d\}\}$. (Note that, as usual, we don't include "loops" in undirected graphs.)

The terminology used to describe paths as simple, or elementary, is exactly the same as we used with undirected graphs, except that this terminology can be applied to either undirected paths or directed paths. In Section 6.4 we officially introduce the concept of connectivity as related to directed graphs, but for the moment we leave it open to the reader's imagination as to how that might be defined.

Many of the questions raised in Example 6.1 are related to existence and nonexistence of paths. As we progress through Chapter 6, we will be able to answer all these questions, and in many cases, we will be able to discuss algorithms for supplying the answers.

Incidence and Degree

DEFINITIONS

A vertex is **incident** to an edge if and only if it is one of the entries of the ordered pair.

The **outdegree** of a vertex is the number of edges to which the vertex is incident in which that vertex is the first entry of the ordered pair.

The **indegree** of a vertex is the number of edges to which the vertex is incident in which the vertex is the second entry of the ordered pair.

In terms of a representation of a digraph, the outdegree of a vertex is the number of arrows leading away from the vertex while the indegree is the number of arrows leading into the vertex. These ideas obviously relate to the number of paths leading into and out of a vertex. In Example 6.2, vertex a has outdegree 1 and indegree 0, b has outdegree 2 and indegree 2, c has outdegree 1 and indegree 1, and d has indegree 1 and outdegree 0.

The following theorem is closely related to Theorem 5.1.

THEOREM 6.1

For any directed graph, the sum of the outdegrees of the vertices is equal to the sum of the indegrees of the vertices.

PROOF
The proof of this theorem is very similar to that of Theorem 5.1, and we leave it as an exercise. ∎

Weighted Graphs

We are also concerned with the concept of a *weighted*, or *labeled*, *graph*. Either an undirected graph or a directed graph may be weighted.

DEFINITIONS

A **weighting function** is a function whose domain is the set of edges in a graph.

A graph is a **weighted graph** if the range of its weighting function is a set of numbers.

A graph is **labeled,** or **colored,** if the range of the weighting function is a set of symbols which is not used as a set of numbers. (If the range is a set of numbers which are used only as symbols, then we usually regard the weighting to be a labeling.) See Fig. 6.3.

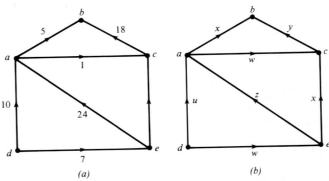

Figure 6.3 Examples of (*a*) weighted and (*b*) labeled graphs.

With a weighting or labeling, we are assigning weights or labels to each edge of a graph. Weighting functions can be used to provide an idea of the length of a path or to provide an alternative way of describing a path. For any path, we can use the sum of the values of the weight function on each of the edges in the path to represent the length of the path, or perhaps the cost of using the path. If the edges of the graph are labeled with symbols, then the sequence of symbols on the edges of the path can be used to describe the path. In either case, we draw the graph by writing the weight or label on the edge. In particular, if each edge has a different label, the sequence uniquely describes each path in the graph. As an illustration of the use of weighted graphs, Section 6.3 provides an algorithm which finds the shortest path between two vertices in a directed graph.

Digraphs and Matrices

In representing a digraph for computer analysis, the adjacency matrix is the method of choice; it works in exactly the same way as the adjacency matrix works for undirected graphs. We have a row and a column for each vertex. We then place a 1 in row i column j if and only if there is an edge from vertex i to vertex j.

EXAMPLE 6.3

Consider the graph from Example 6.2:

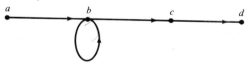

The matrix of this graph is

$$
\begin{array}{c}
 \\
a \\
b \\
c \\
d
\end{array}
\begin{array}{cccc}
a & b & c & d \\
\begin{bmatrix}
0 & 1 & 0 & 0 \\
0 & 1 & 1 & 0 \\
0 & 0 & 0 & 1 \\
0 & 0 & 0 & 0
\end{bmatrix}
\end{array}
$$

∎

Note that this matrix, unlike the adjacency matrix for undirected graphs, is *not* a symmetric matrix. We can use this matrix to determine many important facts about the graph. Since the row labeled a represent edges *from a*, the sum of the entries in row a will represent the outdegree of vertex a. In a similar fashion, the sum of the entries in column a represents the indegree of vertex a. The adjacency matrix provides a unique description of the graph. If we are given the adjacency matrix for a graph, the graph represented by that adjacency matrix is unique.

EXAMPLE 6.4

Determine the digraph whose adjacency matrix is

$$
\begin{array}{c c}
 & \begin{array}{c c c} a & b & c \end{array} \\
\begin{array}{c} a \\ b \\ c \end{array} &
\left[\begin{array}{c c c}
1 & 1 & 0 \\
0 & 0 & 1 \\
0 & 1 & 0
\end{array}\right]
\end{array}
$$

The graph has three vertices, a, b, and c. We need to construct edges from a to b, b to c, and c to b. We also need a loop at vertex a. The graph then looks like the following:

Use of the adjacency matrix is the key to developing a very efficient algorithm for determining whether a path exists between vertices in a directed graph, and the same algorithm can be used in undirected graphs just as well.

Weighted Graphs and Matrices

The adjacency matrix can be modified to describe a weighted graph by placing in row i, column j the weight of the edge joining i and j. In doing this, it becomes necessary to use some symbol other than 0 to indicate the lack of an edge. We use a dash to indicate no value assigned, which we might think of as an infinity sign to indicate an edge of infinite length. The latter is particularly useful in algorithms which determine paths of minimum length or subgraphs of minimum total weight.

EXAMPLE 6.5

Construct an adjacency matrix to represent this weighted graph:

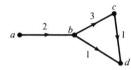

The adjacency matrix for this graph can be described by noting that the edge from b to c has weight 3, so that entry in the matrix (row b, column c) will be a 3. In the same way we can put 1s in row c, column d and row b, column d. Finally we place a 2 in row a, column b. The remaining edges do not exist.

This leaves us with the matrix

$$
\begin{array}{c}
\\
a \\
b \\
c \\
d
\end{array}
\begin{array}{cccc}
a & b & c & d \\
\end{array}
\left[
\begin{array}{cccc}
- & 2 & - & - \\
- & - & 3 & 1 \\
- & - & - & 1 \\
- & - & - & -
\end{array}
\right]
$$

■

It should be easy to see in Example 6.5 that this kind of matrix also uniquely determines a graph. Our next order of business is to explain how to make use of these tools.

Problem Set 2

1. Let $G = (V, E)$ be a graph in which $V = \{a, b, c, d, e\}$ and $E = \{(a, b), (b, a), (a, c), (a, d), (b, c), (d, e)\}$.
   (a) Draw a representation of G.
   (b) Find the adjacency matrix for G.
   (c) Determine the indegree and outdegree of each vertex.
   (d) Find (if possible) a directed path in G from a to e.

2. Let $G = (V, E)$ be a graph in which $V = \{1, 2, 3, 4, 5, 6\}$ and $E = \{(1, 3), (2, 4), (3, 5), (5, 3), (4, 6), (4, 3), (2, 2)\}$.
   (a) Draw a representation of G.
   (b) Find the adjacency matrix for G.
   (c) Determine the indegree and outdegree of each vertex.
   (d) Find (if possible) a directed path in G from 1 to 6.

3. Repeat the instructions for Problem 2 with the graph $G = (V, E)$ if $V = \{1, 2, 3, 4, 5, 6, 7\}$ and $E = \{(1, 3), (2, 4), (3, 5), (4, 6), (6, 4)\}$.

4. Repeat the instructions for Problem 1 with the graph $G = (V, E)$ if $V = \{a, b, c, d, e, f\}$ and $E = \{(a, b), (b, c), (c, f), (e, f), (d, e), (a, d)\}$.

In Problems 5 to 8, find adjacency matrices for the graphs.

5.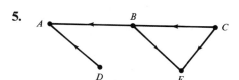

6.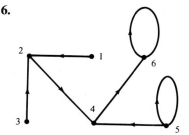

7.

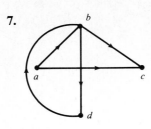

8.

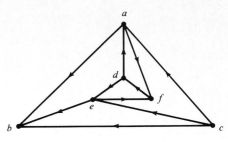

9. Consider this airline fare table for four cities:

	Boston	Chicago	Kansas City	Seattle
Boston	—	195	—	300
Chicago	187	—	65	200
Kansas City	—	75	—	180
Seattle	300	170	180	—

(a) Represent the information in this table by a labeled directed graph.

(b) There is no direct fare given from Boston to Kansas City. Determine the cheapest fare that can be found by putting together two direct flights.

10. The following graph represents distances between some cities in eastern Nebraska. Use the graph to produce a matrix to serve as a mileage table for the cities in the graph. (If no direct route is indicated, indicate "no connection" with a dash.)

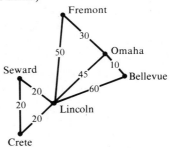

11. Several trucking companies have made bids on carrying a shipment of lumber all or part of the way from Seattle, Washington, to Minneapolis, Minnesota. Only company A can carry the load all the way, but the shipment can be transferred between lines at any point along the way. Here is a table of the bids received:

Company	Start	Seattle	Yakima	Missoula	Billings	Fargo	Minneapolis
A	Seattle	0	21	27	35	46	64
B	Yakima	—	—	21	27	35	46
C	Missoula	—	—	—	22	28	36
D	Billings	—	—	—	—	22	28
E	Fargo	—	—	—	—	—	23

Represent this information in a digraph.

*12. Determine the least expensive way of shipping the lumber from Seattle to Minneapolis.

13. Propose a definition of the associated undirected graph for a directed graph.

*14. Prove Theorem 6.1.

15. In the graph below we say that a "word" is "accepted" if there is a path from x_0 to x_5 in which the labels corresponding to the edges match the "letters" in the word. Test the acceptability of the following words.
(a) *abbacab*
(b) *ab*
(c) *abca*
(d) *aaa*
(e) *acab*

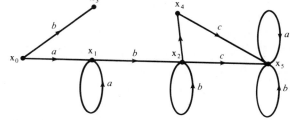

*16. Explain why the flowcharts often employed by **BASIC** and **FORTRAN** programmers are actually implementations of directed graphs.

*17. Phrase the questions asked in Example 6.1 in the terms used in graph theory.

P18. Write and execute a computer program to solve Problem 9 from Section 6.1.

6.3 DIJKSTRA'S ALGORITHM

Optimization Problems

In Example 6.1c we asked you to find the fastest route from 1st and A to 4th and D. Clearly there are many different ways to approach this particular problem. But since we may wish to find the shortest path through very large directed graphs, it is important to be able to find an efficient algorithm to solve problems like this. This particular problem is one of many mathematical problems of *optimization*, in which we are concerned with finding a solution which minimizes a quantity that we would like to have as small as possible or which maximizes a quantity that we would like to have as large as possible. In particular, in graph theory we are often interested in finding the shortest path between two points, the cheapest spanning tree for a graph, and solutions to other similar problems.

Because of their frequent occurrence, it is important for us to be able to answer such questions. Thus the computer scientist and the mathematician must be interested in finding efficient algorithms for solving optimization problems. There has been considerable investigation into problems of this

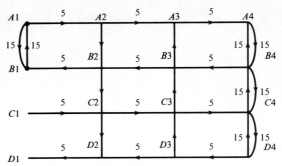

Figure 6.4 Weighted, directed graph of downtown street system.

nature, and in some cases the best-known algorithms are not very efficient in terms of execution time (in some cases this is because the problem turns out to be one of the "hard" or NP-complete problems). But in the case of the shortest-path problem, a very efficient solution was found very early in the computer age. We discuss this now because it involves the use of weighted directed graphs (see Fig. 6.4), and it is particularly clever. To understand why this algorithm is so efficient, we first look at a very natural, but extremely inefficient, solution.

The Algorithm SHORTPATH

A first pass at constructing an algorithm for finding the path of least weight would involve using an exhaustive algorithm and simply choosing the smallest of all possible weights.

ALGORITHM SHORTPATH

{Infinity represents a very large constant.}
{First is the first vertex in the desired path, and
last is the vertex which we are trying to reach.}
{n represents the number of vertices in the graph.}
{The subalgorithm CONSTRUCT_A_PATH constructs a path from a given
vertex to another. We assume that the results of
CONSTRUCT_A_PATH are stored in the variable "path".}
begin
min: = infinity;
repeat
   CONSTRUCT_A_PATH (first, last);
   if length (path) < **min then**
                 min: = length (path);
                 route: = path
   fi;
until all paths from first to last with fewer than n edges are constructed
end.

In analyzing this algorithm, one point is particularly clear: The sub-algorithm CONSTRUCT_A_PATH is likely to be extremely inefficient. We limited the algorithm to consider paths with fewer than n edges, and if the graph has e edges, just considering the number of possible paths to be explored with exactly $n - 1$ edges will require e^{n-1} possible sequences of edges as possible paths. (This goes back to our counting principles from Chapter 2.) Thus the time required for this algorithm will be at least an exponential function with the number of vertices as the variable. Algorithms whose efficiency is given by an exponential function are called *exponential time algorithms*. Exponential time algorithms are considered particularly undesirable because of their extremely rapid growth rate, which can mean that small problems can be practically solved with such an algorithm, but the amount of time required to solve a moderate-size problem makes it quite infeasible to use. (If the function were, for example, 2^n and time were measured in seconds, a problem with 4 vertices would be solved in 16 seconds, while a problem with 20 vertices would require 2^{20} seconds, which is approximately 12 *days*.)

We have presented a very inefficient algorithm in order to verify that the problem in question does have an algorithmic solution, no matter how bad it may be, which is *not* always the case with every problem.

It is useful to consider a specific example to verify the inefficiencies which are built into this algorithm and thus to see whether it is possible to remedy some of those inefficiencies in a new algorithm.

EXAMPLE 6.6

Find the shortest path from A to E in the following graph:

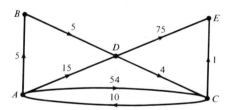

Two wasteful things occur when we use an exhaustive algorithm to solve this problem. First, edges (A, C) and (D, E), which are inordinately long, are considered with the same frequency as the edge (C, E), even though it would seem that while (C, E) is *quite* likely to be included in the final path because of its short length, edges (A, C) and (D, E) appear to have very little chance of being included in the shortest path. Second, edge (C, A) is considered with the same frequency as (A, C). Since (C, A) leads back to A, any path which includes (C, A) and goes from A to E can be replaced with a shorter path to E by simply not using the part of the path which caused us to get back to A in the first place. Although it has no great practical value, SHORTPATH does illustrate the inefficiencies which are built into exhaustive algorithms. ∎

Dijkstra's Algorithm

Dijkstra's algorithm was first posed by E. Dijkstra in 1959, and it represents a considerable advance over SHORTPATH.

The idea behind Dijkstra's algorithm is quite clever. The algorithm keeps track of two sets of vertices. The set S consists of all vertices to which a shortest path from the given initial vertex has been found. The set T consists of all the rest of the vertices. When the algorithm begins, S consists of only the initial vertex, and with each iteration of the algorithm, a vertex is removed from T and placed in S. When the desired final vertex is placed in S, the process stops. The algorithm keeps track of the path used to reach each vertex and maintains data on the shortest actual distance (for vertices in S) as well as an estimate of the shortest distance to each vertex in T.

The estimates of the distances for each vertex in T are based on paths which include only vertices in S, except for the last vertex. It can be shown (and we do so later) that the shortest estimated distance to any vertex in T is indeed correct, so that the vertex in T with the shortest estimated distance will be the one added to S (and removed from T). Once the new vertex is added to S, the estimated distances to the vertices remaining in T are updated, and the process is repeated.

We explore Dijkstra's algorithm by applying it to the graph in Example 6.6. This discussion illustrates the process and indicates how each of the steps in the algorithm is accomplished.

We begin by letting $S = \{A\}$, and thus $T = \{B, C, D, E\}$. The distance to A is known (it is zero), and we estimate the distances to B, C, and D by using the length of the edges from A to each of these vertices. The distance to E is regarded as infinite, because as yet we have not found a path to E. The shortest estimated distance is the one to B. We thus remove B from T and place it in S. We also retain the information that the paths to each of B, C, and D *end* with an edge from A.

The next step is to revise the estimated distances for the vertices still remaining in T. Our estimates are based on using paths remaining in S, except for the final edge. To determine whether a better path could be found, we only need to determine whether a *better* path can be found which passes through the vertex (B) that was just added to S. Since the distance to A was 5, we simply add 5 to the distance from A to each of C, D, and E. If this result is smaller, then the new best path to that vertex would be to go to A and to follow the edge from A to that vertex. This does cause us to change the estimated distance to the vertex D to the value 10 (now the path to D is by way of B). We now find that the vertex in T with the smallest estimated distance is D, with a distance of 10. We remove D from T and put it in S. At this point $S = \{A, B, D\}$, and $T = \{C, E\}$.

The process continues as we revise the estimated distances again. Now the estimates give the distance for C of 14 and for E of 85. The process finishes with C going to the set S, a revised value of 15 being calculated for E, and finally E being placed in S. At the same time we retained the fact that (after revisions) the last vertex preceding E was C and preceding C was D. Since D was preceded by B and B by A, the final path is $ABDCE$, with a length of 15.

A table can be used efficiently to track the progress of the algorithm. In the notation of the tables, vertices with actual distance values are in S, and those with estimated distances are in T.

Vertex	Actual Distance	Estimated Distance	Previous
A	0	—	—
B	X	5	A
C	X	54	A
D	X	15	A
E	X	infinite	X

(The dash indicates that it does not apply, and the X indicates that the value has not been calculated yet. The column "Previous" contains the vertex in S which precedes the vertex on the best path.) The vertex in T with the smallest estimated distance becomes the next vertex to move into S. Its actual distance is the old estimated distance. The estimated distances to the other vertices in T are recalculated. The algorithm proceeds in this fashion until E is placed in the set S, and an actual distance is assigned to it. The first iteration of the algorithm gives the following table:

Vertex	Actual Distance	Estimated Distance	Previous
A	0	—	—
B	5	—	A
C	X	54	A
D	X	10	B
E	X	infinite	X

We continue, using vertex D as our new focus:

Vertex	Actual Distance	Estimated Distance	Previous
A	0	—	—
B	5	—	A
C	X	14	D
D	10	—	B
E	X	85	D

Now we use vertex C as the focus:

Vertex	Actual Distance	Estimated Distance	Previous
A	0	—	—
B	5	—	A
C	14	—	D
D	10	—	B
E	X	15	C

And finally we use vertex E:

Vertex	Actual Distance	Estimated Distance	Previous
A	0	—	—
B	5	—	A
C	14	—	D
D	10	—	B
E	15	—	C

We get to E from C, and to C from D, and to D from B, and to B from A, so the shortest path is $ABDCE$.

In counting the number of steps required for this algorithm, we need only note the following: If n is the number of vertices, we will have at most n iterations of the algorithm. At each step, we need to do no more than n additions and comparisons (to compute new estimates for the vertices still in T) and n additional comparisons (to determine which vertex to move into S). Thus we can use no more than $3n$ steps with each iteration, and thus by the fundamental counting principle we need a total of $n * 3n = 3n^2$ steps to complete the algorithm. This is a tremendous improvement over SHORTPATH, and it can be proved that this is, in some sense, the best that we can expect for such an algorithm. We present a partial description of the algorithm insofar as it applies to the process of finding the shortest distance from the initial vertex to the terminal vertex. We leave out the parts of the algorithm which allow the shortest path to be listed, and we encourage readers to determine for themselves how that task should be carried out.

ALGORITHM PARTIAL_DIJKSTRA

{Initial is the starting vertex, terminal is the final vertex, and all the vertices are referred to as integers. The distance matrix is called M.}

```
begin
V := [all vertices]; {V is the set of all vertices.}
S := [initial]; {S is the set of "known" vertices.}
T := V − S; {T is the set of vertices at unknown distances.}
D[initial] := 0; {D[x] is the value for distance from the initial vertex—it is
        evaluated only for vertices in set S.}
for vertex := 1 to number_of_vertices do Temp[vertex] := 9999 od;
{Temp represents the current estimate of the distance—it is initially "infinite."}
focus := initial; {we start with the initial vertex as "focus".}
while terminal in T do
min := 9999;
for vertex := 1 to number_of_vertices do
    if vertex in T then
        if D[focus] + M[focus, vertex] < Temp[vertex] then
            Temp[vertex] := D[focus] + M[focus, vertex] fi;
                {If it is shorter to the vertex by way of focus than is given by
                the current estimate, change the estimate.}
```

```
              if Temp[vertex] < min then
                 min := Temp[vertex];
                 minvert := vertex fi
                 {Keep track of smallest Temp value for this
                 iteration of the process.}
              fi
        fi
     od;
     D[minvert] := Temp[minvert];
     T := T - [minvert];
     S := S + [minvert];
     focus := minvert;
  od
  distance := D[terminal]
end.
```

We also note that Dijkstra's algorithm can be used with weighted undirected graphs just as easily as it can be used with digraphs. In fact, as we shall see in the next section, if we use the appropriate weights, Dijkstra's algorithm can be used to calculate distances and diameters in nonweighted undirected graphs.

The algorithm can be summarized in the following way: We maintain two sets of vertices, S and T. Set S represents the set of vertices whose exact distance has been calculated, and T represents those vertices whose distances have not been calculated. The vertex most recently added to S is chosen as a focus, and the estimated distance to each vertex in T is compared to the sum of the actual distance to the focus vertex and the distance from the focus to the given vertex. If this sum is smaller, then this represents a shorter path to the vertex and the estimated distance is replaced with this new value. After this is done for all vertices in T, the smallest estimate can be proved to be an exact value; so the vertex with the smallest estimate is moved from T to S and becomes the new focus. As soon as the terminal vertex is placed in S, the process halts. The key to the success of this algorithm is the fact that the smallest of the estimated distances to vertices in T is exact, and thus we are able to move the vertex from T to S. The proof that this is actually the case provides an interesting application of mathematical induction, and we present it here:

PROOF
For all vertices in the set S, we prove that $D[v]$ is the minimum distance from initial to v, and we proceed by induction on $|S|$.

Basis   When $|S| = 1$, $S = \{initial\}$ and $D[initial] = 0$, as required.

Induction   We suppose that the statement is true for $|S| = k$, and then we show that it follows for $|S| = k + 1$.

So we can assume that we have followed the algorithm to the point at which $|S| = k$ and that $D[x]$ is exact for all $x \in S$. The algorithm will add a new vertex b to S. That vertex is chosen in such a way that Temp$[b]$ is a minimum for all vertices remaining in T. The value of Temp$[b]$ represents the length of a path which travels through S to vertex c and then proceeds by a single edge to vertex b. Temp$[b]$ is the minimum value of all such paths in the graph, since for each vertex x in T, Temp$[x]$ is the minimum of the lengths of such paths for that vertex, and b is chosen so that Temp$[b]$ is the smallest of all values of Temp. Suppose that there is a shorter path to b. That path must be of the form

initial $x_1\, x_2 \ldots x_n\, b$

This path must leave the set S at some point *before* vertex b; otherwise, we would have calculated Temp$[b]$ as being the length of this path. Let x_i be the first vertex in the path not in S. In this case, Temp$[x_i]$ would have to be no larger than the sum of the lengths of (initial, x_1), (x_1, x_2), ..., (x_{i-1}, x_i). This means, however that Temp$[x_i] <$ Temp$[b]$, which is a contradiction. ■

The Traveling Salesman Problem

The famous "traveling salesman" problem appears to be related to Dijkstra's algorithm in that it asks a related question, namely, how to find the length of the shortest hamiltonian cycle through a directed graph. It is called the *traveling salesman problem* because it is often posed as a problem in which a traveling salesperson is asked to find the shortest route which will cause her or him to visit each city in her or his territory exactly one time and then return to home base. Dijkstra's algorithm is useless in this problem; in fact, the traveling salesman problem is another of those vexing NP-complete problems which we discussed earlier. The best attacks on this problem have not succeeded in doing significantly better than the obvious exhaustive algorithm.

Problem Set 3

Use Dijkstra's algorithm to find the shortest path from A to Z in the following digraphs.

1.

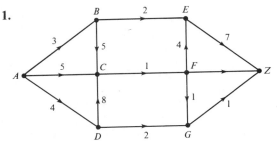

2.

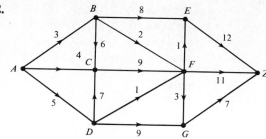

3.

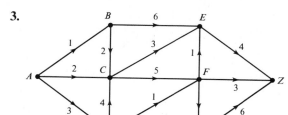

4.

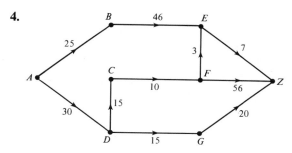

5.

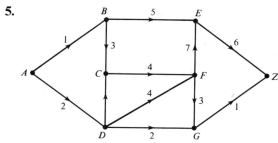

6.

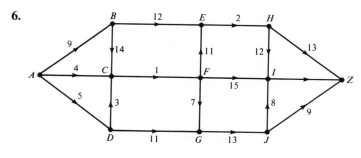

*7. Complete Dijkstra's algorithm in our algorithm language so as to be able to list the vertices in the minimum path. (*Hint*: Recall in our informal discussion of the algorithm we used an array "Previous" to keep track of which vertices were to precede a given vertex in the shortest path to that vertex.)

*P8. Write a Pascal program to implement Dijkstra's algorithm and use it to solve Problems 1 to 6.

*9. Describe how to use Dijkstra's algorithm to find the distance between two vertices in an undirected, nonweighted graph.

*10. The diameter of an undirected graph was defined in Chapter 5. Describe how to use Dijkstra's algorithm to find the diameter of an undirected graph.

*11. Explain why Dijkstra's algorithm is of no use in solving the traveling salesman problem.

12. Use an exhaustive algorithm to solve the traveling salesman problem for the graph in Fig. 6.5.

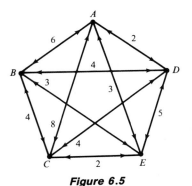

Figure 6.5

*P13. Write a Pascal program to solve the traveling salesman problem exhaustively for graphs with six or fewer vertices, and use that algorithm to solve the problem for both the graph in Fig. 6.5 and the one in Fig. 6.6.

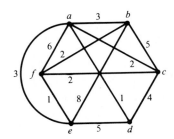

Figure 6.6

14. The "nearest neighbor" heuristic algorithm produces a path that approximately solves the traveling salesman problem by beginning at an arbitrary vertex and selecting the next vertex by using the nearest available vertex. (The algorithm rules out selecting any vertex already selected and cannot return to the original vertex until *all* other vertices have been chosen.) Use the nearest-neighbor algorithm on the graphs in Figs. 6.5 and 6.6, starting with vertex A in each case.

*15. Explain why the nearest-neighbor algorithm could fail if the graph is not complete.

*P16. Write a Pascal program to implement the nearest-neighbor algorithm, and apply it to the graphs in Figs. 6.5 and 6.6.

17. Use combinatorial techniques from Chapter 2 to verify that SHORT-PATH must consider at least e^{n-1} different paths.

18. Verify by combinatorial techniques that Dijkstra's algorithm will need no more than $3n^2$ steps.

6.4 CONNECTIVITY

Kinds of Connectivity

As we mentioned in Section 6.2, there are two kinds of paths to be considered in dealing with digraphs, and as a consequence we are led to deal with two kinds of connectivity. The first kind is known as *strong connectivity*.

> **DEFINITION**
>
> A graph is **strongly connected** if and only if for any ordered pair of vertices there is a (directed) path connecting the two vertices.

EXAMPLE 6.7

In the following pair of graphs, G is strongly connected while H is not.

G

H

Since $ABCA$ is a cycle in G, it is possible to go from any vertex to any other by means of a path with one or two edges. However, H is not strongly connected because vertex C has outdegree 0, and as a consequence it is not possible for *any* directed path to begin at C. ∎

Example 6.7 also illustrates the possibility of another kind of connectivity. The representation for *H* certainly seems to be in "one piece," which suggests that it should be connected according to *some* definition. In fact, we note that in graph *H*, for any pair of vertices, there is an undirected path which joins them.

EXAMPLE 6.8

In the figure below, we see two other digraphs:

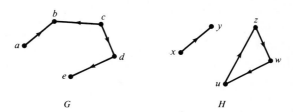

G   H

In graph *G* we see again that for any two vertices which we choose there is an *undirected* path joining them. For example, although *abcde* is not a directed path, it is a path which does join vertices *a* and *e*. However, in graph *H* there is not even an undirected path between vertices *x* and *z*. Graph *G* seems to be the kind of graph we would like to call connected, while *H* is not connected in any sense of the word. ∎

We offer the following definitions:

> **DEFINITIONS**
>
> A directed graph is **connected** if and only if for every pair of vertices in the graph there is an undirected path joining the vertices.
>
> A directed graph is **weakly connected** if it is connected but not strongly connected. (It is clear that a graph will be connected if it is strongly connected.)

Theorem 6.2 relates the connectivity for directed graphs to the connectivity for undirected graphs.

THEOREM 6.2

A digraph is connected if and only if its associated undirected graph is connected.

PROOF

This is an if-and-only-if theorem, and so the proof requires two parts. We must prove that (1) if a graph *G* is connected, then its associated undirected graph is connected and (2) if the associated undirected graph is connected, then the directed graph is connected. We prove the first statement and leave the second as an exercise.

To prove the first statement, we assume that G is connected and show that between any two vertices in the associated undirected graph there is a path in the associated undirected graph. If a and z are vertices of G, then they are also vertices of the undirected graph. We need to be able to construct a path in the associated undirected graph between a and z. Since G is connected, we know that there is an undirected path $av_1v_2 \cdots v_n z$ from a to z. Since this path is an undirected path, for all i, either (v_i, v_{i+1}) or (v_{i+1}, v_i) is an edge of G. In either case, $\{v_i, v_{i+1}\}$ is an edge of the associated undirected graph, and hence $av_1 \cdots v_n z$ is a path in the associated undirected graph. ∎

A consequence of Theorem 6.2 is that all the theorems which we dealt with in Chapter 5 regarding connectivity of undirected graphs can be re-phrased in terms of connectivity of a directed graph. Although these theorems do not apply to strong connectivity, they will apply to the concept of connectivity. To apply these theorems, we simply use them with the associated undirected graph.

We might add that some authors deal with a third concept of connectivity called *unilateral connectivity*. This is defined by the property that for any two vertices a and z in the graph, there is either a directed path from a to z or a directed path from z to a. Theorem 6.3 delineates the relationship between the three kinds of connectivity.

THEOREM 6.3

If G is strongly connected, then it is unilaterally connected. If G is unilaterally connected, then it is connected.

PROOF
This is left as an exercise. ∎

Facts from *Un*directed Graph Theory

A number of facts about undirected graphs apply without change to directed graphs. Theorem 5.10, for example, is true without modification or use of any reference to the associated undirected graph. It is also just as convenient to represent directed graphs by the adjacency matrix and to use it in dealing with directed graphs as it was used in dealing with undirected graphs. You should verify that we can make the same uses of the adjacency matrix for testing the existence of directed paths as we did in dealing with the existence of paths in undirected graphs. In the problems at the end of this section, you are asked to verify that Theorems 5.10, 5.15, and 5.16 are true for directed paths in directed graphs. This means that the algorithms which we discussed for testing connectivity of undirected graphs can be used without change to test for strong connectivity of directed graphs. Although this is convenient, we hasten to point out that there are better algorithms to do this task, and that is the focus of the remainder of this section.

EXAMPLE 6.9

Use matrix multiplication to show that the graph in Fig. 6.7 is strongly connected.

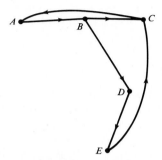

Figure 6.7 Directed graph.

We write the adjacency matrix of the digraph and then compute the first four powers of the matrix and the sums of those powers.

$$M = \begin{bmatrix} 0 & 1 & 0 & 0 & 0 \\ 0 & 0 & 1 & 1 & 0 \\ 1 & 0 & 0 & 0 & 0 \\ 0 & 0 & 0 & 0 & 1 \\ 0 & 0 & 1 & 0 & 0 \end{bmatrix}$$

(Note that this adjacency matrix, unlike the adjacency matrix for an undirected graph is *not* a symmetric matrix.)

$$M^2 = \begin{bmatrix} 0 & 0 & 1 & 1 & 0 \\ 1 & 0 & 0 & 0 & 1 \\ 0 & 1 & 0 & 0 & 0 \\ 0 & 0 & 1 & 0 & 0 \\ 1 & 0 & 0 & 0 & 0 \end{bmatrix} \qquad M + M^2 = \begin{bmatrix} 0 & 1 & 1 & 1 & 0 \\ 1 & 0 & 1 & 1 & 1 \\ 1 & 1 & 0 & 0 & 0 \\ 0 & 0 & 1 & 0 & 1 \\ 1 & 0 & 1 & 0 & 0 \end{bmatrix}$$

$$M^3 = \begin{bmatrix} 1 & 0 & 0 & 0 & 1 \\ 0 & 1 & 1 & 0 & 0 \\ 0 & 0 & 1 & 1 & 0 \\ 1 & 0 & 0 & 0 & 0 \\ 0 & 1 & 0 & 0 & 0 \end{bmatrix} \qquad M + M^2 + M^3 = \begin{bmatrix} 1 & 1 & 1 & 1 & 1 \\ 1 & 1 & 2 & 1 & 1 \\ 1 & 1 & 1 & 1 & 0 \\ 1 & 0 & 1 & 0 & 1 \\ 1 & 1 & 1 & 0 & 0 \end{bmatrix}$$

$$M^4 = \begin{bmatrix} 0 & 1 & 1 & 0 & 0 \\ 1 & 0 & 1 & 1 & 0 \\ 1 & 0 & 0 & 0 & 1 \\ 0 & 1 & 0 & 0 & 0 \\ 0 & 0 & 1 & 1 & 0 \end{bmatrix} \qquad M + M^2 + M^3 + M^4 = \begin{bmatrix} 1 & 2 & 2 & 1 & 1 \\ 2 & 1 & 3 & 2 & 1 \\ 2 & 1 & 1 & 1 & 1 \\ 1 & 1 & 1 & 0 & 1 \\ 1 & 1 & 2 & 1 & 0 \end{bmatrix}$$

Since all the entries of the final sum are nonzero except for the entries on the diagonal of the matrix, the graph is strongly connected. ■

We use the graph in this example in discussing other algorithms for testing connectivity of directed graphs.

EXAMPLE 6.10

Determine if the following graph is connected or strongly connected by use of matrix algorithms.

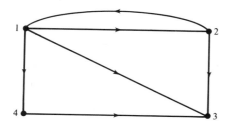

By examination, we determine that the graph is not strongly connected since the outdegree of vertex 3 is 0. To test for connectivity, we apply our algorithms to the adjacency matrix for the associated undirected graph, which can be obtained by using the OR operation on the adjacency matrix and its transpose. (See the problems.)

$$M = \begin{bmatrix} 0 & 1 & 1 & 1 \\ 1 & 0 & 1 & 0 \\ 0 & 0 & 0 & 0 \\ 0 & 0 & 1 & 0 \end{bmatrix} \qquad M^T = \begin{bmatrix} 0 & 1 & 0 & 0 \\ 1 & 0 & 0 & 0 \\ 1 & 1 & 0 & 1 \\ 1 & 0 & 0 & 0 \end{bmatrix}$$

$$M \vee M^T = \begin{bmatrix} 0 & 1 & 1 & 1 \\ 1 & 0 & 1 & 0 \\ 1 & 1 & 0 & 1 \\ 1 & 0 & 1 & 0 \end{bmatrix}$$

The square of this matrix is

$$\begin{bmatrix} 3 & 1 & 2 & 1 \\ 1 & 2 & 1 & 2 \\ 2 & 1 & 3 & 1 \\ 1 & 2 & 1 & 2 \end{bmatrix}$$

The fact that the square of the adjacency matrix of the undirected graph has no 0s is sufficient to show that the undirected graph is connected, and as a consequence, the directed graph is connected. We conclude that this graph is connected, but not strongly connected. ■

Improved (Faster) Connectivity Tests

As we mentioned in Chapter 5, if n is the number of vertices in the graph, these algorithms require approximately a multiple of n^4 operations to determine whether a graph is connected. Dijkstra's algorithm is an efficient algorithm which finds paths through directed graphs. So it might be worthwhile to examine the algorithm to see whether it might provide a clue as to how to improve the matrix multiplication algorithm.

If we weight an unweighted graph by assigning a weight of 1 to each edge, Dijkstra's algorithm *will* provide us with a simple test for the existence of a directed path from vertex a to vertex z. If Dijkstra's algorithm provides us with a finite value for the distance between the vertices, there must be a directed path between the vertices in question. In fact, if at any stage of Dijkstra's algorithm a finite *estimated* distance is assigned, there is a path connecting the vertex with the initial vertex.

EXAMPLE 6.11

Use Dijkstra's algorithm to show that there is a path from A to E in the graph in Fig. 6.7.

The matrix for Dijkstra's algorithm is

$$\begin{bmatrix} - & 1 & - & - & - \\ - & - & 1 & 1 & - \\ 1 & - & - & - & - \\ - & - & - & - & 1 \\ - & - & 1 & - & - \end{bmatrix}$$

Using Dijkstra's algorithm, we have the following steps:

Vertex	Actual Distance	Estimated Distance	Previous
A	0	—	—
B	X	1	A
C	X	X	X
D	X	X	X
E	X	X	X

Continuing with the algorithm, we find

Vertex	Actual Distance	Estimated Distance	Previous
A	0	—	—
B	1	—	A
C	X	2	B
D	X	2	B
E	X	X	X

Next we have

Vertex	Actual Distance	Estimated Distance	Previous
A	0	—	—
B	1	—	A
C	2	—	B
D	X	2	B
E	X	X	X

Then we get

Vertex	Actual Distance	Estimated Distance	Previous
A	0	—	—
B	1	—	A
C	2	—	B
D	2	—	B
E	X	3	D

and finally

Vertex	Actual Distance	Estimated Distance	Previous
A	0	—	—
B	1	—	A
C	2	—	B
D	2	—	B
E	3	—	D

■

Some work can be saved in Example 6.11 if we recall that if a finite estimated distance is computed for a vertex, there is at least one path from the initial vertex to that vertex. With this in mind, we could simplify the information from Dijkstra's algorithm for our current purpose in the following way:

	A	B	C	D	E
Step 1	1	1	0	0	0
Step 2	1	1	1	1	0
Step 3	1	1	1	1	0
Step 4	1	1	1	1	1
Step 5	1	1	1	1	1

The first row is obtained by placing a 1 for each vertex for which an actual distance or a finite estimated distance has been computed in step 1 of Dijkstra's algorithm. The remaining rows are similarly constructed. These rows can actually be constructed as follows from the original matrix:

$$\begin{bmatrix} 0 & 1 & 0 & 0 & 0 \\ 0 & 0 & 1 & 1 & 0 \\ 1 & 0 & 0 & 0 & 0 \\ 0 & 0 & 0 & 0 & 1 \\ 0 & 0 & 1 & 0 & 0 \end{bmatrix}$$

The first row of the table represents the fact that a path exists between A and B and the fact that (of course) A is connected to A. To obtain the second row, we recall that in Dijkstra's algorithm we computed estimated distances for all vertices to which B was connected. We can obtain the same information by noting that initially there is a 1 in column 2 of our table, and so we know that B can be reached from A. *Any* vertex to which B is connected can be reached from A. This information can be obtained by ORing the contents of row 2 of the incidence matrix into our results, producing row 2 of the table.

To obtain row 3, we note that the new second row has a 1 in column C, and thus we OR the contents of row 3 of the matrix. We continue and note that the entire resulting row consists of 1s after vertex D is dealt with. This means that there is a path from vertex A to each of the other vertices. This process could be repeated for each vertex, and then we would be able to determine whether directed paths exist from every vertex to every other vertex. Approximately n^2 operations will be required to test a vertex as the initial vertex, so carrying the algorithm to its completion will require n^3 steps. For an *un*directed graph, we need only determine which vertices can be reached from a *single* vertex, since if paths exist from A to C and from A to E, we can follow a path from C to A (which must exist in an undirected graph) and then go from A to E to produce a path from C to E.

Warshall's Algorithm

A bit of fine tuning on the algorithm described above will produce a way of testing a directed graph for connectivity and existence of paths. This new algorithm is known as *Warshall's algorithm*, and its efficiency is on the order of n^3. At each step this algorithm focuses on finding vertices which can be reached either directly or by going through a given set of vertices. This set of vertices will be increased from the first vertex, to the first and second vertex, etc., until it includes all the vertices in the graph. At that point, if the final matrix includes a 1 in row i column j, a directed path exists from vertex i to vertex j.

Informally, we describe the algorithm as follows:

1. OR the matrix with the identity matrix. This represents the fact that a (trivial) directed path exists between a vertex and itself, and each edge represents a directed path from its initial vertex to its terminal vertex. Call this matrix N.

2. With the matrix N, proceed down column 1. For each row j in which a 1 is found in row j, column 1, we know that there is an edge from vertex j to vertex 1. Thus there must be a path from vertex j to each of the vertices for which an edge exists from vertex 1. To record this fact, create a new row j which consists of the old row j ORed with row 1. The matrix N now records information about the paths which consist of single edges and those which pass through vertex 1.

3. To obtain information about direct connections, and those which may pass through the set of vertices $\{1, 2\}$, repeat step 2, using column 2 in the role played by column 1, and use row 2 in the way that row 1 was used in step 2.

4. Continue this process with the remaining columns of matrix N.

The nth iteration of the algorithm will produce a 1 in the matrix entry i, j if and only if there is an edge from vertex i to vertex j *or* if there is a path from i to j which passes through the set of vertices $\{1, 2, \ldots, n\}$.

If when the process is finished matrix N has all 1s in it, the graph in question is strongly connected; otherwise, it is not. In any case, if the final result has a 1 in row i, column j, there is a directed path from vertex i to vertex j.

Before we write Warshall's algorithm formally, let us use it on the graph from Example 6.9.

EXAMPLE 6.12

Use Warshall's algorithm to show that the graph in Example 6.9 is strongly connected.

SOLUTION
The original matrix is

$$
\begin{bmatrix}
0 & 1 & 0 & 0 & 0 \\
0 & 0 & 1 & 1 & 0 \\
1 & 0 & 0 & 0 & 0 \\
0 & 0 & 0 & 0 & 1 \\
0 & 0 & 1 & 0 & 0
\end{bmatrix}
$$

ORing this matrix with the identity yields

$$
\begin{bmatrix}
1 & 1 & 0 & 0 & 0 \\
0 & 1 & 1 & 1 & 0 \\
1 & 0 & 1 & 0 & 0 \\
0 & 0 & 0 & 1 & 1 \\
0 & 0 & 1 & 0 & 1
\end{bmatrix}
$$

Proceeding down column 1, we see that there are 1s in row 1 and in row 3. We need not OR row 1 with itself, but ORing row 1 into row 3 gives us

$$\begin{bmatrix} 1 & 1 & 0 & 0 & 0 \\ 0 & 1 & 1 & 1 & 0 \\ 1 & 1 & 1 & 0 & 0 \\ 0 & 0 & 0 & 1 & 1 \\ 0 & 0 & 1 & 0 & 1 \end{bmatrix}$$

Next we move to column 2. We find 1s in rows 1, 2, and 3. This means that row 2 should be ORed into rows 1 and 3. When this is done, the new matrix is

$$\begin{bmatrix} 1 & 1 & 1 & 1 & 0 \\ 0 & 1 & 1 & 1 & 0 \\ 1 & 1 & 1 & 1 & 0 \\ 0 & 0 & 0 & 1 & 1 \\ 0 & 0 & 1 & 0 & 1 \end{bmatrix}$$

Moving to column 3, we find 1s in all rows except row 4. We must OR row 3 with rows 1, 2, and 5. When this is done, the matrix is

$$\begin{bmatrix} 1 & 1 & 1 & 1 & 0 \\ 1 & 1 & 1 & 1 & 0 \\ 1 & 1 & 1 & 1 & 0 \\ 0 & 0 & 0 & 1 & 1 \\ 1 & 1 & 1 & 1 & 1 \end{bmatrix}$$

Next, we use column 4. Finding 1s in every row, we know that row 4 must be ORed into each row (except for row 4). The matrix which results will contain all 1s except for row 4. The final step with column 5 produces a matrix of all 1s. Since the final matrix contains only 1s, the graph is strongly connected. ∎

Now we write out Warshall's algorithm formally.

WARSHALL'S ALGORITHM

{*M* is an *n* × *n* matrix initially representing the adjacency matrix of a digraph.}
begin
for i := 1 **to** n **do**
    M[i,i] := 1 **od**;
            {This has the effect of ORing *M* with the identity matrix.}
for i := 1 **to** n **do**
      for j := 1 **to** n **do**
          if M[j, i] = 1 **then**
              for k := 1 **to** n **do**

$$M[j,k] := M[j,k] \text{ or } M[i,k]$$
od fi
  od
 od
 end.

∎

From the nesting of the loops in the algorithm it is clear that Warshall's algorithm will require a maximum of n^3 steps to complete its work. The outer two loops require that the **if** statement be executed n^2 times. Each time the condition in the **if** part is true, n steps are executed, thus leading to a worst case of n^3 steps. Since the number and location of 1s in the original matrix actually determine how many times the **if** statement is true, and since this density in general will be less than 1, this algorithm should have an efficiency of slightly better than n^3.

If a person wishes to test a directed graph for connectivity, there are two possibilities: Use the modified Dijkstra's algorithm for any selected vertex, or use Warshall's algorithm. In either case the adjacency matrix will be that of the associated undirected graph. You should be able to explain why, in most cases, the use of Dijkstra's algorithm is preferable. If, however, the question is one of strong connectivity, or if information relating to the components is needed, then Warshall's algorithm is the most appropriate to use.

Problem Set 4

Use the matrix multiplication algorithm to test the following digraphs for connectivity.

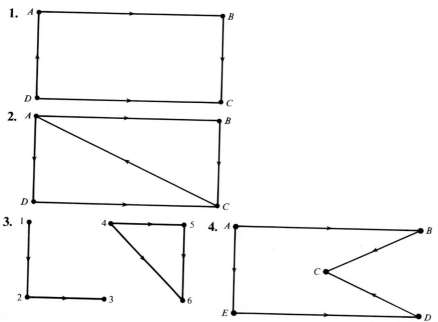

Apply Warshall's algorithm to the following graphs, and interpret the meaning of the final matrix.

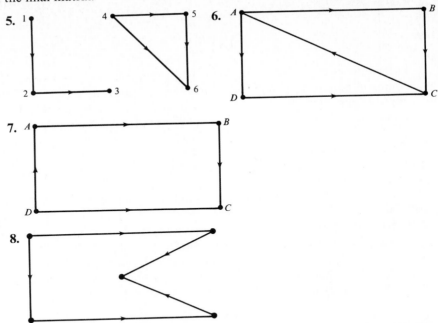

5.

6.

7.

8.

Use Dijkstra's algorithm as modified in the text to test the following undirected graphs for connectivity.

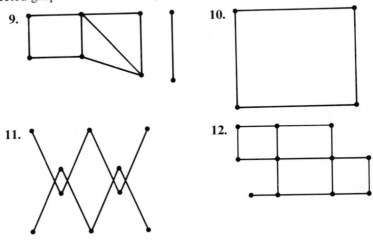

9.

10.

11.

12.

13. Verify the statement made in the text that if the adjacency matrix of the digraph G is M, then the adjacency matrix of the associated undirected graph is $M \vee M^T$. [*Hint:* If (i, j) is an edge in G, then $\{i, j\}$ will be an edge of the associated undirected graph and in the matrix of the associated undirected graph $A_{i, j}$ and $A_{j, i}$; thus must both be 1.]

*14. Write a formal description of the modification of Dijkstra's algorithm to determine connectivity of an undirected graph.

15. Explain how the results of Warshall's algorithm can be used to determine the components of an undirected graph.

*16. Prove that the statement made in the description of Warshall's algorithm is correct, namely that the nth iteration of the algorithm will produce a 1 in the matrix entry i, j if and only if there is an edge from vertex i to vertex j or if there is a path from i to j which passes through the set of vertices $\{1, 2, \ldots, n\}$.

*17. Complete the proof of Theorem 6.2.

*18. Prove Theorem 6.3.

*19. Prove that Theorem 5.10 is true for directed paths in directed graphs. (*Hint:* Look at the proof of Theorem 5.10 and see what changes, if any, need to be made.)

*20. Verify that Theorems 5.15 and 5.16 apply to directed paths in directed graphs.

*21. Modify Warshall's algorithm to find a matrix indicating the shortest distance between any two vertices in a directed graph, and apply your algorithm on the graph in Problem 1, Section 6.3.

P22. Write a Pascal program to implement the modified Dijkstra's algorithm, and use it on the graphs in Problems 9 to 12.

P23. Write a Pascal program to implement Warshall's algorithm, and use it on the graphs in Problems 5 to 8.

6.5   DIRECTED TREES

EXAMPLE 6.13

How many different lines are required in a truth table with three basic propositions?

SOLUTION

We have already solved this problem once, but it is useful to use a directed graph to describe another way of looking at the same problem. We construct a directed graph in which each possible path from a fixed starting vertex to a vertex of outdegree 0 describes a line in a truth table. We start with a single vertex A, and we give it outdegree 2. One of the edges leading out of this vertex will be labeled T and the other F. Using the edge labeled T corresponds to choosing the value T (true) for the first proposition, and using the edge labeled F corresponds to choosing the value F (false) for the first proposition.

The first stage of construction gives us the following:

From each of the vertices B and C, we can do the same thing as we did with A and thus obtain a graph with four vertices of outdegree 0. There is a unique path from A to each of the four vertices D, E, F, and G, and each of the paths corresponds to one of the four possible combinations of T and F for the first two propositions in the truth table.

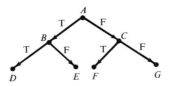

We finish the process by repeating it once more at each of the four vertices D, E, F, and G. The resulting graph has eight vertices of outdegree 0, and there is a unique path from A to each of them. Each sequence of labels on the edges of a path corresponds to a line in a truth table with three propositions. The first label corresponds to the value of the first proposition, the second label to the value of the second proposition, and the third label to the value of the third proposition.

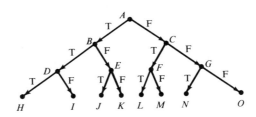

Since there are eight possible paths in the graph from vertex A to one of the vertices of outdegree 0, there must be eight lines in a truth table with three propositions. ∎

The completed graph in Example 6.13 is typical of the kind of graph which we would like to call a directed tree. Two features stand out:

1. The associated undirected graph is a tree.

2. There is only one vertex with indegree 0.

We use those two features to define a directed tree, or a "rooted" tree.

A **directed** or **rooted tree** is a directed graph for which the associated undirected graph is a tree and for which there is a unique vertex of indegree 0.

The **root** of a directed tree is the vertex of indegree 0.

EXAMPLE 6.14

The following graphs are directed trees.

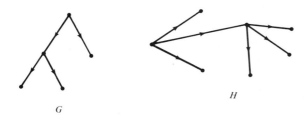

G H

The following graphs are *not* directed trees.

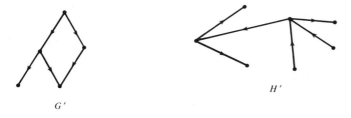

G' H'

Graphs G and H both clearly satisfy the definition of a directed tree; G' fails to be a directed tree because the associated undirected graph contains a cycle, and H' fails because there are too many vertices of indegree 0. ∎

Terminology

The **leaves** or **terminal vertices** of a directed tree are those vertices of outdegree 0.

The **interior vertices** of a tree are those vertices which are not leaves.

In dealing with trees, as with other graphs, the terms "node" and "vertex" are used interchangeably. Edges of a tree are also often called *branches*.

For the remainder of the chapter, we do not use the adjective "directed" when referring to a directed tree unless we need to distinguish a directed tree from an undirected one. In drawing representations of trees, it is traditional to put the root at the top and direct all edges down, so that it becomes unnecessary to indicate direction with arrows.

In many instances trees arise in situations in which there are no obvious connections to graph theory. In Chapter 2, for example, we introduced the tree as a device to help clarify the ideas involved with the development of the multiplication principle of counting, even though we had not discussed *any* graph theory at that point. Trees are useful in describing a large number of processes in mathematics. Even though trees *can* be discussed without reference to digraphs, we find it convenient to discuss them in this context, especially owing to the convenience of having the language and theorems of graph theory available to us. This convenience becomes clear in the proof of Theorem 6.4.

THEOREM 6.4

Every vertex of a tree except for the root has indegree 1.

PROOF

If a tree has v vertices, it has $v - 1$ edges, and so the sum of the indegrees must be $v - 1$. There are $v - 1$ vertices which are not roots, and each must have indegree of at least 1. This can happen only if each vertex has indegree of exactly 1. ∎

DEFINITION

An ***n*-ary tree** is a tree in which each of its vertices has outdegree n or less.

Traditionally, we use "binary" to refer to 2-ary trees and "ternary" for 3-ary trees.

Clearly much of the terminology (root, leaves, branches) for dealing with trees comes from forestry, but some also comes from genealogy. The genealogy connection occurs because of the fact that in tracing one's family heritage, the graph which results is (except in very unusual circumstances) a tree. With this motivation, if a is a vertex in a directed tree and (a, b) is an edge in the tree, we call b an *offspring* of a. (Some authors use "son" while others use "daughter." We remain neutral in that controversy.) If there is a directed path from a to d, we say that a is an *ancestor* of d and that d is a *descendant* of a.

Of course, directed trees are connected because the associated undirected graph is an undirected tree, and undirected trees are connected, but directed trees are *not* strongly connected. Only for the root is it true that there is a directed path to each of the other vertices in the tree.

THEOREM 6.5

If v is any vertex in a tree T, there is a unique directed path from the root r to the vertex v.

PROOF

Since T is connected, there is an elementary undirected path from r to v. Since a tree has no cycles, this path is unique. If the path is $r a_1 \cdots b_k v$ and r has indegree 0, then (r, a_1) must be an edge of T. Since a_1 has indegree 1, it follows that (a_1, a_2) is an edge of T. (Why is a_2 not the vertex r?) By the same logic, in general (a_i, a_{i+1}) is an edge of T, as is (a_k, v); hence our undirected path is also a directed path. (In the problems you are asked to formalize this informal argument as a proof by induction on the length of the elementary undirected path.) ∎

DEFINITIONS

The **depth of the vertex** v is the length of the directed path from the root to v.

The **depth** (sometimes **height**) **of a tree** is the maximum depth of all its vertices.

THEOREM 6.6

A vertex v and all its descendants form a subgraph of a tree which is also a tree, and the root of this tree is v.

PROOF

The associated undirected graph for this subgraph is a tree since it is connected (there is a path from v to every vertex in the subgraph) and has no cycles (the original graph had no cycles, and this is a subgraph). In this subgraph v has indegree 0 and is the only such vertex. ∎

The subgraph described in Theorem 6.6 is called the *subtree determined by v*.

Applications

Trees have a number of valuable uses in computer science and mathematics. One use of the tree in mathematics is illustrated in Example 6.13. Trees can be used to count the number of ways that a process can occur. The beginning of a process is associated with the root of the tree, and one branch is created for each possible way the first step of the process can be carried out. This process is repeated at the vertices at depth 1, with a new collection of vertices at depth 2 added. This continues as long as there are steps to be described in the process. The end result is one leaf for each way that the event in question can occur. This use of a tree is quite common in discrete probability problems. Other uses of trees will occur with alarming frequency in the rest of this text. Example 6.15 illustrates one way in which the idea of a tree has been applied to the very real world of computer operating systems.

EXAMPLE 6.15

Several computer operating systems (including, for example, UNIX and VAX/ VMS) use a tree to describe the way in which files and directories are organized by the system. In each of these systems, there are two kinds of files: directory files, which serve as interior nodes and contain information as to where to find other files, and ordinary data files. Users of the system are given a directory which they access every time they start to use the system. There is a system directory, known as the *root*, which functions in the file system exactly as we might expect. The root directory belongs to the system manager. Connected to the root are several other files, some of which are directory files and others of which are ordinary files. On a simple system, the file structure might look like Fig. 6.8, where the squares represent directory files and the circles represent ordinary files.

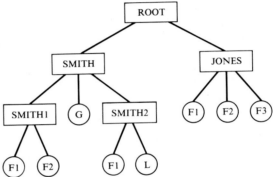

Figure 6.8 Typical tree-structured file system. ■

A structure such as that in Example 6.15 has a large number of advantages:

1. There is a unique path from the root to any file, and this provides a simple way of completely specifying files. In Fig. 6.8, file F1, in the directory SMITH1, can be uniquely described by listing the path which reaches it: ROOT.SMITH.SMITH1.F1. (We have used periods to separate vertex labels from each other.) If a file named F1 occurs in some other directory, its specification is different from that of this F1.

2. Each user of the system is assigned a main directory, but each user can create directories of his or her own. That is what has happened in Fig. 6.8 in the directory ROOT.SMITH. The person using the directory ROOT.SMITH has access to the entire subtree determined by the vertex SMITH.

3. A person who is accessing or using a particular directory is located in the subtree determined by its vertex, and the naming of files can take place within that context. Suppose someone is using SMITH. Then to refer to the file F1, that person can list the path which reaches it, namely, SMITH1.F1. (In systems like this it is not necessary to list the current directory in describing the path, since the current directory automatically is assumed to be the first vertex.) If SMITH1 is the current directory, the file is accessed by simply referring to F1.

4. Different hierarchies of accounts could be set up. Every directory is in the subtree determined by the root, but two users reporting to the same supervisor could be given accounts with main directories in the subtree determined by their supervisor's main directory.

5. Users who are working on several projects can create within their own directories subtrees of files in which each subtree is related to a different project.

Universal Address Method

In the application of trees outside of computer science, often a convenient labeling of the vertices is not possible, and yet a tree structure is still convenient. If we are fortunate enough to be able to order each of the descendants from each interior vertex, we can use a scheme which is very similar to that of Example 6.15 to describe vertices. A tree in which each node's offspring are ordered is called an *ordered tree*. In an ordered tree, we can describe any path from the root (and hence any vertex) by listing the number of the offspring taken at each step. Thus path 1.2.3 would mean to take the first offspring of the root, then take the second offspring of that vertex, and stop with the third offspring of that vertex. In drawing an ordered tree for each vertex, we place its first offspring farthest left, its second offspring next to it, and so on.

In the tree in Fig. 6.9, a path to vertex H is 1.1.2, and to vertex J the path is 1.3.2. This way of describing paths to vertices is called the *universal address method*. From the address of a vertex we can also determine its depth. In an ordered tree we refer to the subtree determined by the smallest (or leftmost) offspring of a vertex as the *leftmost* subtree of the vertex. The second subtree, third subtree, etc., are defined in a similar way. In a binary tree, each vertex has at most a left subtree and a right subtree.

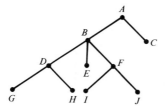

Figure 6.9 Ordered tree.

Problem Set 5

Use trees to solve Problems 1 to 6.

1. In how many ways can the elements of the set $\{a, b, c\}$ be written to form a "word"?

2. Answer Problem 1 if a and c are not permitted to be adjacent to each other (abc is OK, but acb and bca are not allowed).

3. Repeat Problem 1 with the set $\{a, b, c, d\}$.

4. Repeat Problem 2 with the set $\{a, b, c, d\}$.

5. Suppose that we develop a three-valued logic in which propositions can be valued as T, F, or M (maybe). How many lines would a truth table have if there were two basic propositions?

6. There are three flights a day from Boston, Massachusetts, to Burlington, Vermont, and two connecting flights from Burlington to Albany, New York. There are also six flights a day from Boston to New York City and three connecting flights per day from New York City to Albany. How many total combinations can one use to get from Boston to Albany?

7. Define the relation R on the vertices of a directed tree T in the following way: aRb if and only if $a = b$ or there is a directed path from a to b. Prove that R is a partial order on the vertices of T.

8. Prove that if x and y are vertices of T, with R defined as above, then x and y have a "least common ancestor" a, which has the properties (1) aRx and aRy and (2) if d is any vertex such that dRx and dRy, then dRa.

9. Find the address of each vertex in the following ordered tree.

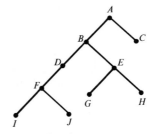

10. Repeat Problem 9 for the following tree.

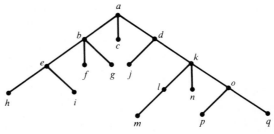

11. The Beaver Crossing Golf Club has four members and designed its Club Match Play Tournament in the following way. A played against B, and C played against D, with the winners to meet in the championship. Notice that the following binary tree describes the tournament.

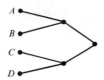

(a) Show that the number of interior nodes in the tree represents the number of matches to be played.

(b) Show that the depth of a leaf represents the number of matches that a person must play to win the title.

*12. Tournaments as described in Problem 11 are regarded as balanced if the difference in height of any two leaves is no greater than 1. For the following situations, find a graph which represents a balanced tournament, and determine the height and number of interior nodes in the tree.

(a) Cozad Community Country Club—32 members.

(b) Broken Bow Birdie Brigade—55 members.

*13. State and prove a theorem which will provide general answers to the questions asked in Problem 12.

14. Formalize the argument used in the proof of Theorem 6.5 by making use of mathematical induction.

*15. In a tree T we can define the following operation, which we denote by $\wedge : x \wedge y =$ the root of the smallest subtree containing x and y.

(a) Prove that the $\wedge$ operation is commutative and associative.

(b) Show that the $\wedge$ operation is the same as the operation of finding the least common ancestor described in Problem 8.

16. In reference to Problem 15, prove that $x \wedge y = y$ if and only if x is a descendant of y.

17. In reference to Problem 15, is there an element which has the property that $x \wedge i = i \wedge x = x$ for all x? Explain your answer.

18. A binary tree could be represented in a Pascal program by means of an array in the following way: The root is $T[1]$. For each vertex $T[i]$, its left offspring is $T[2*i]$, and its right offspring is $T[2*i+1]$. All unused entries in the array are indicated by the use of a "distinguished" value, such as -1 if the vertices are labeled with numbers, or a distinguished symbol, such as @ if the vertices are labeled with symbols. Represent the tree in Problem 9 in this manner.

What is the minimum array size needed to store a tree with 10 vertices such as this one? What is the maximum array size that could be required?

19. How could a method similar to that of Problem 18 be used to store a ternary tree? Represent the tree in Problem 10 in this fashion. What is the minimum array size needed to store a ternary tree with 10 vertices? What is the maximum size that might be required?

P20. Using the storage method for a binary tree described in Problem 18 and a maximum array size of 1024, write a program to determine the depth and address of each vertex in the tree.

6.6 SEARCH TREES

Directed trees are useful for representing information for computer storage and retrieval. In Section 6.5 we saw an illustration of the use of a directed tree to represent the file storage system on a computer. There are many other applications of value as well, and we focus on one in this section.

Special Features of Binary Trees

The trees which we use in this section are all binary, and as such, binary trees have some particular advantages and terminology. In a binary tree, each node has at most two offspring, so it is particularly easy to consider a binary tree to be ordered. One offspring of a node is considered to be the left offspring and the other to be the right offspring, as discussed in Section 6.5. We can thus consider the left and right subtree of each vertex.

> DEFINITION
>
> A **full binary tree** is one in which each interior vertex has outdegree 2 and all the leaves are at the same depth. (A full n-ary tree is defined in similar fashion.)

Full binary trees are particularly good for representation of data because they have the largest possible number of nodes for a given height tree. If a binary tree is not full, it is always possible to add an edge and vertex to the tree without changing the height of the tree. Theorem 6.7 describes the relationship between the number of vertices in a full binary tree and the height of the tree.

THEOREM 6.7

A full binary tree of height h has 2^h leaves and $2^{h+1} - 1$ vertices.

PROOF
We proceed by induction on the height of the tree, letting $P(h)$ be the statement which we wish to prove.

Basis Step   A tree of height 0 must have a single vertex. In that case both 2^0 and $2^{0+1} - 1$ are equal to 1, as required.

Induction Step   Let T be a full binary tree of height $k + 1$. Consider the subgraph T' consisting of all vertices in the tree of height k or less. This is a full binary tree of height k. If we use as the induction hypothesis that $P(k)$ is true, then T' contains 2^k leaves and $2^{k+1} - 1$ vertices since T' has height k. The tree T is obtained from T' by adding two offspring at each of the leaves of T'. These new offspring are the leaves of T, and there must be $2(2^k) = 2^{k+1}$ of them. Each vertex in T' is an interior node of T, and the total number of vertices in T is equal to the number of vertices in T' plus the number of leaves. This amounts to $2^{k+1} - 1 + 2^{k+1}$ which is equal to $2(2^{k+1}) - 1 = 2^{k+2} - 1$. ∎

COROLLARY 6.1

A full binary tree with n vertices has height $\log_2 (n + 1) - 1$.

PROOF
This is left as an exercise.   ∎

Not every tree which we wish to use will have exactly the number of vertices that allows it to be a full binary tree, but we often wish to construct trees with a given number of vertices which come as close as possible to being a full tree. As a consequence, it is necessary to provide some measure of "closeness" to being a full tree. In Fig. 6.10 we present two binary trees; one is reasonably close to being a full binary tree, and the other is much less like a full tree, even though both trees have the same number of vertices.

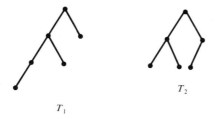

T_1 T_2

Figure 6.10 Two binary trees.

DEFINITION

A binary tree of height k is **almost full** if and only if the subgraph induced by all its vertices at height $k - 1$ or less is a full binary tree.

In Fig. 6.10, T_1 is an almost full binary tree, and T_2 is not.

We state without proof the following theorem about almost full binary trees, which is closely related to Corollary 6.1.

THEOREM 6.8

The height of an almost full binary tree with n vertices is equal to the greatest integer less than or equal to $\log_2 n$. ∎

In dealing with algorithms which use trees for information storage and retrieval, it is helpful to use the idea of a *balanced* binary tree. The property of balance for binary trees is easier to deal with algorithmically, and although this property is weaker than the property of being almost full, the depth of a balanced binary tree does not vary significantly from the depth of an almost full binary tree with the same number of vertices. Every almost full binary tree is balanced, but the converse of that statement is not true.

> **DEFINITIONS**
>
> A vertex is **in balance** if the heights of its left subtree and its right subtree differ by no more than 1.
>
> A binary tree is **balanced** if all its vertices are in balance.

It is often useful to consider a process in which we store information at the vertices of a tree and then later wish to retrieve this information for some use. This kind of arrangement is commonly used for storing information in a computer. It is extremely valuable to find a way of representing this information so that the cost of recovering it is as small as possible. If we place information in a binary tree and are able to retrieve the information by following a single path, we never need to exceed the depth of the tree in terms of comparisons required to search for the information. If the tree is almost full, or even balanced, the depth of the search is quite small in relation to the total amount of information which can be stored in the tree.

The discussion above is the motivation behind what is known as a *binary search tree*. In a binary search tree, information is stored in the vertices of a binary tree. This information must have some natural order. The information

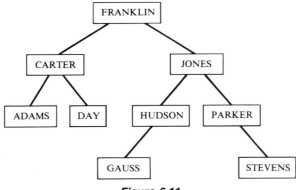

Figure 6.11

may be numerical, in which case we can use ordinary numerical ordering; it could be alphabetical, in which case we could use alphabetical order; or there could be some other special ordering associated with the data. In the search tree, for each vertex, all information stored in its left subtree is less than or equal to the information stored in the vertex, and all information stored in the right subtree is greater than the information stored in the vertex. Figure 6.11 shows a search tree.

An algorithm for locating information in a search tree is easy to describe:

ALGORITHM SEEK(ITEM)

begin
repeat
  if item < current_vertex
      then current_vertex := left(current_vertex)
      else current_vertex := right(current_vertex)
      fi
until current_vertex is empty **or** current_vertex = item;
if current_vertex is empty **then** print('not in tree')
                         else print('found in tree')
end.

Usually, the data on which a tree is built actually consist of "keys" that each identify a much larger collection of data. The key may be a social security number, and the vertex may actually include a considerable amount of information about the person to whom that social security number belongs. The key may be a person's name, and considerable other information may be stored with it. In cases like this, the search tree is used to locate the information on file about a particular person.

If the process ends in failure, the next step may be to add data to the tree at the point at which it should be located. Considerable care is taken in such an event to make sure that the tree is balanced after the insertion takes place. The number of comparisons required by this algorithm is no more than the depth of the tree. If the tree remains balanced with n vertices, this depth is approximately $\log_2 n$. A linear (item-by-item) search through the same data would require $n/2$ comparisons (on average). The following table illustrates the difference between the number of comparisons required:

n	$\log_2 n$	$n/2$
2	1	1
4	2	2
8	3	4
16	4	8
100	7	50
1,000	10	500
1,000,000	20	500,000

For small values of n, the values of $\log_2 n$ and $n/2$ are quite similar (identical, in fact, at $n = 2$ and $n = 4$); but as n becomes large, the difference between $n/2$ and $\log_2 n$ becomes quite striking. This is why a search with a balanced binary search tree is an efficient way of representing information in a large database.

The situation in which data are sought but not found serves to further reinforce this concept, since if an item is *not* stored in the search tree, this fact can be discovered when a leaf of the tree is reached. Again, this lack of information can thus be discovered in time proportional to $\log_2 n$, whereas if the items are unordered, one needs n observations. Even for ordered data an average of $n/2$ observations is needed to verify the absence of an item.

The well-known binary search of an ordered array is related to the idea of a search tree as well. And for exactly the reasons discussed above, it is far more efficient to use a binary search to find data in an ordered array than to search the array in a linear fashion. Problem 21 provides an opportunity to explore the binary search algorithm and its connection to search trees.

Problem Set 6

1. Propose a definition of a full n-ary tree based on our definition of a full binary tree.

2. Propose a definition of an almost full n-ary tree based on our definition of an almost full binary tree.

*3. Prove that a full n-ary tree of height h has n^h leaves and $(n^{h+1})/(n-1)$ total vertices.

4. Based on the result of Problem 3, prove that a full n-ary tree with k vertices has height of about $\log_n (k)$.

5. Discuss Problems 11 and 12 in Section 6.5 in light of the information contained in Theorems 6.7 and 6.8.

6. Consider the search tree in Fig. 6.12.

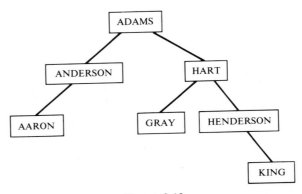

Figure 6.12

(a) Is the tree balanced? Explain.

(b) Calculate the number of comparisons needed to locate or to determine the absence of each of the following keys: (i) Adams, (ii) Gordon, (iii) Hart, (iv) Henderson, (v) Aaron.

7. The search tree in Fig. 6.13 is essentially a linear list. Answer the same questions about this tree as were asked in Problem 6.

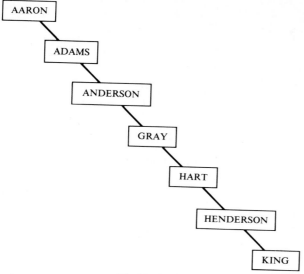

Figure 6.13

8. Construct an almost full tree to represent the data in Problems 6 and 7, and then answer the questions posed in Problem 6 for your tree.

9. Consider the search tree in Fig. 6.14.

(a) Is the tree balanced?

(b) Calculate the number of comparisons needed to locate or determine the absence of each of the following keys: (i) 9, (ii) 18, (iii) 19, (iv) 4.

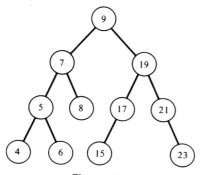

Figure 6.14

10. The same data as in Problem 9 can be represented by the search tree shown in Fig. 6.15. Answer the same questions as in Problem 9.

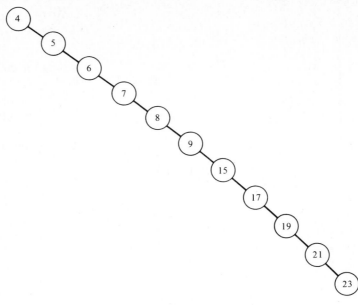

Figure 6.15

11. Represent the keys stored in Problems 9 and 10 in an almost full tree, and answer the questions in Problem 9.

12. Construct an almost full search tree to store the following list of keys: 9, 52, 17, 21, 83, 42, 1, 71, 25, 5. Is your tree balanced?

13. Construct an almost full tree to store the keys 0, 1, 3, 5, 7, 8, 10, 12, 16, 18, 19, 20.

14. In Problem 18 of Section 6.5, we discussed a method for using an array to represent a binary tree in Pascal. Determine how the trees in Problems 6, 7, and 8 would be represented in this array. Does this suggest any other advantages for an almost full tree?

15. Repeat Problem 14, using the trees from Problems 9, 10, and 11.

*16. Make a table to compare the function $\log_2 n$ with kn for $k = 1, 0.25,$ 0.125, 0.0625. Then prove that for any positive value of k, there is a value N such that if $n > N$, then $\log_2 n < kn$. This is what we mean when we say that $\log_2 n$ is eventually smaller than any linear function.

17. Prove Corollary 6.1. (*Hint:* Use $\log_2$ with the statement of Theorem 6.7.)

18. Prove that if a binary tree is not full, it is always possible to add a new leaf to the tree without changing its height.

19. Prove that if a tree is almost full, it is balanced.

*20. Prove that the converse of Problem 19 is false by constructing a balanced binary tree which is not almost full. (*Hint:* Try to construct as "unbalanced" a balanced tree as you can, and make the depth of your tree 4.)

*21. The *binary search* is a well-known algorithm for searching for data in an ordered array. The process can be informally described in the following way:

Suppose that the array has subscripts from 1 to *n*. *Top* represents the largest subscript at which the data can be found, and *Bottom* represents the smallest subscript at which the data can be found. Initially *Top* is set equal to *n*, and *Bottom* is set equal to 1.

We repeat the following process until either the data being sought are located in the array or *Bottom* becomes larger than *Top*.

Make *Middle* equal to (*Top* + *Bottom*) DIV 2. If A[*Middle*] is the information sought, we stop. If A[*Middle*] is less than the item sought, let *Bottom* = *Middle* + 1. If A[*Middle*] is greater than the item sought, let *Top* = *Middle* − 1.

(a) Formalize this as an algorithm.
(b) Observe the array locations tested in seeking each of the data points 1, 3, 5, 7, 9, 11, 13, 15, 19. Show that the process is identical to a search of the search tree shown in Fig. 6.16.

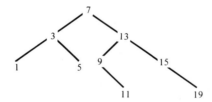

Figure 6.16

(c) Attempt to generalize this algorithm, and make relevant comparisons to the standard linear search.

P22. Write a program to implement the binary search described in Problem 21. You may assume that the size of the array does not exceed 1023.

P23. Write a program to implement a search through a search tree. Use the representation of a tree introduced in Section 6.5. Run your program, using the data in Problem 9.

P24. Write a program to determine whether a binary tree is balanced. Use the representation of a tree introduced in Section 6.5. You may assume that the size of the array does not exceed 1023.

6.7 TREE TRAVERSALS

Expression Trees

In Section 6.6, we discussed a way of using trees to represent data for computer storage. Another use of a tree which has interesting consequences in both mathematics and computer science is a tree representing the computation of an arithmetic expression. We can illustrate this idea with the construction of a tree to represent the computation of the expression $2 * 3 + [4 * (2 - 1) + 5]$, as in Example 6.16.

EXAMPLE 6.16

The tree shown represents the expression $2 * 3 + [4 * (2 - 1) + 5]$. In this tree, each of the leaves is labeled with a value, and the interior nodes are labeled with arithmetic operators. The vertices labeled with operators are

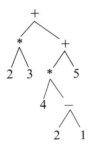

intended to represent the application of that operator to the results contained in the subtrees of that vertex. We can "work our way up" to compute the value represented by the tree. ∎

EXAMPLE 6.17

Construct a tree to represent the result of the computation of the expression $(2 + 3) * 4 - 1$.

SOLUTION

We can construct the tree in a "top-down" fashion by noting that the final calculation will be subtracting 1 from the result of $(2 + 3) * 4$. We place a minus sign at the root of our tree; we make its right subtree consist of the number 1, and its left subtree must be a calculation of $(2 + 3) * 4$. After this first step, the tree is

$$(2 + 3) * 4 \qquad 1$$

Next we must expand the expression $(2 + 3) * 4$. With this expression, the final calculation is to multiply the result of $2 + 3$ by 4. Proceeding as before, we label the root of the subtree with $*$, and we label its two subtrees as $2 + 3$ and 4:

The final step is to express $2 + 3$ as a subtree with the root labeled as $+$ and leaves 2 and 3.

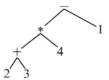

The most obvious feature of trees constructed in this fashion is the fact that each subtree represents a computation, and the entire computation represented by the tree can be completed by systematically completing the calculations indicated by the subtrees.

In this section we look at algorithms which describe the process of expanding the vertices of a binary tree. We use the tree in Fig. 6.17 to illustrate the application of each of the algorithms, which are known as *tree traversal algorithms*.

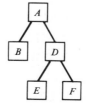

Figure 6.17 Sample binary tree.

Inorder Traversal

We begin with the traversal algorithm which seems to be the most natural, the one known as *inorder traversal*. In inorder traversal, each subtree of a binary tree is traversed by

1. Traversing the left subtree of the root
2. Expanding the root of the subtree
3. Traversing the right subtree of the root

Expanding the root may involve performing some computation or following some instruction listed at the vertex, or it may simply mean listing the label stored there. In Fig. 6.17, we must first traverse the left subtree of the root (A). In the case of the graph in Fig. 6.17, this is easy to do, because the subtree consists of a single vertex B. Next we expand vertex A which is the root of the tree. Next we must traverse the right subtree of A, which is the subtree with root D. To traverse this subtree, we must in turn traverse D's left subtree, expand D, and traverse the right subtree. This means expanding vertices in the order E, then D, and finally F. An inorder traversal of the tree expands the vertices in the order B, A, E, D, F. We can write a formal algorithm for the process as follows:

ALGORITHM INORDER(ROOT)

{Left (a) and right (a) indicate the left and right offspring of vertex a.}
begin
if left(root) exists **then** INORDER(left(root)) **fi**;
expand(root);
if right(root) exists **then** INORDER(right(root)) **fi**
end.

Note that this algorithm is recursive, so to complete its work, it must call upon itself. This is feasible. And the fact that the calls to INORDER are not made if left(root) or right(root) does not exist allows the process to eventually terminate as it reaches the leaves of the tree.

Applying inorder traversal to search trees and arithmetic expression trees yields some interesting results. If we do an inorder traversal of the search tree shown in Fig 6.11, the vertices are expanded in the following order: Adams, Carter, Day, Franklin, Gauss, Hudson, Jones, Parker, Stevens. The vertices are listed exactly in alphabetical order! In fact, we can prove that for a search tree, *in*order traversal will always expand the vertices in *in*creasing order.

THEOREM 6.9

If T is a search tree and a and b are data values stored in the tree with $a < b$, an inorder traversal of T will expand a before b.

PROOF
This is a good example of a theorem whose proof is handled by considering cases. If a and b are vertices of a tree, one of the following three conditions must be true:

1. Vertex a is a descendant of b.

2. Vertex b is a descendant of a.

3. There is a vertex c such that a is in one left subtree of c and b is in the other subtree.

(The proof that one of these three must be true is left as an exercise.) We show that no matter which of the three cases holds, a will be expanded before b.

If a is a descendant of b, then a is in the left subtree of b (why?). And in an inorder traversal, the left subtree is traversed before the root is expanded, so a is expanded before b.

If b is a descendant of a, then b is in the right subtree of a (why?). Hence b must be expanded after a.

If a and b have a common ancestor c, then a is in c's left subtree and b is in c's right subtree (why?). In traversing the tree, a and b will be expanded as part of the traversal of the subtree determined by c. Since a is in the left subtree, it will be expanded before any vertices in the right subtree, and hence a will be expanded before b. ■

Inorder Traversals and Arithmetic Expressions

If inorder traversal is applied to a tree representing an arithmetic expression, an interesting result occurs. The vertices are listed in the order in which they are written in ordinary arithmetic notation. An inorder traversal of the tree in Example 6.16 expands the vertices in the order $2 * 3 + 4 * 2 - 1 + 5$, and an inorder traversal of the last tree shown in Example 6.17 expands the vertices in the order $2 + 3 * 4 - 1$. Inorder traversal of an expression tree does not recover all the information stored in the tree. Parentheses were needed in the expression to set up the tree, because they indicate which operations are associated with which subexpressions. The relationships indicated by the parentheses determine the subtree structure. However, the location of the parentheses is not provided by the inorder listing of the vertices. Inorder traversal listing of the tree provides an ambiguous description of the tree because more than one expression tree has the same inorder listing. Figure 6.18 shows two expression trees with inorder traversal $2 * 3 + 4 * 2 - 1 + 5$.

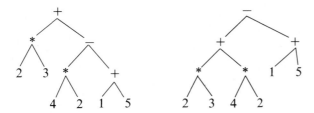

Figure 6.18 Two expression trees with inorder traversal

$2 * 3 + 4 * 2 - 1 + 5$.

The usual notation for arithmetic expressions is called *infix notation*, and the "in" in infix indicates its close relationship with the inorder traversal of an expression tree.

Postorder Traversal and Reverse Polish Notation

Many pocket calculators use infix notation, but other calculators use another kind of notation, known as *reverse Polish notation* (RPN), or *postfix notation*. In using an RPN calculator, to enter the calculation 2 + 3, we key in the keys 2 (enter) 3 +. RPN arises from a different traversal order for a binary tree. A tree for 2 + 3 looks like this:

If we write 2 3 + as a traversal of the tree, we have expanded the tree by taking first the left subtree of the root, then the right subtree, and finally the root. This is the traversal algorithm known as *postorder traversal*. The algorithm for postorder traversal is described in our algorithm language as follows:

ALGORITHM POSTORDER(ROOT)

begin
if left(root) exists **then** POSTORDER(left(root)) **fi** ;
if right(root) exists **then** POSTORDER(right(root)) **fi** ;
expand(root)
end.

A postorder traversal of the tree in Fig. 6.17 expands the vertices in the order *B*, *E*, *F*, *D*, *A*. In traversing binary trees, postorder traversal has an advantage over inorder traversal because as long as we know which vertices are interior vertices and their outdegrees, the postorder listing contains all the information needed to completely reconstruct the tree. In postorder traversal, the root of each subtree stands out. Suppose that *D* and *A* are interior nodes of degree 2 in a binary tree with postorder traversal *B*, *E*, *F*, *D*, *A*. Example 6.18 shows how the tree can be reconstructed uniquely from that information.

EXAMPLE 6.18

Reconstruct a tree whose postorder traversal is *B*, *E*, *F*, *D*, *A* if the interior nodes are *D* and *A* and each has outdegree 2.

SOLUTION
Node *B* occurs first in the list and is a leaf. Since *B* occurred first, it must be in the left subtree of some interior vertex. Node *E* occurs next. We cannot tell at this point if *E* and *B* have a common parent; but if they did, it would be the

next vertex to occur in a postorder traversal because the vertex E would be its entire right subtree and B would be its entire left subtree. Both of the parent's subtrees would have been traversed, and thus the root (the parent) would be the next vertex traversed. Next we encounter F. Node F is also a leaf. The fact that F was not an interior vertex now tells us that E and B do not have a common parent; but it is possible that E and F do have a common parent. We can record the information we have at this point:

Since postorder traversal does proceed from left to right, we can be sure that in the final figure B, E, and F should be arranged in that order from left to right. Now we reach an interior node F of degree 2. In a postorder traversal, a vertex can be expanded only after both its left and its right subtrees have been completely traversed. Node D has outdegree 2, so its successors must be among the vertices B, E, and F, and in fact they must be E and F. This information is recorded by

Finally we encounter the interior node A. Since this is the last vertex, it must be the root of the entire tree, and its subtrees must be the single vertex B and the tree constructed at the previous step.

Postorder traversal of the tree in Example 6.16 yields the traversal order 2 3 * 4 2 1 − * + +. In an arithmetic expression tree, we know that the numbers represent leaves, and the operation symbols represent interior vertices of outdegree 2. The tree can be reconstructed by using the same techniques as were applied in Example 6.18. Since the tree can be reconstructed and the tree contains all information needed to perform the computation, we should be able to perform the calculation represented by the tree directly from the postorder traversal listing.

This is the idea behind postfix notation, which is the description of a calculation by giving a postorder traversal of its tree. Postfix notation is, as we mentioned before, also called RPN and is used by some calculator makers, most notably Hewlett-Packard. Postfix notation allows us to describe calculations without parentheses, a fact which is most convenient in the design of pocket calculators. For calculators with infix notation either keys for parentheses must be provided to allow computation of expressions such as $(2 + 3) * 4$, or else the intermediate results must be stored elsewhere. On an RPN calculator, however, the keystrokes 2 (enter) $3 + 4 *$ perform the calculation. (The main problem with RPN calculators is making allowances for unary operators, which correspond to interior vertices with outdegree 1. The unary minus must have a key different from the subtraction key, for example.)

Let us trace the process involved in performing the calculation $2 \ 3 * 4 \ 2 \ 1 - * 5 + +$. In doing this we need only keep in mind the process involved in reconstructing the tree. Encountering first a 2 and then a 3, we recall that these are leaves of the tree. Finding next an asterisk (an interior node), we conclude that the 2 and 3 are the left and right successors, respectively, of the asterisk; thus the operation $2 * 3$ must be performed. Since the result of a calculation is to be placed at the root of its subtree, we replace the asterisk with a 6, and the expression is now $6 \ 4 \ 2 \ 1 - * 5 + +$. The next interior vertex successor is the minus. The minus must apply to the 2 and 1, to compute $2 - 1 = 1$. The expression becomes $6 \ 4 \ 1 * 5 + +$. Next, $4 * 1$ is evaluated, and the new expression is $6 \ 4 \ 5 + +$. All that is left to do is compute $4 + 5$ and then $6 + 9$, for a final result of 15.

RPN and Compilers

Most computers achieve arithmetic computations by making use of special registers within the processor. Suppose, for example, 2 is to be added to 3. The number 2 is loaded into the arithmetic register, and an instruction is then executed which adds 3 to the contents of the arithmetic register. Assembler-language instructions for this sequence of operations would typically be of the form

```
LDAM 2

ADM 3
```

These instructions load the value 2 into register A and then add the value 3 to the register.

If we were to perform the computation $(2 + 2) * 4$, the sequence of instructions would be performed in the following way:

```
LDAM 2

ADM 2

MULM 4
```

This sequence means to load the value 2 into register A, add the value 2 to it, and then multiply the new contents of the register by the value 4. (We have used "immediate" instructions here which operate on their operands as values rather than as addresses.) This sequence of instructions is exactly the same as the sequence we would use to perform the same calculation on an RPN calculator. The process is facilitated, as it is on an RPN calculator, by the availability of a "stack" for storing intermediate results of calculations. In writing machine-language (or assembler-language) instructions it is convenient to express the calculations to be performed in RPN. For this reason, one of the first steps taken by a compiler in setting up the machine-language code for an arithmetic computation is to translate the expression into RPN from the given infix notation. One way of achieving this, although not necessarily the most efficient, would be to take the infix notation, set up the tree representing the computation, and perform a postorder traversal of that tree. Although in practice compilers do not perform in exactly this fashion, the net effect is exactly the same.

It is also interesting that the language FORTH, which has become fairly popular for systems development on microcomputers, is oriented around the use of RPN for the description of arithmetic computations.

Preorder Traversal and Polish Prefix Notation

Preorder traversal is similar to postorder traversal except that now the root is expanded first. The algorithm for preorder traversal of a tree is as follows:

ALGORITHM PREORDER(ROOT)

begin
expand(root);
if left(root) exists **then** PREORDER(left(root)) **fi**;
if right(root) exists **then** PREORDER(right(root)) **fi**
end.

Applying preorder traversal to the tree in Fig. 6.17 results in the traversal order A, B, D, E, F. Students sometimes think that preorder and postorder traversals produce exactly reversed traversal orders, but that is *not* the case. Preorder traversal has many of the same properties as postorder traversal, the most important being that arithmetic expressions can be evaluated from a preorder traversal of the tree in a manner similar to that used for a postorder traversal. The notation which results from this is called *Polish prefix notation* (PPN), or just *prefix notation*. Prefix notation is often useful for theoretical discussions of operations in mathematical logic.

One final note is in order. Obviously postorder or preorder traversal of a search tree provides little interesting information about the data in the tree. This is reasonable, since the search tree was designed for use with inorder traversal.

Problem Set 7

Construct trees to represent the expressions in Problems 1 to 4.

   1. $(2 * 3) + (4 * 6)$

   2. $(-7 + 6) * 3 - (4/2) + 5$ (Remember the usual rules of precedence of operations!)

   3. $(3 * 4 * 5 * 6) - (7 + 3 * 5)$

   4. $(-3 * -5)/[2 + (-3 * -8) + (6 - 2 * -1)]$

In Problems 5 to 7 show the results of each of the traversal orders of the tree.

5.
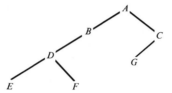

6.
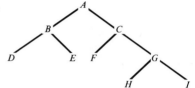

7.
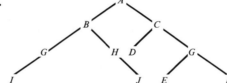

   8. Apply postorder and preorder traversal to the search tree shown in Fig. 6.11, and thus verify the comment at the end of this section.

Evaluate each of the RPN expressions, and draw the tree representation for the expressions in Problems 9 to 12.

   9. $2\ 3\ +\ 4\ *$

   10. $-2\ 4\ /\ 3\ 5\ +\ 2\ +\ +$

   11. $1\ 2\ 3\ 4\ 5\ +\ *\ -\ /$

   12. $1\ 3\ -\ 4\ +\ 2\ 3\ 4\ 1\ -\ *\ -\ +$

   13. Find PPN representations of the expressions in Problems 9 to 12.

   ***14.** Describe an algorithm for converting an expression from RPN to PPN.

   15. Show the ambiguity of infix notation by constructing two expression trees which have inorder traversal $2\ +\ 3\ *\ 5\ -\ 1\ +\ 2$.

   16. Repeat Problem 15 with the expression $-1\ +\ 4\ *\ 3\ -\ 2\ *\ -\ 3\ +\ 5\ *\ 2$.

Evaluate the following PPN expressions, and produce the tree representing the expressions in Problems 17 to 20:

17. $- + 2\ 3 * - 5\ 1\ 2$

18. $- + * 2\ 3 - 5\ 1\ 2$

19. $+ * - 4\ 3 - * 2\ 3 + * 5\ 2 - 1 - 2$

20. $+ * - - * + * 4\ 3\ 2\ 3\ 5\ 2 - 1 - 2$

21. Prove the statement made in the proof of Theorem 6.9 that if a and b are vertices of a binary tree, one of the following must be true: (1) a is a descendant of b; (2) b is a descendant of a; (3) there is a vertex c such that a is in one left subtree of c and b is in the other subtree.

22. Answer the three why's that appear in the proof of Theorem 6.9.

6.8 RELATIONS AND FUNCTIONS

In Chapter 2 we introduced the ideas of relations and functions as sets of ordered pairs of elements. Functions are special kinds of relations and are usually viewed or used in a different context from relations in general. But the techniques that we discuss in this section apply to both functions and more general relations.

If R is a relation on A, then R is a subset of $A \times A$. If G is a digraph (V, E) with $V = A$, then E is a relation on A. Every directed graph defines a relation, and every relation on a set A defines a directed graph with vertex set A. For this reason it is valuable to examine the relationships between properties of relations and those of digraphs.

EXAMPLE 6.19

Let $A = \{1, 2, 3\}$, and let $R = \{(1, 2), (1, 3), (2, 3)\}$. Represent the digraph with vertices A and edges R.

SOLUTION

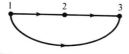

■

For finite relations, digraphs provide a very neat pictorial representation of the relation, and we should not be surprised that some of our algorithms and theorems for dealing with directed graphs have very nice applications to relations as well. In fact, the only real difference between the two objects lies more in the point of view than in mathematical substance. As we have noted before, it is a particularly gratifying triumph for mathematics when the same theory can be made to apply to apparently disparate topics.

We also need to deal with the situation in which a relation is not a relation on a single set but rather a relation on a pair of sets. If R is a relation on A and B, at first glance it might seem as though we could *not* produce a digraph to deal with R, since we don't have a single vertex set. However, if we let the vertex set V be $A \cup B$, $A \times B$ is a subset of $V \times V$, and the elements of R can be edges of a graph with vertex set V. The graph produced by this process will be a bipartite graph which has the additional property that all the edges are directed *from* set A *to* set B.

EXAMPLE 6.20

Let $A = \{1, 2, 3\}$ and $B = \{x, y, z\}$. Let R be a relation on A and B defined by $R = \{(1, x), (1, y), (2, y), (3, z)\}$. In this case $V = A \cup B = \{1, 2, 3, x, y, z\}$, and we obtain the following picture of the graph of R:

We make a couple of quick remarks which you are asked to verify in the problems. If (V, E) represents a relation, the domain of the relation is the set of all vertices with outdegree greater than 0, and the range of the relation is the set of all vertices with indegree greater than 0. A relation is a function if and only if the outdegree of each vertex in the domain is exactly 1.

In studying relations defined on a single set, we found that several properties of relations were particularly interesting in mathematics. We now consider those properties again, this time in relation to the theory of directed graphs.

Closures of Relations

One of the easiest properties of a relation to consider is reflexivity. A relation R on A is *reflexive* if and only if $\forall x(xRx)$. In terms of graphs, this means that for all vertices x there is an edge from x to x. In graph theory terms this means that there is a loop of each vertex.

For many relations, mathematicians find it useful to form the closure of the relation with respect to some property.

DEFINITION

The **reflexive closure** of relation R on A is a relation R' such that (1) $R \subset R'$, (2) R' is reflexive, and (3) if R'' is a reflexive relation containing R, then $R' \subset R''$.

The reflexive closure of R on A is the smallest subset of $A \times A$ which contains R and which is a reflexive relation.

Almost identical definitions are made for the symmetric and transitive closures of a relation. Essentially, we are asking to make R reflexive (or symmetric, or transitive) by "messing it up" as little as possible.

Often we are given only a limited amount of information about a relation, and still we know that the relation is reflexive, or symmetric, or transitive. Under those circumstances, the prudent approach to studying the relation is to find some relation known to be a subset of the relation that we are dealing with and examine its properties. Any pairs of elements that we find related in the closure will be guaranteed to be related in the actual relation.

EXAMPLE 6.21

If $A = \{1, 2, 3\}$ and R is the relation $\{(1, 2), (1, 3), (2, 3)\}$, represent the reflexive closure of R as a digraph.

SOLUTION

In Example 6.20, we represented R as a digraph. All we need to do to the relation to make it reflexive is to add a loop at each vertex. This results in the following graph:

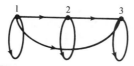

One question goes begging when we say, as we did in Example 6.21, that we were constructing *the* reflexive closure of R. The reflexive closure of R is a unique relation, and we would be remiss in not proving that fact. (Constructing the proof also forces us to recall from Chapter 2 how to prove that two sets are equal.)

THEOREM 6.10

If R is a relation on A and R' and R'' are both reflexive closures of the relation R, then $R' = R''$.

PROOF

We can prove $R' = R''$ if we can show that $R' \subset R''$ and $R'' \subset R'$. Since R' is a reflexive closure of R, we know R' contains R and is reflexive. According to the definition of reflexive closure and the fact R'' is a reflexive closure of R, it follows that $R'' \subset R'$. A similar argument shows that $R' \subset R''$. This completes the proof. ∎

From this argument, clearly similar statements can be made about the symmetric and transitive closures of a relation.

We recall that a relation R on set A is symmetric if and only if, for all elements a and b in A, aRb only if bRa. In graph theory terms, this means that if (a, b) is an edge in G, then (b, a) must also be an edge. The graph in Example 6.19 is not symmetric, but it can be turned into a symmetric graph by adding edges running in the opposite direction to each of the existing edges. Figure 6.19 is the graph of the symmetric closure of the relation.

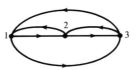

Figure 6.19 Graph of the symmetric closure.

Relations can also be studied by considering matrix representations of their graphs. Because of the obvious and extremely close relationship between graphs and relations, the terminology often becomes blurred. When we speak of the transitive closure of digraph G, we obviously mean the transitive closure of the relation defined by G; and if we refer to the matrix of relation R, we mean the adjacency matrix of the graph associated with relation R. In most cases the reference is obvious, and the shorthand does save a lot of words. If the situation is not clear, then it is necessary to use the correct formal terminology.

Simple matrix operations can be used to find the reflexive and symmetric closures of relations. To find the matrix of the reflexive closure of R, we just OR the matrix of R with the identity matrix. To find the matrix of the symmetric closure of R, we can OR the matrix of R with its own transpose.

Graphs and the Transitive Property

The question of transitivity poses a greater challenge for graph theory. We recall that the definition of a transitive relation R on set A was that for all a, b, and c in set A, if aRb and bRc, then aRc. Translating that idea to graph theory, we say that if (a, b) is an edge and (b, c) is an edge, then (a, c) is an edge. Thus if there is a path of length 2 between a and c, there must be an edge between a and c. This provides a simple test for transitivity. If M is the matrix for a digraph G, we remember that M^2 is a matrix which has a nonzero value in row i, column j if and only if there is a path of length 2 from vertex i to vertex j. The following algorithm is a transitivity test:

ALGORITHM TRANSITIVE

begin
{M is the adjacency matrix of R.}
matrix N $:= M^2$
trans $:=$ true
for i $:= 1$ **to** n **do**
    for j $:= 1$ **to** n **do**
        trans $:=$ trans **and (not** (($N[i, j] < > 0$) **and** $M[i, j] = 0$))
    od
od
end.

The statement in the loop remains true as long as with each i and j, $M_{i,j}$ is not zero when $M_{i,j}^2$ is not, because the matrix M^2 will be nonzero in position i, j exactly when there is a path of length 2 joining vertices i and j. This algorithm makes sure that we never have a path of length 2 without an edge already existing in the graph. In the case of the graph in Example 6.19,

$$M = \begin{bmatrix} 0 & 1 & 1 \\ 0 & 0 & 1 \\ 0 & 0 & 0 \end{bmatrix} \quad \text{and} \quad M^2 = \begin{bmatrix} 0 & 0 & 1 \\ 0 & 0 & 0 \\ 0 & 0 & 0 \end{bmatrix}$$

Matrix M has a nonzero entry in the only position in which M^2 is not zero.

Constructing the Transitive Closure

The test for transitivity is easy, but the process of constructing the transitive closure is not so easy. A reasonable guess would be for any nontransitive relation, insert a 1 in its adjacency matrix in those positions in which the square of the adjacency matrix has a nonzero entry but the original matrix does not have a nonzero entry. Unfortunately this does not work. This idea corresponds in the graph to inserting an edge for every pair (a, b) for which a path of length 2 exists.

EXAMPLE 6.22

Let $A = \{1, 2, 3, 4\}$ and R be a relation defined on A by $1R2$, $2R3$, and $3R4$. Find the transitive closure of R.

SOLUTION

Rather than using matrix multiplication, we draw this graph and look for paths of length 2:

There is a path of length 2 from a to c and from b to d. We simply add edges (a, c) and (b, d). If our conjecture had been correct, this would have given us the transitive closure:

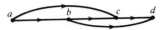

This is not the graph of a transitive relation. In this new graph, there is a path of length 2 from a to d. (In fact, there are two such paths.) If R is to be transitive, we must also add that edge to the graph. It is useful to note here that there originally was a path of length 3 from a to d. ∎

Theorem 6.11 clarifies the situation regarding the relationship between paths in a directed graph and the transitive property.

THEOREM 6.11

If G is the digraph of a transitive relation and a and z are any vertices of G such that there is a directed path from a to z in G, then there must be an edge from a to z.

PROOF

The proof is a simple induction on the length of a path. We let $P(n)$ be the statement that if there is a path of length n connecting two vertices, then there must be an edge from the initial vertex of the path to the terminal vertex.

Basis Step   This is easy (as it usually is): $P(1)$ is true because a path of length 1 is an edge.

Induction Step   The induction hypothesis is that if there is a path of length k from a to z, there must be an edge; and we must show that if there is a path of length $k + 1$ from a to z, there must be an edge. To this end, suppose that $av_1 \cdots v_k z$ is a path in G. The length of this path is $k + 1$. The path $av_1 \cdots v_k$ is a path of length k joining a and v_k. By the induction hypothesis, (a, v_k) must be an edge. But (v_k, z) is also an edge (it is part of the path), and R is transitive. Thus (a, z) must belong to relation R and hence be an edge of graph G. ∎

This theorem does all the work that we need to construct a neat algorithm for finding the transitive closure of a relation. Warshall's algorithm finds all pairs of vertices such that there is a path of any length between them; as a consequence, the final matrix from Warshall's algorithm must be the matrix of the transitive closure of the relation.

Although we have dealt with only reflexive, symmetric, and transitive properties of relations in this section, a number of other facts about relations become clear from their graphs.

THEOREM 6.12

The following facts are true about relations considered as graphs:
(a) A relation is *antisymmetric* if whenever a and b are distinct vertices in its graph and there is an edge from a to b, there is no edge from b to a.
(b) A relation is *asymmetric* if it is antisymmetric and has no loops.
(c) Let G be the graph of relation R, and let S be the set of vertices with outdegree not zero in G. Then R is a function if and only if each vertex in S has outdegree 1.

PROOF
See the problems. ∎

Problem Set 8

1. Let $A = \{1, 2, 3, 4, 5, 6\}$ and R be defined by aRb if $b = a + 2$. Draw the graph of R.

2. Let $A = \{1, 2, 3, 4, 5, 6\}$ and R be defined by aRb if a is a divisor of b. Draw the graph of R.

3. Let $A = \{1, 2, 3\}$ and $B = \{a, b, c, d\}$, and let R be the relation on A and B defined by $R = \{(1, a), (2, a), (3, a)\}$. Draw the graph of R.

4. Let $A = \{x, y, z\}$ and $B = \{4, 5\}$, and let R be the relation on A and B defined by $R = A \times B$. Draw the graph of R.

5. Use Theorem 6.12 to determine which of the relations in Problems 1 to 4 are also functions.

6. Prove the remark made following Example 6.20 that the domain of a relation is the set of vertices in its graph with outdegree greater than 0, and that the range is the set of vertices with indegree greater than 0.

7. Find the domains and ranges of the relations in Problems 1 to 4 by using the remark mentioned in Problem 6.

8. Find the reflexive and symmetric closures of the relations in Problems 1 and 2.

9. Use Warshall's algorithm to find the transitive closures of the relations in Problems 1 and 2.

10. Formally define the symmetric and transitive closures of a relation (we skipped that, to leave it to you), and prove that they are unique.

In Problems 11 to 14, draw the graphs of the relations whose adjacency matrices are given, and find the indicated closures. In each case draw the graph of the closures requested, don't just give the matrix.

11. $\begin{bmatrix} 1 & 0 & 1 \\ 0 & 0 & 1 \\ 0 & 1 & 0 \end{bmatrix}$ Find the reflexive, symmetric, and transitive closures.

12. $\begin{bmatrix} 0 & 1 & 0 & 0 \\ 0 & 0 & 1 & 0 \\ 0 & 0 & 0 & 1 \\ 0 & 0 & 0 & 0 \end{bmatrix}$ Find the reflexive, symmetric, and transitive closures.

13. $\begin{bmatrix} 0 & 1 & 0 & 0 & 0 \\ 1 & 0 & 0 & 0 & 0 \\ 0 & 0 & 0 & 1 & 0 \\ 0 & 0 & 0 & 0 & 1 \\ 0 & 0 & 0 & 0 & 0 \end{bmatrix}$ Find the symmetric and transitive closures.

14. $\begin{bmatrix} 1 & 1 & 1 & 1 & 1 & 1 \\ 0 & 1 & 0 & 1 & 0 & 0 \\ 0 & 0 & 1 & 0 & 0 & 1 \\ 0 & 0 & 0 & 1 & 0 & 0 \\ 0 & 0 & 0 & 0 & 1 & 0 \\ 1 & 0 & 0 & 0 & 0 & 0 \end{bmatrix}$ Find the transitive closure.

15. Suppose that R is an equivalence relation. Describe its graph. In particular, what do the components of the graph of R look like?

In the following problems we use the notation $r(R)$ to indicate the reflexive closure of the relation R, $s(R)$ for the symmetric closure, and $t(R)$ for the transitive closure. Thus $rs(R)$ is the reflexive closure of the symmetric closure of R, etc. In Problems 16 to 20 determine whether the given statement is true or false for all relations R on a set A. If it is true, prove the statement; but if it is false, provide an example showing it to be false.

16. $rs(R) = sr(R)$ **17.** $st(R) = ts(R)$

18. $tr(R) = rt(R)$ **19.** $rst(R) = tsr(R)$

20. $tsr(R)$ is an equivalence relation on A.

21. Prove that R is transitive if and only if $R = t(R)$.

22. Prove that if the adjacency matrix of R is M, then the adjacency matrix of $s(R)$ is $M \vee M^T$.

23. Prove Theorem 6.12.

In this chapter we continued our development of graph theory by introducing the concept of a directed graph. In many ways directed graphs resemble the undirected graphs from Chapter 5, except for the fact that each edge in a directed graph carries with it a sense of direction. We described the appropriate concepts for connectivity of directed graphs, and we introduced some algorithms to assist in dealing with the properties of directed graphs.

The directed or rooted tree is a special case of the directed graph, and we explored in some detail the concept of the directed tree. Directed trees are especially important because of the numerous applications of trees in computer science.

KEY TERMS

We list some of the important ideas introduced in this chapter (section numbers are found within parentheses).

adjacency matrix (6.2)
almost complete binary tree (6.6)
associated undirected graph (6.2)
balanced tree (6.6)
binary tree (6.5)
colored graph (6.2)
connected graph (6.4)
depth of a vertex (6.5)
digraph (6.2)
Dijkstra's algorithm (6.3)
directed graph (6.2)
directed path (6.2)
directed tree (6.5)
expression tree (6.7)
full binary tree (6.6)
incident (6.2)
indegree (6.2)
infix notation (6.7)
initial vertex (6.2)
inorder traversal (6.7)
labeled graph (6.2)

leaf (6.5)
n-ary tree (6.5)
ordered tree (6.5)
outdegree (6.2)
postfix notation (6.7)
postorder traversal (6.7)
PPN (6.7)
prefix notation (6.7)
preorder traversal (6.7)
rooted tree (6.5)
search tree (6.6)
strongly connected (6.4)
subtree determined by v (6.5)
terminal vertex (6.2)
traveling salesman problem (6.3)
traversal (6.7)
undirected graph (6.2)
undirected path (6.2)
universal address method (6.5)
Warshall's algorithm (6.4)
weighted graph (6.2)

7

Mathematical "Machines"

7.1 BINARY NUMBERS AND PARITY CHECKING

Magnetic tape stores data coded in binary (base 2) form. Typically a segment of tape is coded magnetically in such a way to represent a series of 0s and 1s in an organization called a *frame*. Each frame is a vertical strip of tape on which a fixed number of the 0s and 1s (or *bits*) can be represented. The number of bits represented in each frame is usually 9, with 8 bits being used to represent each character stored on the tape and the ninth bit used as an error-checking device known as a *check bit*.

In Fig. 7.1, the frame illustrated represents the binary number 01101101. The check bit is *not* used in determining the value of the frame. As we recall, the number represented by this frame is $0 \times 128 + 1 \times 64 + 1 \times 32 + 0 \times 16 + 1 \times 8 + 1 \times 4 + 0 \times 2 + 1 \times 1 = 109$. Each bit represents the number of times (either 0 or 1) that the power of 2 represented by its position occurs in the number, with the rightmost bit representing 2^0 and each bit to the left representing the next higher power of 2. This frame thus may represent the number 109 or perhaps the ASCII character number 109 (m), or it may be part of a longer binary representation of a number. The exact meaning of the frame depends on how the tape is being used. In any case, the binary system is very useful in computing because of the bipolar nature (on or off, high or low voltage, magnetic field present or not present) of the storage media used by

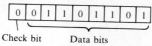

Check bit   Data bits

Figure 7.1 One frame of a tape with bits labeled.

computers. In the problems at the end of this section, we provide the opportunity to review the processes of converting back and forth between binary and our usual base-10 notation system.

Of special significance to the contents of this chapter is the fact that the information in each frame of the tape is parity-checked. This is a system which is designed to catch many of the errors which can occur during reading from or writing to magnetic tape. The parity check system is designed so that the number of bits in a frame which are 1s will always be odd. (This is known as *odd parity checking*. It is also possible to use even parity checking.) If the tape reader senses an even number of 1s in a frame, then we know that either the reader has misread one of the bits or one of the bits was sent incorrectly when the tape was originally written. In either case something is not right, and an error has occurred. Note that this scheme will only detect an error if an *odd* number of bits have been written or read incorrectly. (Can you explain why?)

The scheme for testing parity will serve as a good example of the mathematical model of a computer known as a *finite automaton*.

In checking parity, we must test each bit in the frame one by one. Each time a 1 is encountered, we can change our parity count from odd to even or even to odd. When the last bit is examined, if the number of 1 bits is even, then an error has occurred, and if the number of 1 bits is odd, then all is regarded as correct. We can set up an algorithm to test a frame of a tape for odd parity.

ALGORITHM PARITY

```
begin
odd := false;
for I := 1 to 9 do
    if frame[I] = 1 then
        if odd then
                odd := false
        else
                odd := true
        fi
    fi;
od;
parity := odd
end.
```

When we examine this algorithm, we can begin to get an idea of what is happening. There are really two states that the algorithm takes on: one when the number of 1s found so far is even, and the other when the number of 1s found is odd. When a 0 is encountered, the state remains the same; but when a 1 is found, the state changes. We start with the state as being even, and we regard the process as succeeding if we end in the odd state.

A directed graph serves to illustrate the process very nicely. See Fig. 7.2. The meaning of the graph is that the process starts in the even state and then changes state when a 1 is found. The double circle around the odd state indicates that computations which end in odd are acceptable results. The absence of a double circle around the even state indicates that ending in an even state is an unacceptable result.

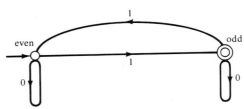

Figure 7.2 Directed graph description of parity checking.

This process can be modified easily to determine the value of the check bit when data are to be written onto a tape. We simply allow the 8 bits used to represent a value to be stored on the tape to be processed by the parity checker. If the result is odd, then the parity bit will be sent as a 0; if the result is even, the parity bit will be sent as a 1.

This is a simple example of what is known as a *finite automaton*. This is the simplest model of the workings of a computer program. Finite automatons are frequently used to model processes which involve the determination of whether a particular "string" is acceptable for a specific purpose. Although they are not the best model of a computer executing a program, finite automatons make a good start in that direction. In later chapters we look at more sophisticated models. The first step in studying finite automatons is to carefully consider the concepts of strings.

Problem Set 1

1. Convert the following numbers from binary to base-10 notation:
   (a) 1   (b) 101   (c) 11011   (d) 10001   (e) 1100111

2. Convert the following numbers from binary to base-10 notation:
   (a) 10   (b) 1001   (c) 100100   (d) 1111100   (e) 10011000

*3. Prove that the number which is represented in binary notation by a sequence of n consecutive 1s is $2^n - 1$. (*Hint:* Try induction.)

*4. Explain why parity checking can detect an error only if an odd number of bits have been incorrectly transmitted.

5. The following provides a technique for converting numbers written in base 10 to binary representation:

Divide the number by 2, noting the quotient and remainder. Repeat this process, dividing the new quotient by 2 at each step. Stop when the quotient becomes 0. Writing the remainders in reverse order from which they occurred gives a binary representation of the original number.

For example,

10 divided by 2 has a quotient of 5 and a remainder of 0

5 divided by 2 has a quotient of 2 and a remainder of 1

2 divided by 2 has a quotient of 1 and a remainder of 0

1 divided by 2 has a quotient of 0 and a remainder of 1

The remainders in reverse order are 1010, which is a binary representation of the number 10.

Find binary representations for the following numbers:
(a) 9   (b) 35   (c) 62   (d) 113   (e) 951

6. Find binary representations of the following numbers:
(a) 6   (b) 27   (c) 109   (d) 197   (e) 1000

*7. Express the process described in Problem 5 as an algorithm.

*8. Express the process of determining the appropriate parity bit for writing a binary number to tape as an algorithm.

7.2   STRINGS

The basic building block of programming languages—and, for that matter, for languages used by people (usually called *natural languages*)—are *character strings*. A character string is a sequence of symbols taken from a set of characters. Computer languages are usually sequences of characters taken from the set of ASCII characters (or, in the case of some machines, from the set of EBCDIC characters). In any language, natural or programming, some strings are part of the language and others are not. Often languages are built from *words*, which are particular strings. In a later chapter we look in more detail at ways of defining languages, but in this chapter we study the process of recognizing whether strings belong to a given language. This is one of the important functions which a compiler must perform.

In this chapter we develop a mathematical model of a "machine" which can process strings to determine whether they belong to a language. This model will not be the final one, but it is a good way to begin the process. The kind of device we develop acts as a computer executing a program in the same manner as the example in Section 7.1. To do this, we first must spend some time discussing strings and the operations that can be performed with strings.

> **DEFINITION**
>
> A finite sequence of elements from a set Σ is called a **string** on the **alphabet** Σ. The set of all strings on Σ is denoted by Σ^*.

For notational convenience, we indicate strings by setting them off with single quotes, as in '*abcd*'.

Character Strings and BASIC

EXAMPLE 7.1

In BASIC, character strings are sequences of ASCII characters. They are named by variable names which end in a $, and they are denoted by enclosing them in double quotes. Thus we can use "ABCD" as a character string in BASIC. A statement such as

    10 LET X$= "ABCD"

illustrates the use of character strings in BASIC.   ∎

Note that character string operations available in BASIC include most of the important manipulations which one wishes to perform on strings. We provide examples of this later on.

In this section we make extensive use of BASIC, because it is the commonly known language which provides the most convenient facilities for manipulation of strings. Although the operations we discuss in BASIC *can* be performed in FORTRAN or Pascal, it is not terribly convenient to do so. Thus one either must have a collection of string manipulation procedures available or must write such a collection to perform many of the string operations which are supplied with many versions of BASIC. The problems at the end of the section provide an opportunity to develop Pascal procedures to do some fundamental manipulations with strings. Of course, there are also some sophisticated (and less widely used) languages, such as SNOBOL, which are specifically designed for dealing with strings.

EXAMPLE 7.2

English is a language which consists of words. Valid English words are strings on the alphabet {a, b, c, d, e, f, ..., z, space, -, '}. Valid English sentences are strings of words, spaces, and punctuation marks. This book is a string of sentences.   ∎

EXAMPLE 7.3

The notation of binary numbers is provided by using the set of strings on the alphabet {0, 1}. ∎

Syntax and Semantics

In Examples 7.1 and 7.2 we have conveniently ignored some of the important issues, such as which strings belong to the language and what the meaning is for a particular string.

> **DEFINITIONS**
>
> The **syntax** of a language is the set of rules which determines whether a particular string belongs to the language.
>
> The **semantics** of a language is the set of rules which associates meaning to the strings belonging to the language.

In the construction of a programming language, the issues of both syntax and semantics are crucial—a translator (either a compiler or an interpreter) for a language must check the syntax of the strings that a programmer writes in the language, and the translator must determine the machine operations needed to carry out the meaning (semantics) of the statements.

Operations on Strings

Certain operations are often used in processing strings. Two of the most important are concatenation and testing for substrings.

> **DEFINITION**
>
> The **concatenation** of strings A and B, denoted AB, is the sequence which consists of the sequence A followed by the sequence B.

EXAMPLE 7.4

Suppose that string A is 'protest' and string B is 'march'. The string AB is then 'protestmarch', and the string BA is 'marchprotest'. Notice that concatenation of strings is most definitely not a commutative operation. ∎

EXAMPLE 7.5

In many versions of BASIC, string concatenation is provided as an operation for character strings and is denoted with the operator $+$. If A\$ = "CAT" and B\$ = "DOG", then A\$ + B\$ = "CATDOG" and B\$ + A\$ = "DOGCAT". ∎

> **DEFINITION**
>
> *A* is said to be a **substring** of *B* provided that the entire sequence of symbols in *A* occurs as a subsequence of *B*.

EXAMPLE 7.6

'*abcd*' is a substring of '*cd**abcd**cdxy*'.
'*xyab*' is not a substring of '*cdabcdcdxy*'. ∎

Next we provide a sequence of definitions which will be useful as we formalize the ideas of strings and languages.

> **DEFINITIONS**
>
> The **length** of a string is the number of symbols in the string. The length of string *A* is often denoted by $\lambda(A)$.
>
> The string whose length is 0 is called the **empty string**. The empty string is denoted by Λ.

In the examples above, $\lambda(\text{'protest'}) = 7$, and $\lambda(\text{'march'}) = 5$. Note that the length of the concatenation of the two strings is 12, which is the sum of their lengths.

Λ is a substring of any string. In BASIC, Λ is denoted by " " (two consecutive double quotes with nothing between them).

> **DEFINITION**
>
> A **language** on Σ is any nonempty subset of Σ^*.

BASIC and Substrings

BASIC provides a number of capabilities for dealing with substrings. Although the exact notation varies from one version to another, most provide the capability both of testing whether one string is a substring of another and of extracting substrings from a given string. For example, in one version of BASIC the function INSTR tests whether one string is a substring of another. INSTR(A$,B$,N) returns a value in the following way:

1. The result is 0 if A$ is *not* a substring of the substring of B$ which consists of all characters in B$ from position *N* on.

2. The result is the value *k* if A$ *is* found as a substring of that substring of B$, and the first time that A$ occurs as a substring begins at character *k* in B$.

If A\$ is the string "ABC" and B\$ is the string "DXABCWYABCXXXX",
then INSTR(A\$,B\$,1) = 3, INSTR(A\$,B\$,4) = 8, and INSTR(A\$,B\$,10) = 0,
because "ABC" first occurs as a substring of B\$ beginning at position 3. But if
we look at only the substring of B\$ which begins in position 4, the first
occurrence of "ABC" is not until position 8 (this is the position in the *original*
string), and "ABC" does not occur as a substring of the substring of B\$ which
begins at position 10.

In the same version of BASIC, the functions LEFT\$, RIGHT\$, and
MID\$ extract substrings from a given string. The function LEN reports the
length of a string.

EXAMPLE 7.7

If A\$ = "ABCDEFGHIJKL" and B\$ = "EFGHI", then
(a) INSTR(A\$,B\$,1) = 0 because A\$ is *not* a substring of B\$.
(b) INSTR(B\$,A\$,1) = 5 because B\$ occurs as a substring of A\$ starting at
character number 5.
(c) INSTR(B\$,A\$,6) = 0 because B\$ does not occur as a substring of the sub-
string of A\$ which begins at position 6.
(d) LEFT\$(A\$,3) = "ABC", the leftmost three characters of A\$.
(e) RIGHT\$(A\$,9) = "IJKL", the substring of A\$ which begins at character
number 9.
(f) MID\$(A\$,4,2) = "DE", the substring starting at character number 4 which
has length 2.
(g) LEN(B\$) = 5. ∎

The set of strings on an alphabet together with the operation of concatenation reveals a number of interesting properties.

Properties of the Strings and Concatenation

THEOREM 7.1

The following facts are true about Σ^* for any alphabet Σ:
(a) The set Σ^* is closed under the operation of concatenation. That is, concatenation of any two strings in Σ^* yields another string in Σ^*.
(b) The operation of concatenation is associative. That is, $A(BC) = (AB)C$ for
any A, B, C in Σ^*.
(c) The empty string Λ serves as an identity since $A\Lambda = \Lambda A = A$ for all strings
in Σ^*.

PROOF
This is left as an exercise. ∎

> **DEFINITION**
>
> A **semigroup** is a set together with an operation which satisfies the closure, associative, and identity properties.

Thus for any set of symbols Σ, Σ^* is a semigroup. As we noted before, in mathematics one often proves theorems about all structures of a certain kind, such as semigroups, in order to find statements which are true and useful in all instances of that structure. There are also some facts which are specifically true about strings of characters.

THEOREM 7.2

Let A and B be strings on alphabet Σ.
(a) $\lambda(AB) = \lambda(A) + \lambda(B)$.
(b) A is a substring of AB.
(c) If A is a substring of B and B is a substring of A, then $A = B$.
(d) If $AB = \Lambda$, then $A = \Lambda$ and $B = \Lambda$.

PROOF
This is left as an exercise. ∎

Strings are the medium in which languages are created. In the next section, we look at a device which will allow us to test strings for membership in *some* languages.

Problem Set 2

For Problems 1 to 9, use the following definitions:

$A = \text{'thisisa'} \qquad B = \text{'test'} \qquad C = \text{'hisisa'} \qquad D = \text{'history'}$

  1. Find AB, BA, BC, BD, and DC.

  2. Find CD, DB, AD, DA, and CB.

  3. Which of the strings are substrings of the others?

  4. Compute $(AB)C$ and $A(BC)$, indicating intermediate steps.

  5. Compute $(BC)D$ and $B(CD)$, indicating intermediate steps.

  6. Find $\lambda(A)$, $\lambda(B)$, and $\lambda(AB)$.

  7. Find $\lambda(C)$, $\lambda(D)$, and $\lambda(CD)$.

P8. Write a program in BASIC to find the following:
- (a) $\lambda(BC)$
- (b) The position (if one exists) at which A occurs as a substring of BC
- (c) The substring of A consisting of the leftmost three characters
- (d) The substring of D consisting of all characters to the right of character number 3
- (e) The substring of length 4, starting at position 7 of $ABCD$

P9. Write a program in BASIC to find the following:
- (a) $\lambda(ABC)$
- (b) The position (if one exists) at which C occurs as a substring of ABD
- (c) The substring of CD consisting of the leftmost seven characters
- (d) The substring of BD consisting of all characters to the right of character number 9
- (e) The substring of length 2, starting at position 2 of A

*10. Prove Theorem 7.1.　　　　**11.** Prove Theorem 7.2a.

12. Prove Theorem 7.2b.　　　　***13.** Prove Theorem 7.2c.

*14. Prove Theorem 7.2d. (*Hint:* Use Theorem 7.2a.)

P15. A string in Pascal can be represented by a record with two fields. The first field in the record is an integer which reports the length of the string. The second field in the record is an array of characters whose first n characters comprise the n characters of the string (where n is the length of the string). This array is declared to be larger than one might expect to use, and the excess characters are simply ignored. The declaration for such a type might be something like this:

```
type string = record
            length: integer;
            data: array [1..80] of char
end;
```

Use this representation to write a procedure called CONCAT, which has as input the strings A and B and produces as output the string C which is the concatenation of A and B. (*Hint:* We know the length of C will be the sum of the lengths of A and B, and the data will consist of A followed by B.)

P16. Use your procedure from Problem 15 to write a program to print the results of Problems 1 and 2. (*Hint:* You will also need to create a procedure to output strings.)

P17. Write Pascal procedures to produce results equivalent to those produced by LEFT\$, RIGHT\$, and MID\$ in BASIC.

P18. Use your procedures from Problems 15 and 17 to write programs to solve Problems 8c–e and 9c–e.

A Model of Computation

The finite automaton is the first of our mathematical models of how a computer functions while running a program. It is based on the idea that a computer can be in only a finite number of states and that in proceeding through the processing of information, a computer processes strings. This model of a computation has some drawbacks. In particular, we assume that a finite automaton provides only the answer of yes or no as to whether an input string satisfies certain criteria.

The model of a computation which we considered in Section 7.1 was actually a finite automaton. Before we proceed further, we offer a formal definition of a finite automaton, and then we describe a way to represent finite automatons as directed graphs.

DEFINITION

A **finite automaton** is a 5-tuple (S, Σ, d, s_0, A) which consists of a set S called the **set of states**, a set Σ called the **set of input symbols**, a function d: $S \times \Sigma \to S$ called the **transition function**, an element s_0 of S called the **initial state**, and a subset A of S called the **set of accepting states**.

A finite automaton provides a function from Σ^* into {yes, no} in the following way: Given a string in Σ^* with first symbol a_0, first the value of $d(s_0, a_0) = s_1$ is evaluated. Next, if the second symbol in the string is a_1, the value of $d(s_1, a_1)$ is evaluated, until the last symbol in the string is reached. If the final evaluation of d provides an element of A, then we say that the string is accepted, and we regard the automaton as having returned a value of yes. If the final evaluation of d yields a state which is not in A, then we say that the string in question is rejected, and we regard the automaton as having returned a value of no.

In the example of Section 7.1, we could write the following:

$$S = \{E,O\} \qquad \Sigma = \{0, 1\} \qquad s_0 = E \qquad A = \{O\}$$

and d is defined by $d(E,0) = E$, $d(E, 1) = O$, $d(O, 0) = O$, and $d(O,1) = E$. For convenience, if M is a finite automaton, we can refer to S as States(M), Σ as Symbols(M), d as Instructions(M), s_0 as Start(M), and A as Accepting(M).

Describing Finite Automatons

Often we use a digraph to represent the action of a finite automaton. We indicate the states by vertices of the graph, and we connect states s_i and s_j with an edge labeled by the symbol x provided that $d(s_i, x) = s_j$. The initial state is indicated by an (unlabeled) incoming arrow, and the accepting states are indicated by an extra circle drawn around the vertex. This leads to the picture that we drew in Fig. 7.2.

Another way of describing a finite automaton is by a state table. The *state table* is essentially a matrix, with rows corresponding to states (the first row is used for the initial state) and columns corresponding to elements of Σ. The entry in row s_i, column a is the value of the transition function d as applied to the pair (s_i, a). Accepting states are indicated with asterisks. The top row is used for the initial state. The state table for the example in Section 7.1 is as follows:

	0	1
E	E	O
O*	O	E

EXAMPLE 7.8

Consider the finite automaton whose diagram is drawn here:

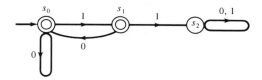

What kinds of strings are accepted by this machine?

SOLUTION

The only rejecting state is s_2. It might also be referred to as a *trapping state*, since once s_2 is entered, there is no way for a computation to leave that state. The strings which are accepted are those which *avoid* s_2. State s_2 can be reached only from state s_1 and then only when the next input is a 1. Rejection can only occur when we have a string of the form $A1X$, where A is a string leading to s_1 and X is any string. In turn, s_1 can only be reached from s_0 with an input of 1. And A must be a string of the form $C1$, where C is a string that leads to s_0. As long as we have not already arrived in s_2, a 0 always leaves us in s_0. We also start in s_0. We conclude that any string of the form $Y11X$ will cause us to reach s_2, where either Y ends in a 0 or Y is Λ. In any case this means that all strings with two consecutive 1s will be rejected, and only strings with two consecutive 1s will be rejected. We leave as an exercise the problem of constructing a state table and a formal description for this finite automaton.

■

An Application

One problem described in Section 7.2 in the processing of strings is that of determining whether one string is a substring of another. It is very easy to construct a finite automaton which tests a string for the occurrence of a particular substring. The process is quite simple to describe: We create a trapping state which is reached if the desired substring is found and a path from the initial state to the trapping state is labeled by the characters in the substring. If we are on that path and the next character in the desired substring is found, then we proceed to the next state. If we do not encounter the desired character and if the character we found does *not* leave us with a partial completion of the substring, then we return to s_0. If the next character is not the next character in the substring but *does* leave the substring partially completed, the next state should correspond to having found that much of the substring. If we are in s_0 and do not find the first character of the string, then we remain in s_0.

EXAMPLE 7.9

Design a finite automaton to test a string to determine whether the string 1101 occurs as a substring.

We need five states: s_0 corresponding to nothing found yet, then s_1, s_2, s_3, and s_4 corresponding to having encountered 1, 11, 110, and 1101 as substrings, respectively. A 1 takes us from s_0 to s_1, another 1 takes us from s_1 to s_2, a 0 takes us from s_2 to s_3, and a 1 takes us from s_3 to s_4. State s_4 is a trapping state. Except for the fact that a 1 encountered in s_2 leaves us in s_2, all other situations cause a return to s_0. The diagram below shows the automaton.

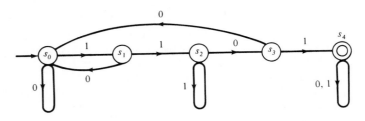

The state table for this automaton is

	0	1
s_0	s_0	s_1
s_1	s_0	s_2
s_2	s_3	s_2
s_3	s_0	s_4
s_4^*	s_4	s_4

From the state table, it is easy to design an algorithm to perform in exactly the same way as the finite automaton. It is easy to see that a sequence of **if** ... **then** statements inside a loop will do the job. You need simply let the counter i run from 1 to the length of the second string, and include within the loop a sequence of statements such as "**if** (state = 0) and (string.data[i] = '1') **then** s := 1", etc. We leave it as an exercise for the student to complete the construction of this algorithm. Although every finite automaton can be realized by an algorithm, the converse is not true. So we need to go further with our development of mathematical machines if we wish to come up with a device that can describe more complicated computer programs.

The basic concepts of finite automatons are especially useful in the design of pattern-matching algorithms in computer science. We ask you to come up with a Pascal version of INSTR in the problems, and you would be well advised to consider how to make use of the design of a finite automaton to expedite the development of your program.

The kind of process that you use in constructing such an algorithm is exactly that which is needed in designing the parts of editors or word processors which search for strings in a text file. The file itself is a very long string, and when you issue a command in a word processor (such as Wordstar, or Select) to seek a string, the program must execute a procedure which determines where your target string occurs as a substring of the file. These kinds of procedures are extremely important in many programming applications.

In many cases people are interested in finite automatons which provide output other than simply a yes or no as to whether the input string is accepted. One common scheme is to output symbols from each state. In this fashion, a finite automaton then accepts input in the form of a string and provides output in the form of a string as well. Such an arrangement is called a *finite-state transducer*. Finite-state transducers provide the opportunity for us to begin exploring the process of computation. The finite-state transducer provides in essence a function from the set of possible input strings to the set of possible output strings. In a sense they are a first attempt at studying the problem of *translating* from one language to another. Finite-state transducers are discussed in Section 7.6.

The set of strings which are accepted by a finite automaton is called the *language* of that automaton. In the next section we consider some properties that are satisfied by the languages of finite automatons.

Problem Set 3

1. Consider the automaton drawn below. Which of the following strings are accepted by this machine?

(a) 101001
(b) 11001010111
(c) 111111
(d) 0000011000

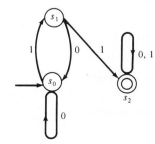

2. What property is required of a string for it to be accepted by the automaton in Problem 1?

3. Consider the automaton drawn below. Which of the following strings are accepted by this machine?

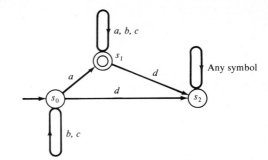

(a) *abcabd*
(b) *abbaccab*
(c) *abababab*
(d) *dabadab*

4. What property must a string have for it to be accepted by the automaton in Problem 3?

5. Construct the state table for the automaton in Problem 1.

6. Construct the state table for the automaton in Problem 3.

7. Construct the state table for the automaton in Example 7.8.

8. Construct the formal description for the automaton in Example 7.9.

9. Construct a finite automaton (by drawing its graph) which will accept all strings on {*a,b,c*} in which all three symbols are used.

10. Construct a finite automaton (by showing its state table) which determines whether the string '*sttsp*' occurs as a substring of a string in {*s, t, p*}*.

11. Write an algorithm to emulate the operation of the finite automaton in Example 7.8.

12. Write an algorithm which will accept the same set of strings as does the automaton in Problem 1.

13. Write an algorithm to accept the same set of strings as the automaton in Problem 3.

P14. Write a Pascal program to emulate the action of the finite automaton in Problem 9.

P15. Write a Pascal program to emulate the action of the finite automaton in Problem 10.

P16. Write a Pascal program to emulate the action of the algorithm in Problem 11.

P17. Write a Pascal program to emulate the action of the algorithm in Problem 12.

*P18. Write a Pascal program to serve the same function with strings as does
the BASIC function INSTR. Apply your program to Problems 8b and
9b in Section 7.2.

[*Warning*: This is harder than it looks. If you are not careful, your
program could return INSTR('in ','iinxx ',1) as 0 instead of 2. Designing
a program for a specific test is easy, but the general one is much more
of a challenge.]

7.4   LANGUAGES AND FINITE AUTOMATONS

Using a Finite Automaton to Define a Language

As we mentioned in Section 7.2, for a given alphabet Σ, a language is any
subset of Σ^*. If M is a finite automaton, then we call the set of all strings
accepted by M the *language* generated by M, denoted by $L(M)$,

---

EXAMPLE 7.10

Consider the finite automaton M illustrated here:

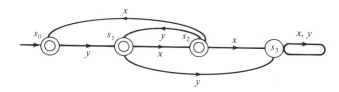

Determine $L(M)$.

SOLUTION

With this automaton clearly there is only one rejecting state. That particular
rejecting state is reached only when one encounters two identical symbols in a
row. The rejecting state is also a trapping state, so once two consecutive identical symbols are found, the string is rejected. The strings come from $\{x,y\}^*$, so
language $L(M)$ is all strings in $\{x,y\}$ which consist of *alternating* x's and y's. ∎

Concatenating *Languages*

It is interesting that it is possible to use the operation of concatenation of
strings to develop an operation which allows us to combine two languages to
create another.

THEOREM 7.3

If Σ is any alphabet, then the set of all languages on Σ with the operation of concatenation forms a semigroup in which the language consisting of only the empty string is the identity.

PROOF

Closure   The concatenation of two languages will produce a set of strings. Now since the concatenation of two strings in Σ^* is also a string in Σ^* (by Theorem 7.1), the set of strings thus produced is a subset of Σ^* and thus is a language on Σ.

Associative   If we consider language $A(BC)$, every element in this language is formed by a string of the form $a(bc)$, where a is a string from A, b is a string from B, and c is a string from C. By Theorem 7.1, we know that $a(bc) = (ab)c$, which is a string from $(AB)C$. This means that $A(BC) \subset (AB)C$. The reverse inclusion is shown true in a similar manner, so $(AB)C = A(BC)$.

Identity   If A is any language and $L = \{\Lambda\}$, since for any string a, $a\Lambda = a$ and $\Lambda a = a$, it follows that $AL = LA = A$. This collection of languages then satisfies the three properties needed for a semigroup.   ∎

DEFINITION

A language is **regular** if there is a finite automaton which accepts it.

It is reasonable to ask whether the concatenation of two regular languages is a regular language. The rather surprising answer is that it is. But to prove that fact, we have to detour through an extension of the idea of a finite automaton to what is known as a *nondeterministic finite automaton*.

Nondeterministic Automatons

The idea of a nondeterministic finite automaton is to provide the possibility of more than one transition for a given state and input symbol. The term "nondeterministic" comes from the idea that given a particular input string, we do not know which state will be reached, whereas with an ordinary (deterministic) automaton we do know exactly what state will be reached.[1]

---

[1] Don't confuse a nondeterministic automaton with a probabilistic automaton. In the latter, some random event will choose the next state. In this case the choice is *not* random.

EXAMPLE 7.11

311

7.4 LANGUAGES AND
FINITE AUTOMATONS

Consider the following nondeterministic finite automaton.

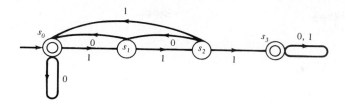

In this example, when one is in state s_2 and the input is 1, the next state may be either s_3 or s_0. The nondeterministic automaton is presumed to choose the next state so as to lead the string to an accepting state, if such is possible.

A string is accepted by a nondeterministic automaton if any of the sequences of states associated with that string ends in an accepting state. It is as though we simultaneously do *all* possible computations associated with the string and regard it as accepted if *any one* of them returns success.

It is sometimes convenient to view a nondeterministic automaton as a machine which has an infinite collection of processors and uses one or more new processors every time a choice needs to be made. A string is accepted if any one of the processors leads to an accepting state. A state table for the nondeterministic automaton in the figure is as follows:

	0	1
s_0^*	s_0	s_1
s_1	s_0	s_2
s_2	s_1	s_0, s_3
s_3^*	s_3	s_3

■

In the automaton above, two paths are associated with the string '1111'. They both begin in s_0 and proceed through transitions to s_1 and s_2. At s_2, the input of '1' yields transitions to either s_0 or s_3. The first path then terminates in s_1, which is a rejecting state, and the second terminates in s_3, which is an accepting state. Since one of the paths associated with the string '1111' terminates in an accepting state, we regard the string as being accepted.

Yet another way to view this process is to note that a string will be accepted by a nondeterministic automaton if and only if there is a path from s_0 to an accepting state with the edges of the path labeled with the symbols in the string being tested.

What we previously called a finite automaton should more properly be called a *deterministic finite automaton* to emphasize the distinction between the two ideas. We supply a formal definition of a nondeterministic finite automaton.

DEFINITION

A **nondeterministic finite automaton** is an ordered 5-tuple $M = (S, \Sigma, d, s_0, A)$, in which S is the set of states, Σ is the set of input symbols, d is a function from $S \times \Sigma$ to $P(S)$ called the transition function, s_0 is an element of S called the initial state, and A is a subset of S called the set of accepting states. A nondeterministic automaton functions in the same way as a deterministic automaton, *except* that when the transition function provides a result with more than one state, the process is regarded as "being in" all the states. A string is accepted by the nondeterministic automaton if one of the states in which it terminates is an accepting state.

In essence, a deterministic finite automaton is a special kind of nondeterministic automaton in which the transition function provides only single-element subsets of S as results.

Even though it might seem that there should be languages which are accepted by nondeterministic automatons and are not accepted by deterministic automatons, that is *not* the case. In fact, given any nondeterministic automaton, it is always possible to construct a deterministic automaton which accepts exactly the same set of strings. The proof of this fact is somewhat technical in its details, but the spirit of the proof is quite easy to comprehend. In constructing a new deterministic automaton M' from a given nondeterministic one M, we proceed in the following way:

1. Symbols(M') = Symbols(M).

2. States(M') = P(States(M)).

3. For each set S in States(M'), $d(S, a) = \{s \in \text{States}(M) \mid s \in d(x,a)$ for some $x \in S\}$.

4. Start(M') = {Start(M)}.

5. Accepting(M') = $\{S \in P(\text{States}(M)) \mid \exists s \in S, s \in \text{Accepting}(M)\}$.

It is clear this is a description of a deterministic finite automaton. Although the details may be somewhat unclear, it also seems reasonable that any string accepted by M is accepted by M', and conversely. The details of proving this are not so much difficult as they are tedious, and thus we leave this fact unproved. We do, however, state it formally.

THEOREM 7.4

If M is any nondeterministic finite automaton, there is a deterministic finite automaton with the same set of input symbols which accepts exactly the same language. ∎

EXAMPLE 7.12

We are given M as a nondeterministic finite automaton:

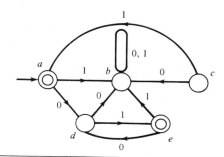

This deterministic finite automaton M' accepts the same language:

We leave as an exercise the effort of constructing state tables for both these automatons, and we ask students to construct a deterministic automaton to accomplish the same purpose as the automaton in Example 7.11.

Combining Automatons

It is now quite easy to describe the process of connecting finite automatons in "parallel." To connect automatons M and N in parallel, from the initial state of a new automaton, we provide two possible transitions for the initial symbol of the string. The first transition takes us from our new initial state to the state in States(M) that would be reached with that symbol as the first input to M. The second possible transition for the same symbol leads us to the state in States(N) that would be reached if that symbol were the first input to N. From that point on, the new automaton "acts as" M or N. The only other adjustment which needs to be made is to allow the initial state to be accepting if *either* machine accepts Λ. This automaton will accept a language which is the union of the languages accepted by M and N. Although the new machine is nondeterministic, the process produces a machine which accepts the union of the languages regardless of whether the old machines were deterministic or nondeterministic.

We summarize the process of connecting automatons A and A' to produce the nondeterministic automaton M:

1. We can assume that States(A) and States(A') are disjoint sets and that none of them is called s_0. (If this is not true, we simply rename the states of the two machines so that the statement is true.)

2. States(M) = $\{s_0\}$ ∪ States(A) ∪ States(A').

3. Start(M) = s_0.

4. If $x \in$ States(A), then

Instructions(M)(s, x) = Instructions(A)(s, x)

5. If $x \in$ States(A'), then

Instructions(M)(s, x) = Instructions(A')(s, x)

6. Instructions(M)(s, x_0) = Instructions(A)(s,Start(A)) ∪

Instructions(A')(s,Start(A')).

7. Accepting(M) = Accepting(A) ∪ Accepting(A') if neither machine accepts Λ and Accepting(M) = Accepting(A) ∪ Accepting(A') ∪ $\{s_0\}$ if either machine accepts Λ.

EXAMPLE 7.13

Connect the following automatons in parallel:

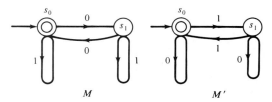

Following the procedure described above, we connect them in parallel to get the following nondeterministic automaton:

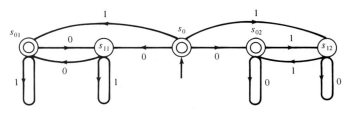

THEOREM 7.5

The union of two regular languages is regular.

PROOF
If L and L' are regular languages, there are finite automatons M and M' which accept them. If, as described above, we connect automatons M and M' in parallel, the language accepted by that nondeterministic automaton is $L \cup L'$. Thus there must be a deterministic automaton which accepts $L \cup L'$ as well. ∎

The preceding discussion of the union of languages shows us the kind of technique that we will use to do the main business of this section—which is to show that the concatenation of two regular languages is also a regular language. The main difference is that we need to describe a way of connecting finite automatons in series. The idea is to provide a nondeterministic automaton which will accept strings that consist of a string in one language followed by a string in the other language. The trick is as follows: Start by using the automaton of the first language, but from each accepting state in that language provide the possibility of multiple transitions—one remains in the first automaton, and the other leads to the state in the second automaton which would be reached from the initial state of the second automaton if the given symbol were the first to be processed by the automaton. Clearly this automaton can accept exactly those strings which are concatenations of strings accepted by the first and second automatons.

The formal description of connecting automatons A and A' in series is as follows:

1. We assume that States(A) and States(A') are disjoint sets. If this is not the case, we simply rename the elements of the two sets.

2. States(M) = States(A) $\cup$ States(A').

3. Start(M) = Start(A).

4. If $x \in$ States (A'), then

$$\text{Instructions}(A')(s, x) = \text{Instructions}(A')(s, x)$$

5. If $x \in$ States (A) and x is *not* an accepting state, then

$$\text{Instructions}(M)(s, x) = \text{Instructions}(A)(s, x)$$

6. If $x \in$ States (A) and x is an accepting state, then

$$\text{Instructions}(M)(s, x) = \text{Instructions}(A)(s, x)$$
$$\cup \text{Instructions}(A')(\text{Start}(A'), x)$$

7. If A' accepts Λ, then

$$\text{Accepting}(M) = \text{Accepting}(A) \cup \text{Accepting}(A')$$

If A' does not accept Λ, then

$$\text{Accepting}(M) = \text{Accepting}(A')$$

EXAMPLE 7.14

Consider the following automatons:

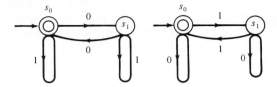

Connecting these two automatons in series yields this automaton:

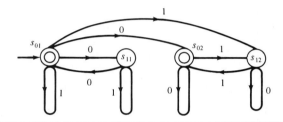

With this much groundwork done, it is very easy to state the following theorem.

THEOREM 7.6

The concatenation of regular languages is a regular language.

PROOF
This is left as an exercise. ∎

With all this out of the way, we can finally state the main theorem of the section:

THEOREM 7.7

The set of regular languages on a given alphabet together with the operation of concatenation forms a semigroup with identity being the language consisting of only the empty string.

PROOF
From Theorem 7.6, we know that the operation of concatenation of regular languages satisfies the closure property. Clearly the associative property holds, and the language consisting of the empty string will serve as an identity. We

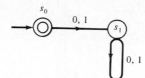

Figure 7.3 Finite automaton accepting only the empty string.

still need to verify that this language is actually regular, but that is clearly seen from the automaton in Fig. 7.3. Thus we have verified that this set does form a semigroup under the operation of concatenation. ∎

Problem Set 4

Determine the language accepted by the following automatons.

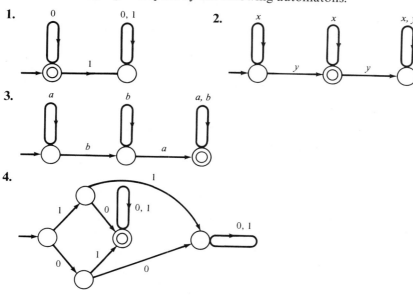

1.

2.

3.

4.

5. Show that the language $A = \{a, ab, ba, b\}$ defined on the alphabet $\{a,b\}$ is regular by designing a deterministic finite automaton to accept it.

6. Show that the language B which consists of all strings in $\{a,b\}^*$ which contain no b's is regular by designing a finite automaton to accept it.

7. Describe the language C which is accepted by the nondeterministic finite automaton of Fig. 7.4.

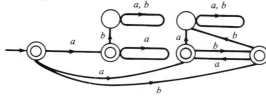

Figure 7.4

8. Find the state table for the automaton in Fig. 7.4.

9. Find a *deterministic* automaton which accepts the same language as in Example 7.11.

10. Construct the state table for the automaton in Example 7.12.

11. Find a nondeterministic finite automaton which accepts the language $A \cup B$, where A and B are as described in Problems 5 and 6.

12. Find a nondeterministic finite automaton which accepts the language AB.

13. Find a deterministic automaton which accepts the language C from Problem 7.

*14. Prove the statement that a string is accepted by a nondeterministic automaton if and only if there is a path in which the labels of the edges form the same string as the input string and which leads from s_0 to an accepting state.

*15. Prove Theorem 7.6. (*Hint:* Construct automatons for each language, and connect them in series. What language is accepted by this machine?)

*16. The language A^n is defined recursively in the following way:

$$A^0 = \{\Lambda\}$$
$$A^{n+1} = A^n A$$

Prove that for all n, if A is regular, so is A^n. (*Hint:* Use induction and the result of Problem 15.)

17. Prove that any finite language is regular. (*Hint:* Construct a finite automaton with an accepting state for each string in the language.)

P18. Write a Pascal program to simulate the operation of the automaton in Problem 1.

7.5   THE SYNTAX DIAGRAMS OF PASCAL

For readers who are familiar with Pascal and who have used one of the standard textbooks on Pascal programming, the preceding sections may have left them with a kind of *déjà vu* feeling. The state diagrams for finite automatons bear a striking resemblance to those somewhat mysterious syntax diagrams that fill parts of Pascal books. In fact, as we shall soon see, some of the syntax diagrams of Pascal can actually be interpreted in terms of finite automatons. Unfortunately, some of the syntax diagrams cannot be dealt with as finite automatons because they require infinitely many states or other features not possible for finite automatons.

The Identifier

EXAMPLE 7.15

The following syntax diagram is used to describe legal identifiers in Pascal.

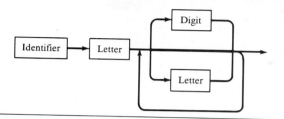

The figure in Example 7.15 is not exactly a state diagram for a finite automaton. We can note the differences: The vertices of the graph shown in the figure are labeled, but the edges are not. There are no states which are labeled, and in particular there is no indication of accepting or rejecting states. It is reasonably easy to interpret the diagram. The idea of the syntax diagram is that a string will be accepted by the diagram if we can get from the start to the end by passing through vertices which list conditions satisfied by successive substrings. Thus, for a string to be accepted as a Pascal identifier, it must begin with a letter and the letter may be followed by one or more characters which are permitted to be either letters or digits. It is easy enough to construct a finite automaton to do exactly the same thing; see Fig. 7.5. (The **else** indicates a transition to be made if the next symbol does not fit the criteria for any of the edges leaving the state.)

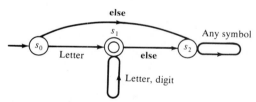

Figure 7.5 Finite automaton for description of Pascal identifier.

It is possible, by a bit of juggling, to obtain the automaton in Fig. 7.6 from the original syntax diagram. The basic task is to make the edges into states and the vertices into edges. We also need to add a rejecting (trapping) state.

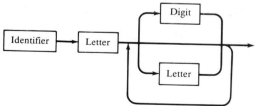

Figure 7.6 Original syntax diagram.

Thus, in this case, we produced s_0 (the starting state) corresponding to the edge leading from the word identifier to the vertex labeled "letter." We need s_1 to correspond to the edge leading from the vertex labeled "letter" to the end of the diagram. The vertex labeled "digit" and the vertex labeled "letter" correspond to edges in the state diagram for the automaton which bear those labels. See Fig. 7.7.

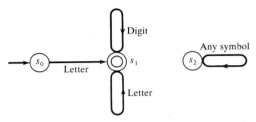

Figure 7.7 Partially completed automaton.

The edge in Fig. 7.6 which leads to the outside corresponds to an accepting state, because the string is accepted if we can follow the diagram and reach the end. We add a rejecting state set up as a trapping state, and we allow transitions to the rejecting state to occur when input occurs which is not provided for by the existing transitions. This produces the final diagram for the automaton (Fig. 7.8).

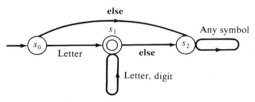

Figure 7.8 Completed automaton.

One of the tricks to this conversion process is determining which edges correspond to states and which are dealt with as though they were parts of vertices in the diagram. In general, a rule of thumb is that for a vertex in the state diagram which is part of a loop, we should regard the edges leading into and out of it as not necessitating the construction of a state in the automaton.

Label Declarations

Let's try the process again with a somewhat more complicated diagram (Fig. 7.9). According to our method, we need four states plus a rejecting state.

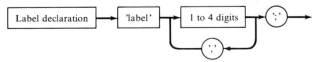

Figure 7.9 Syntax diagram for label.

The four states correspond to the edge leading into 'label', the edge connecting 'label' and the rectangle, the edge connecting the rectangle and the ',' vertex, and the final accepting state. The edges into and out of the ',' vertex are to be regarded as part of the vertex. This leads us to a first pass at a finite automaton, shown in Fig. 7.10.

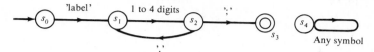

Figure 7.10 First step in automaton for label.

There are two problems with this automaton. First, we really need to insert an automaton to produce the transition called 'label'; second, we need to insert an automaton to produce the transition called 1 to 4 digits. In each case we need to connect in series automatons for recognizing the string 'label' and 1 to 4 digits with the automaton in Fig. 7.10, with the 'label' automaton taking the place of the edge labeled 'label' and the 1 to 4 digits automaton taking the place of the edge with the same label. The automaton for recognizing 'label' is easy to construct. See Fig. 7.11.

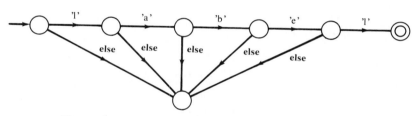

Figure 7.11 Automaton for recognizing the string 'label'.

It is also easy enough to construct an automaton for recognizing 1 to 4 digits. See Fig. 7.12. The final step is to connect the three automatons. We can simplify this a little bit by merging the rejecting states together. See Fig. 7.13.

Unfortunately, not all Pascal syntax diagrams can be converted to finite automatons. This should not be too surprising, since, as we explain later, regular languages are the simplest of the types of languages that we will study.

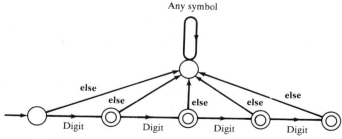

Figure 7.12 Automaton for recognizing 1 to 4 digits.

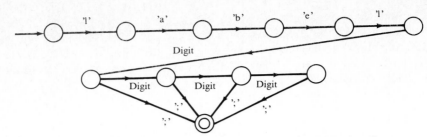

Figure 7.13 Automaton for recognizing Pascal label declaration.
(If symbol is not on edge, go to trapping rejecting state.)

For a number of reasons, particularly convenience of programming, programming languages for a computer are usually more complex than a regular language, in order to facilitate a description of the computations to be performed on the computer.

The depth of this problem arises when we consider the syntax diagrams for compound statement and statement in Figs. 7.14 and 7.15. The problem

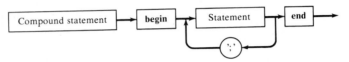

Figure 7.14 Syntax diagram for compound statement.

now becomes serious. If we were to convert Fig. 7.15 to a finite automaton, one of the automatons which would have to be connected in series would be an automaton for compound statement; that, in turn, requires that an automaton for statement be connected in series, which in turn requires .... Since, at least theoretically, the number of times this could be repeated can be any integer, no *fixed* finite number of states can possibly do the job. Some device other than a finite automaton will be needed to check statements in Pascal. (Of course, in practice, the amount of nesting in these circumstances is limited by available memory and other considerations which are not actually part of the language.) The recursive nature of this pair of syntax diagrams creates too many problems for finite automatons to handle. Using the methods in Chapter 9, we could produce a formal proof of this fact, by using the same techniques as those used to prove Theorem 9.5.

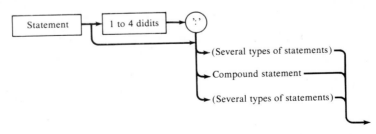

Figure 7.15 Syntax diagram for statement.

Problem Set 5

In Problems 1 to 9, design a finite automaton to recognize the strings described by the Pascal syntax diagram. You may assume that automatons exist (and can be connected in the appropriate fashion) to check for any of the symbols which are not enclosed by single quotes.

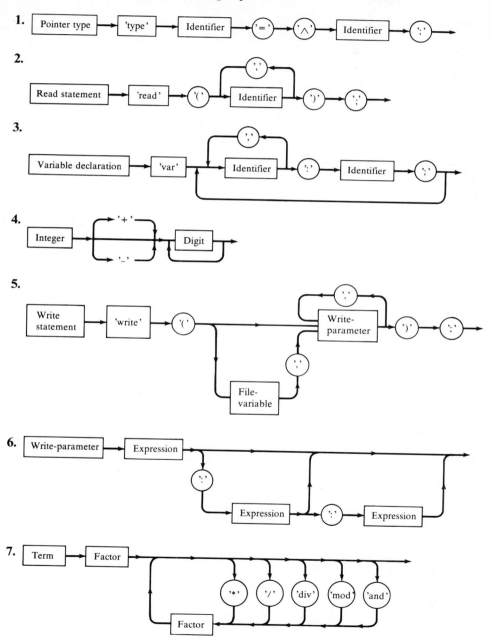

8.

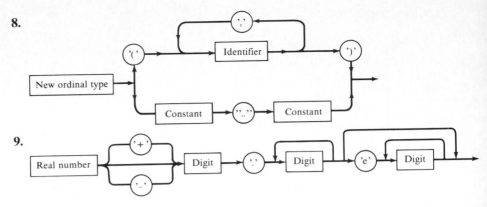

9.

*10. Take a Pascal reference and find the syntax diagram for *expression*. Explain why an expression cannot be described in terms of a finite automaton.

P11. Write a Pascal program to test a string to determine whether it is a valid integer string. Your program should implement a finite automaton equivalent to the syntax diagram in Problem 4.

P12. Write a Pascal program to test a string to determine whether it is a valid real number. Your program should implement a finite automaton equivalent to the syntax diagram in Problem 9.

7.6   FINITE-STATE TRANSDUCERS

Translating from One Language to Another

One of the drawbacks in dealing with finite automatons is the fact that the possible results are quite limited, to say the least—all we get is a yes or no answer to whether the string is accepted by the automaton. Although the finite automaton is a useful tool for beginning to understand the analysis of languages, it does not provide any insight at all into the process of translating from one language into another.

Any compiler or interpreter for a programming language *must* provide a translation between the language in use and the machine language of the computer on which the program is to be run. Although we have already noted that most useful computer languages are not regular, an extension of the finite automaton to provide output for regular languages gives us some insight into the kinds of processes needed for translation. (Actually, with some modification, the model of a finite-state transducer is used to provide parsing and translation on some computers—the utility LIB$TPARSE on the VAX series of minicomputers is described in terms of finite automatons.)

We start by considering a seemingly silly example of a "code language" known as *ob* which is used by teenagers in some parts of the country for communicating "secret" messages. In ob, translation is made from English to ob by inserting the string 'ob' into the word before each vowel. The word "vowel" becomes '*vobowobel*' in ob.

EXAMPLE 7.16

Translate the following words into ob:
(a) example
(b) string
(c) another
(d) job

SOLUTION
(a) *obexobamplobe*
(b) *strobing*
(c) *obanobothober*
(d) *jobob*

We include a few more of these as exercises. The big problem in listening to teenagers using ob is figuring out how they pronounce "words" like these. A person wanting to decode this language back into English needs to filter out the *ob* substrings. The decoder might also want to determine when a word has been used which is not part of the language. Although this example may seem somewhat silly, it contains many of the elements involved in translation from one language to another.

A first step is to note that given a string purported to be an ob string, it is easy to produce a finite automaton which can test whether the string is legal *ob*. See Fig. 7.16.

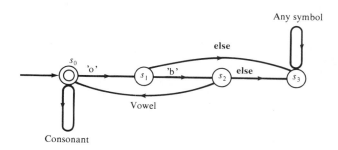

Figure 7.16 Automaton for testing ob strings.

When we use this automaton on the string '*vobowobel*', we begin in state s_0, and encountering a '*v*' as the first symbol, we remain there. Encountering an '*o*' next takes us to state s_1, and the '*b*' moves us into s_2. From s_2, since the next symbol is an '*o*', we return to s_0. The '*w*' leaves us there, and the sequence '*obe*' takes us through states s_1, s_2, and back to s_0. The final symbol in the input string is an '*l*', which causes us to terminate in s_0. Since this is an accepting state, the string '*vobowobel*' is accepted.

The string '*hobby*' is not accepted, however, because the process begins in s_0 and remains there with the '*h*' and the '*ob*' takes the process to s_2. In s_2, the next symbol encountered is another '*b*' which moves the automaton to s_3, where it remains. Since s_3 is rejecting, the string is rejected.

Finite-State Transducers

It would be nice to be able to produce output in English from this automaton. The idea would be to "echo" the input except for the ob parts. We can associate, in addition to a new state, an output string with each transition. The final output is the concatenation of all the output strings. This, in a nutshell, is the idea of what we call a *finite-state transducer*. In drawing a state diagram for such a device, we indicate the input symbols first, followed by a slash, and then the output string associated with that transition. We regard the transducer as having produced an output consisting of the concatenation of the output strings from each transition provided that the final state is an accepting state. See Fig. 7.17.

In describing a finite-state transducer, we expand the ordered 5-tuple of a finite automaton to an ordered 7-tuple by adding a set B, called the *output alphabet*, and a function $f: \Sigma \times S \to B^*$ which provides a string for each combination of input and state. As we mentioned before, the output for a given string is the concatenation of all the results of f. The input is considered to have produced the output, provided that the final state is an accepting state. A finite-state transducer provides a function from the language accepted by the automaton to a language which is a subset of B^*. Since the output function f has strings and not single symbols as values, the length of the output from a given input string does not have to be the same as the length of the input string. (It can be either shorter or longer, since Λ is a possible value of f.)

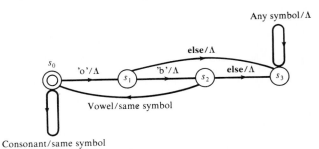

Figure 7.17 Finite-state transducer from ob to English.

In Fig. 7.17 we showed a finite-state transducer which translates from ob to English. We can similarly (in fact, even more easily) produce a transducer (with only one state) to translate from English to ob.

State Tables for Finite-State Transducers

Just as with finite automatons, we can also describe finite-state transducers by state tables. The only difference with the state tables for transducers is that following the listing of the state we place a comma and then list the output associated with the given input and state.

EXAMPLE 7.17

Consider the following state table, and produce a transition diagram for it.

	0	1
s_0	s_1, '0'	s_2, '1'
s_1^*	s_1, '0'	s_2, '0'
s_2^*	s_1, '0'	s_2, '0'

The effect of this transducer is to insert an extra 0 at the end of each sequence of 0s in a binary string. The transition diagram for this automaton is

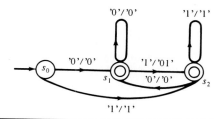

The Sum of Two Integers

It is reasonable to ask whether there are some more realistic problems which can be modeled with finite-state transducers, and the answer is yes. We can create a finite-state transducer which will produce a representation of the sum of two integers. To do this, we have to rearrange the input, but the solution ends up looking *exactly* as we would write out the solution on a piece of paper.

Suppose that we want to add the numbers 5 and 7. In binary, 5 is written as 101 and 7 is written as 111. If we were to add on paper, we would write

$$111$$
$$\underline{101}$$

We then work from right to left, first adding the two 1s. Since the sum of the last two digits is 2, which we represent in binary as 10, we write down a 0 and "carry" the other digit into the next sum.

$$
\begin{array}{r}
1 \\
111 \\
101 \\
\hline
0
\end{array}
$$

The next pair of digits in the sum is a 1 and a 0. Their sum is 1, but with the carry from the previous pair of digits, the sum becomes 2 again. So we must write a 0 in the sum and indicate a carry into the next pair of digits:

$$
\begin{array}{r}
11 \\
111 \\
101 \\
\hline
00
\end{array}
$$

For the next pair of digits, the sum of the digits is 2, and adding the carry makes 3, or 11 in binary notation. Thus we write a 1 and carry:

$$
\begin{array}{r}
111 \\
111 \\
101 \\
\hline
100
\end{array}
$$

There are no digits to which we can add the last carry, so we treat the sum as though it were $0 + 0$ and add the carry to get the final result:

$$
\begin{array}{r}
111 \\
111 \\
101 \\
\hline
1100
\end{array}
$$

We can design a transducer to do exactly the same thing, as long as we can allow the input to the machine to be *pairs* of digits, written from right to left. The states we need for this process will take into account whether there was a carry into the place that is being added. The output will be the last digit of the sum of the two digits or the last digit of the sum with carry. The transition function keeps track of carries. Thus the final result has to be read in reverse, and in order to resolve any final carry (such as we had in $5 + 7$), we send the last pair of digits as $(0,0)$. Since all carries must be resolved, only the "no carry" state will be accepting. The state diagram of Fig. 7.18 shows the result.

You should trace the operation of this transducer with the pairs of digits needed to add 5 and 7 to verify that the transducer operates as advertised.

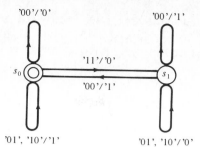

Figure 7.18 Transducer for the sum of two integers.

Changing Bases for Representing Numbers

Another example of a task that can be represented easily with a transducer is the conversion of a number from binary notation to base 4. The trick here is to see that we can group the digits of a binary number in twos and replace the pair 00 by 0, 01 by 1, 10 by 2, and 11 by 3. As long as an even number of digits are presented, the task is easy. In a similar fashion, conversion between binary and octal (base 8) and between binary and hexadecimal (base 16) is easily described by finite-state transducers.

Limitations of Finite-State Transducers

Clearly not all operations can be described with finite-state transducers. The finiteness restriction becomes a real problem. A finite-state transducer cannot find the product of two binary numbers, nor can it raise a number to a power. The problems that arise here are closely related to the limitations of finite automatons and the questions of determining which languages can be accepted by finite automatons. We discuss this topic in much greater detail in Chapter 9. We can find the product of binary numbers with a transducer if we place some restriction on the number of digits in the numbers being multiplied.

Conclusions

We should make a couple of closing comments about mathematical machines. First, it is possible to construct nondeterministic finite-state transducers. But just as with finite automatons, we do not change what can and cannot be done with transducers by considering such a device.

Finally, some comments about finiteness are in order. Computers *are* finite machines, and this fact does influence what can be done on them. Implementations of languages place restrictions on the complexity of expressions and may fail to translate expressions which have too many parentheses in them. Because of word size limitations of computers, we cannot multiply *any* two integers, but instead some "trapping rejecting states" (such as overflow) may occur. We need, however, to know what kinds of things can be done on a computer. And since computers keep getting larger and larger, it is reasonable to want to consider a device in which the finiteness restrictions are not quite so confining. This is what we do when we develop a model known as the Turing machine, which serves as a good model for the modern computer (except that it does allow for infinite memory capacity). With today's ever-increasing computer memory sizes, this does not present serious problems since, relative to the sizes of most problems that we wish to tackle on a computer, memory is effectively limitless.

Problem Set 6

1. Determine the sequence of states entered by the finite automaton which checks for membership in the ob language as seen in Fig. 7.16.
   (a) *jobob*   (b) *obexobamplobe*
   (c) *obanobothober* (d) *strobing*

2. Determine the sequence of states entered by the automaton in Fig. 7.16 as it checks the following strings:
   (a) *troop*   (b) *job*
   (c) *string* (d) *obobject*

3. Construct a state table for the automaton in Fig. 7.16.

4. Follow the progress of the transducer in Fig. 7.17 as it translates the following words from ob to English:
   (a) *jobob*   (b) *obexobamplobe*
   (c) *obanobothober* (d) *strobing*

5. Construct the state table for the transducer in Fig. 7.17.

6. Construct a finite-state transducer to translate from English to ob, and determine the sequence of states which your machine follows as it translates the following words:
   (a) *triple* (b) *drop*
   (c) *hello*   (d) *antidisestablishmentarianism*

*7. What property must a string have in order to be accepted by the ob machine? Will the machine accept words that are not valid in ob? Explain your answer.

*8. A language which is quite popular with elementary school students is "pig Latin." Pig Latin words are formed from English words by taking the initial consonant sound and placing it on the end of the word, followed by "ay." Thus *frog* becomes *ogfray*. Words which begin with a vowel sound are translated into pig Latin by appending "yay." So ink is translated as *inkyay*. Find a finite automaton which tests a string for membership in pig Latin, and test it with the following strings:

(a) *arcay*   (b) *ookbay*
(c) *payday*   (d) *okay*
(The last two may cause trouble.)

*9. Explain why the construction of a finite automaton to translate from pig Latin to English would be difficult to accomplish.

10. A simple substitution cypher is another way of coding information. In this scheme, some of or all the letters of the alphabet are replaced by others. For example, suppose we make the following substitutions:

a—e	o—a
e—i	u—y
i—o	y—u

Then the word "substitution" becomes *sybstotytoan*, "elephant" becomes *iliphent*, and "happy" becomes *heppu*.

Design a transducer to produce these substitutions, and follow its progress on the words "substitution", "elephant", and "happy."

11. Design a transducer to reverse the cypher described in Problem 10.

12. Produce state tables for the transducers in Problems 10 and 11.

13. Construct a finite-state transducer to convert from binary to base 4. (*Hint:* As suggested in the text, you need to group the binary digits into pairs and produce appropriate output for each of the four possible pairs of digits.) Your machine may, if you like, reject strings with an odd number of digits. Test the action of your transducer on the following strings of 0s and 1s:
(a) 0000   (b) 1010   (c) 1101
(d) 0110   (e) 10111

14. Construct a finite-state transducer to convert from binary to base 8. (*Hint:* $000 = 0$, $001 = 1$, $010 = 2$, $011 = 3$, $100 = 4$, $101 = 5$, $110 = 6$, and $111 = 7$.) Try your transducer on the following strings:
(a) 111001   (b) 101100   (c) 001111
(d) 1111111   (e) 110011001

15. Trace the action of the transducer which adds pairs of binary numbers for the sums $5 + 7$ and $8 + 9$.

16. Construct a finite-state transducer to find the *difference* of two binary numbers. (*Hint:* You must be very careful about how the input is specified.) Your machine should reject pairs of numbers for which the result of a subtraction is negative. Test your transducer with the following:

(a) $16 - 8$ (b) $14 - 4$ (c) $12 - 11$ (d) $15 - 16$

P17. Write a Pascal program to implement the ob-to-English transducer.

P18. Write a Pascal program to implement the English-to-ob transducer.

CHAPTER SUMMARY

Chapter 7 introduces the first of our models of computation, that of the finite automaton. Although the finite automaton is, in some ways, too simple to fully describe the process of algorithmic computation, it is a good starting point for our discussion of models of computation, and it paves the way for the introduction of the Turing machine in Chapter 8.

We first used the finite automaton to describe a process for checking whether a given string belonged to a particular language. In doing this we also introduced many of the basic ideas involved in the study of formal languages and some of the operations involved in the processing of strings.

One of the most convenient ways of describing a finite automaton is by means of a labeled digraph known as a transition diagram. The progress of a computation on a finite automaton can be followed by means of labeled paths in its transition diagram.

Section 7.5 discussed the close relationship between finite automatons and the syntax diagrams of Pascal, which should reinforce the idea that the finite automaton is a useful device for testing language syntax.

In the final section we introduced the finite-state transducer as an example of a machine which produces output depending on an input string. Finite-state transducers serve as a simple model of translators; in fact, many of the ideas of finite-state transducers are implemented in the construction of compilers for programming languages.

KEY TERMS

The following are some of the important terms discussed in this chapter:

accepting state (7.3)
alphabet (7.2)
character string (7.2)
concatenation (7.2)
empty string (7.2)
finite automaton (7.3)
finite-state transducer (7.6)
language (7.2)
nondeterministic automaton (7.4)

regular language (7.4)
rejecting state (7.3)
semantics (7.2)
semigroup (7.2)
state table (7.3)
substring (7.2)
syntax (7.2)
word (7.2)

8

Turing Machines

8.1 WHAT IS A TURING MACHINE?

In Chapter 7 we introduced the idea of mathematical machines by using the finite automaton. Finite automatons are an example of the kind of mathematical model that we use to illustrate what can be done with a computer program. There are, unfortunately, features of the finite automaton model which limit the kinds of computations possible, so the finite automaton does not provide a complete model of modern computation. The number of states in a finite automaton must be fixed before the input is read, and the actions taken can be based only on the state. The input can be read only once, and there is no provision for storing the input for future reference or for storing intermediate results of a procedure. When we try to check a string of parentheses to see if it is balanced, these limitations really get in the way; in fact, that problem cannot be solved with a finite automaton. (For a proof of this fact, see Chapter 9.) These limitations on finite automatons also restrict the functions which can be computed with a finite-state transducer. We need a model which is more similar to the way in which a modern computer executes a program.

The Pushdown Automaton

One model of computation which has more flexibility is the pushdown automaton (abbreviated pda). A *pushdown automaton* is a device equipped with a limited kind of storage feature known as a *stack*, or pushdown store. Information may be placed and retrieved from a stack in last-stored, first-retrieved fashion. The most recently stored information is referred to as the *top* of the stack. The operation of a stack is analogous to the way that dishes are put on a stack and removed. The pda's use the stack for storing input and intermediate results. Calculators which use reverse Polish notation (RPN) have a stack that they use in exactly the same fashion, and RPN calculators can even be viewed as kind of a prototypical pda.

Although we did mention the operation of an RPN calculator in Chapter 6, it is worth reviewing here, with emphasis on the operation of the stack. The RPN notation for the expression $2 + [(3 + 4) * 5]$ is $2\ 3\ 4 + 5 * +$. When we use an RPN calculator to evaluate that expression, the keystrokes are 2 (enter) 3 (enter) $4 + 5 * +$. Items are placed on the stack as they are keyed in, and the enter key is used to separate consecutive items and place them on the stack [2 (enter) 3 distinguishes the storage of a 2 and a 3 on the stack from the storage of the number 23]. An operation key causes that operation to be evaluated by removing first the top of the stack to serve as the second operand and then the value preceding it on the stack to serve as the first operand. The result of the operation is then placed on the stack. The sequence of keys listed above first places 2, 3, and 4 on the stack and then performs the addition of 3 and 4. The result is 7, so a 7 is placed on the stack as the top entry. The stack now contains 2 and 7. The sequence of keys 5 * causes the 7 to be removed from the stack and multiplied by 5. The resulting 35 is put on the stack, leaving it with 2 and 35. The final + causes both values on the stack to be added, resulting in a final value of 37.

Pushdown automatons operate with a stack in a very similar manner, except that pda's do not have built-in arithmetic operations, and calculators have stacks with limited size. Symbols are placed (*pushed*) on the stack or removed (*popped*) from the stack according to the instructions of the automaton. A pda operates as a nondeterministic automaton, except that the instructions require an ordered 3-tuple consisting of the state, input symbol, and the contents of the top of the stack (or Λ if the stack is empty). The result of an instruction consists of a new state and a stack operation. A pda is called *deterministic* if each ordered triple produces only a single operation. A computation is successful if the machine ends in an accepting state with no symbols left on the stack. Languages which are accepted by pda's are called *pushdown languages*, and languages which are accepted by deterministic pda's are called *deterministic pushdown languages*.

Example 8.1 shows how a deterministic pda can be used to accept a language which cannot be decided by an ordinary finite automation.

EXAMPLE 8.1

The following deterministic pda accepts the language of all balanced strings of parentheses (read from left to right):

States(M) = $\{s_0, s_1\}$
Symbols(M) = $\{\ )\ ,\ (\ \}$

In the ordered pair giving the results of an instruction, the stack operations will be given as '*pop*', '*push*', or '*no*'. If the stack operation is '*pop*', the stack is popped. If the stack operation is '*push*', the input symbol is pushed onto the stack. If the stack operation is '*no*', nothing is done to the stack. For convenience we use an asterisk in the input to an instruction to indicate the result of that instruction does not depend on the value of the entry.

Instructions$(s_0,), ( )$ $= (s_0, \text{pop})$
Instructions$(s_0, (, *)$ $= (s_0, \text{push})$
Instructions$(s_0,), \Lambda)$ $= (s_1, \text{no})$
Instructions$(s_1, *, *)$ $= (s_1, \text{no})$
Accepting(M) $= s_0$

If the input string is '$(()())$', then as the input is read, the first two left parentheses are pushed onto the stack and the first right parenthesis causes one of the left parentheses to be popped. The next input symbol is another left parenthesis, which is again pushed, and the following two right parentheses cause the stack to be emptied. The state remains as s_0 throughout the operation.

If the input string is $())()$, again the first left parenthesis is pushed onto the stack, and the first right parenthesis causes the stack to be popped and be empty. The next right parenthesis causes the state to change to s_1 because the stack is now empty. The remaining input leaves the machine in s_1, and thus the string is rejected. ∎

In view of our results about nondeterministic finite automatons, it is somewhat surprising that some pushdown languages are not deterministic. We leave it to the reader to provide formal mathematical definitions of pushdown automatons and nondeterministic pushdown automatons.

Even though pushdown automatons are capable of recognizing sets of balanced parentheses and some other kinds of languages not recognized by finite automatons, this model does not have all the power that we need. We will not go any further into the concepts of pda's, other than to note that the languages which they determine are more complex than those accepted by finite automatons but are still not complex enough to represent all the kinds of languages that can be processed with a computer. For that reason, we introduce the Turing machine.

The Turing Machine—The Basic Idea

A Turing machine is a deterministic finite-state machine which has a two-way tape. The tape functions as the source of input to the machine, a place to store intermediate results, and the device to which output is sent. The tape is an infinite string of symbols, which consists of a finite string preceded and followed by infinite strings of blank characters. We usually view the tape as consisting of the finite nonblank portion of the tape, and we allow blank symbols to be appended in front or behind the nonblank portion of the tape as necessary. The machine is equipped with a head which allows it to scan one symbol on the tape at a time. The instruction set for the machine allows the machine to change state, alter the current symbol on the tape, and move the head. The action taken depends on the state of the machine and the symbol scanned by the machine. The symbol scanned by the machine is called the *current symbol*. The tape head is allowed to be moved one position to the right or left with each instruction, or the tape head may remain stationary. The machine halts when it encounters a state and symbol pair for which no instruction is given. At that point, we check to see whether the state is an accepting state. A Turing machine can function as either a transducer or an acceptor, and both ideas are useful in the study of computability.

The Turing Machine—A Formal Definition

> **DEFINITION**
>
> A **Turing machine** is an ordered 6-tuple $M = (S, \Sigma, d, s_0, B, A)$, where S is a set of states, Σ is an alphabet, d is a function having as a domain a subset of $S \times \Sigma$ and as a range a subset of $S \times \Sigma \times \{L,R,H\}$, $B \in \Sigma$, and $A \subset S$. Here d is called the **set of instructions** for M, and B is called the **blank symbol**; also A is the **set of accepting states**.

The machine operates in much the same manner as a finite automaton. The initial configuration puts the machine in the start state and places the head at the leftmost nonblank symbol of the input string. When the instruction function d is evaluated, it is interpreted in the following way: The second entry in the value of the instruction is a symbol from the alphabet of the machine. This symbol replaces the currently scanned symbol on the tape. The third entry gives the direction of motion of the head. If it is an L, the head moves left; if it is an R, then the head moves right; and if the last entry is an H, then the head holds its position and does not move. The head movement instructions move the *head*, and *not* the tape. If the ordered pair of current state and current symbol is not in the domain of d, we say that the machine has *halted*. If the machine halts in an accepting state, the initial string is said to have been *accepted*; and if that state is not an accepting state, then the string is said to have been *rejected*. If a string is accepted and the head is on the leftmost nonblank symbol, then the string which is on the finite portion of the tape is regarded as the output.

We use the following notation for the various parts of the Turing machine: If $M = (S, \Sigma, d, s_0, B, A)$, we refer to S as States(M), Σ as Symbols(M), d as Instructions(M), s_0 as Start(M), B as Blank(M), and A as Accepting(M). In all our examples, Start(M) is denoted by s_0, Blank(M) is #, and Accepting(M) = $\{h\}$.

EXAMPLE 8.2

337

8.1 WHAT IS
A TURING MACHINE?

The following Turing machine takes a string of 0s and 1s and erases the right-most symbol. (By erasing a symbol we mean replacing it with a blank.)

States(M) $= \{s_0, s_1, s_2, s_3, h\}$
Symbols(M) $= \{ \#, 0, 1\}$
Instructions(M):

$d(s_0, 1) = (s_1, 1, R)$	{If the leftmost symbol is a 0 or a 1,
$d(s_0, 0) = (s_1, 0, R)$	move to state s_1 and search for the
	right end of the input string.}
$d(s_1, 1) = (s_1, 1, R)$	{If in state s_1 and a nonblank is
$d(s_1, 0) = (s_1, 0, R)$	found, move right and continue.}
$d(s_1, \#) = (s_2, \#, L)$	{If in state s_1 and a blank is found,
	the right end has been reached, so
	change state and begin moving left.}
$d(s_2, 1) = (s_3, \#, L)$	{s_2 asks for erasure of the symbol
$d(s_2, 0) = (s_3, \#, L)$	found, then begins search for left
	end of string.}
$d(s_3, 1) = (s_3, 1, L)$	{s_3 searches for $\#$ marking left end
$d(s_3, 0) = (s_3, 0, L)$	of string and halts when it is
$d(s_3, \#) = (h, \#, R)$	located.}

The statements in braces are comments designed to help explain the purpose of each instruction or group of instructions. They are provided for your convenience and are not part of the machine.

We can follow the progress of a Turing machine on an input string by the following device: We list the string, underlining the current character; then we insert a slash and list the current state. The leading and trailing blanks are not listed unless they become the current character. We take the string 110 as an example:

Step 1: $\underline{1}10/s_0$

Step 2: $1\underline{1}0/s_1$

Step 3: $11\underline{0}/s_1$

Step 4: $110\underline{\#}/s_1$

Step 5: $11\underline{0}/s_2$

Step 6: $1\underline{1}/s_3$

Step 7: $\underline{1}1/s_3$

Step 8: $\underline{\#}11/s_3$

Step 9: $\underline{1}1/h$

The machine halts with output 11. You might note that this machine will not accept an empty string, since there is no instruction of the form $d(s_0, \#)$. ■

We will refer to Example 8.2 when we use it to describe the relationship between Turing machines and algorithms as well as in our discussion of the efficiency of a Turing machine.

Turing machines do not always halt for every string. The machine in Example 8.2 halts for every string, but many do not. It is easy to describe a Turing machine which will accept any string of 0s and 1s that begins with a 1 and that fails to halt when the string starts with a 0.

EXAMPLE 8.3

The following Turing machine N does not halt for all strings.

States(N) $= \{s_0, s_1, h\}$
Symbols$(N) = \{0, 1\}$
Instructions(N):

$d(s_0, 1) = (h, 1, H)$ {If the leftmost character is a 1, halt.}

$d(s_0, 0) = (s_1, 0, R)$ {If the leftmost character is a 0, enter s_1 and move right.}

$d(s_1, 0) = (s_1, 0, R)$ {If in s_1, regardless of current
$d(s_1, 1) = (s_1, 0, R)$ character, write a 0, and move
$d(s_1, \#) = (s_1, 0, R)$ right on the tape.}

If the input string begins with a 1, N halts and accepts the string; but if the string begins with a 0, N does not halt and instead creates an infinitely long string of 0s. ∎

This example is a warning that Turing machines are much more complex devices than finite automatons and pda's, because those machines must halt when the machine runs out of input. The Turing machine never runs out of input, and so it is possible for it to continue operating indefinitely.

Problem Set 1

1. Propose a formal definition for a pda and for a nondeterministic pda.

2. What features of a pda allow the design of a machine like the one in Example 8.1, which will accept all balanced strings of parentheses, when a finite automaton cannot be designed to do that task?

3. Determine the keystrokes necessary and the use of the stack in an RPN calculator in the evaluation of the following expressions:
   (a) 2 3 + 4 ∗   (b) 1 2 3 4 + ∗ +
   (c) 4 3 2 1 − + −   (d) 6 5 2 + 11 − /

4. Trace the operation of the pda in Example 8.1 on the following strings:
   (a) ((())()()   (b) (()()())   (c) (())))   (d))()((e) Λ

*5. Design a pda to test a string to determine whether a string of a's and b's has the same number of a's as b's. (*Hint*: Design your pda to put a's on the stack if the first input symbol following an empty stack is an 'a' and to put b's on the stack if the first symbol following an empty stack is a 'b'. Every time the opposite symbol is encountered, pop the stack. Use the state to keep track of which symbols are being put on the stack.)

6. Trace the operation of the Turing machine M from Example 8.2 on the following strings:
   (a) 1001   (b) 11101   (c) 0110   (d) 1100101   (e) Λ

7. Consider the following Turing machine O:

   States(O) = $\{s_0, s_1, h\}$
   Symbols(O) = $\{\#, a, b\}$
   Instructions(O):
      $d(s_0, a)$ = (s_1, a, R)
      $d(s_0, b)$ = (s_0, b, R)
      $d(s_1, a)$ = (s_0, a, R)
      $d(s_1, b)$ = (s_1, b, R)
      $d(s_1, \#)$ = $(h, \#, H)$

   (a) Trace the execution of O on the string '$abbaab$'.
   (b) Trace the execution of O on the string '$abbbabb$'.
   (c) What property must a string have to be accepted by O?
   (d) What output is provided by O?

8. Consider the Turing machine Q.

   States(Q) = (s_0, s_1, h)
   Symbols(Q) = $(\#, a, b)$
   Instructions(Q):
      $d(s_0, a)$ = (s_0, b, R)
      $d(s_0, b)$ = (s_0, a, R)
      $d(s_0, \#)$ = $(s_1, \#, L)$
      $d(s_1, a)$ = (s_1, a, L)
      $d(s_1, b)$ = (s_1, b, L)
      $d(s_1, \#)$ = $(h, \#, R)$

   Trace the execution of Q on the strings in parts (a) to (c).
   (a) *aabbba*
   (b) *ababab*
   (c) Λ
   (d) What does this Turing machine actually do?

9. Design a Turing machine to take a string of a's and erase all the entries in the string.

10. Design a Turing machine to take a string of a's and b's and erase all the b's, leaving a string of a's separated by blanks.

8.2 TURING MACHINES AND LANGUAGE RECOGNITION

In Section 7.4, we examined the use of finite automatons as a device to determine whether a particular string belongs to a particular language. We paid little attention to how the language was actually defined. Indeed, one reasonable way to define a language would be to define the language $L(M)$ as the set of all strings accepted by the automaton M. We can do the same thing with pda's and Turing machines. In Chapter 9, we discuss an alternative way of describing a language, and we focus on the issue of what kinds of machines will accept what kinds of languages.

In this section we wish to consider the processes involved in using a Turing machine to recognize a language. If M is a Turing machine, then $L(M)$ is the set of all strings accepted by M. We say that $L(M)$ is the language *enumerated* by M. There are actually two different ways that M can determine $L(M)$. One way is that machine M halts for every string and thus explicitly accepts or rejects all candidates for membership in $L(M)$. The other possibility is that for some strings the machine may not halt. If the first condition holds, we say that language $L(M)$ is *decided* by M.

DEFINITIONS

A language L is **recursively enumerable** if and only if there is a Turing machine M such that $L = L(M)$.

L is **recursive** if M decides L.

Obviously some relationships exist among the various kinds of languages, and we detail them in Theorem 8.1.

THEOREM 8.1

Every regular language is a pushdown language. Every pushdown language is recursive. Every recursive language is recursively enumerable.

PROOF

Most of the theorem is pretty obvious. Regular languages are those which are accepted by finite automatons. A finite automaton can be viewed as a (deterministic) pda in which the stack is never used. The transition from pda's to Turing machines is a bit more tedious, but it is possible to construct a Turing machine that produces exactly the same results as a given pda. The relationship between recursive and recursively enumerable languages is obvious. ∎

The Venn diagram in Fig. 8.1 illustrates the relationships described in Theorem 8.1.

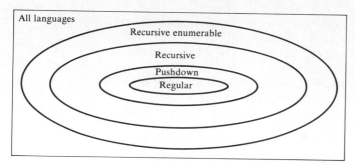

Figure 8.1 Relationship between types of languages.

Turing machines simulate the actions of computers very closely. Since this is the case, any computer programming language must, at the very least, be a recursive language. If the language were recursively enumerable but not recursive, and a statement were given to a computer to check which was not part of the language, the computer program which tested the statement might not halt. Obviously this situation should not occur for a programming language. We need compilers to check statements for membership in a language in a way which provides a yes or no in finite time.

In Example 8.1 we illustrated how a pda can be designed to test a string for balanced parentheses, and now we can use that same problem to illustrate how a Turing machine can store intermediate results on its tape.

The key to designing this Turing machine is the fact that a balanced string of parentheses must have the same number of left parentheses as right parentheses, and at no point in a left-to-right scan of the string should there be more of one than of the other. (The discerning student will note that the Turing machine seems more complex than the pda which we described in Example 8.1, but this example does illustrate some of the ways in which a Turing machine operates.)

A Turing Machine to Test for Balanced Parentheses

EXAMPLE 8.4

The following Turing machine tests a string of parentheses to see if it is balanced. If it is, it halts and accepts the string. If the string is not balanced, the machine halts in a nonaccepting state.

The strategy is straightforward: the machine keeps track of all the left parentheses that it has found and looks for right parentheses to form pairs of parentheses. Every time a right parenthesis is found, the machine erases it, backs up to find the left parenthesis that it matches, and erases the matching left parenthesis. If a matching left parenthesis is not found, or if too many right parentheses are found, the machine halts in a nonaccepting state. If all parentheses are erased, the machine halts in an accepting state.

To facilitate the process, we introduce an extra symbol into Σ which we call a *dirty blank*. The dirty blank is used in the same way as a blank, except that finding a dirty blank on the tape tells us that a symbol was present on the tape in that position at some time. Instead of replacing an erased character with a blank, we replace it with a dirty blank. The use of the dirty blank assures us that we do not find a "real" blank except when we reach the blanks which mark the ends of the string. We use the symbol & as a dirty blank. Here is the machine:

States(M) $= \{s_0, s_1, s_2, h\}$
Symbols(M) $= \{ (,), \&, \# \}$
Instructions(M):

$d(s_0, () = (s_0, (, R)$	{If the symbol is a left parenthesis keep moving right.}
$d(s_0,)) = (s_1, \&, L)$	{If the symbol is a right parenthesis, erase it and reverse direction to find a matching left.}
$d(s_0, \&) = (s_0, \&, R)$	{If the symbol is &, continue the search.}
$d(s_0, \#) = (s_2, L)$	{If the symbol is #, we have reached the right end of string, so now reverse to make sure only dirty blanks remain.}
$d(s_1, \&) = (s_1, \&, L)$	{If the symbol is &, keep seeking.}
$d(s_1, () = (s_0, \&, R)$	{If the symbol is a left parenthesis, erase it and reverse direction to seek another.}
$d(s_2, \&) = (s_2, \#, L)$	{If there is a dirty blank erase, continue moving left.}
$d(s_2, \#) = (h, \#, H)$	{If there is a real blank, stop; we are done.}

The states describe what the machine is doing. State s_0 means the machine is looking to the right for a right parenthesis. State s_1 means the machine is looking right for a left to match the right parenthesis already located. The machine halts in s_1 if it encounters the blank which marks the left end of the input, because that means for some substring there were too many right parentheses. State s_2 is the state for a reverse search to make sure that all symbols have been turned into dirty blanks. The machine halts in a failed condition if it encounters a nonblank symbol in s_2 and halts in an accepting state if it encounters the "real" blank which marks the left end of the string. We present two sample strings. First there is $((((()(:$

Step 1: $\underline{(}((()(/s_0$

Step 2: $(\underline{(}(()(/s_0$

Step 3: $((\underline{(}()(/s_0$

Step 4: $((((\underline{(}(/s_0$

Step 5: $((((\underline{(}(/s_0$

Step 6: $((((\underline{\&}(/s_1$

Step 7: $(((\&\underline{\&}(/s_0$

Step 8: $(((\&\&\underline{(}/s_0$

Step 9: $(((\&\&(\underline{\#}/s_0$

Step 10: $(((\&\&(\underline{\#}/s_2$

The machine halts at step 10, and the string is not accepted.
The second set is $()(())$:

Step 1: $\underline{(})(())/s_0$

Step 2: $(\underline{)}(())/s_0$

Step 3: $\underline{(}\&(())/s_1$

Step 4: $\&\underline{\&}(())/s_0$

Step 5: $\&\&\underline{(}())/s_0$

Step 6: $\&\&(\underline{(})/s_0$

Step 7: $\&\&((\underline{)})/s_0$

Step 8: $\&\&((\underline{\&})/s_1$

Step 9: $\&\&(\underline{\&}\&)/s_0$

Step 10: $\&\&(\&\&\underline{)}/s_0$

Step 11: $\&\&(\&\underline{\&}\&/s_1$

Step 12: $\&\&(\underline{\&}\&\&/s_1$

Step 13: $\&\&\underline{(}\&\&\&/s_1$

Step 14: $\&\&\underline{\&}\&\&\&/s_0$

From this point on, the head moves to the right until the right-hand blank is found, and the machine enters s_2. It proceeds to erase the dirty blanks until it finds the blank at the left end of the string. When this blank is found, the machine enters state h and halts. ∎

Turing Machines and Other Problems

Language recognition techniques can be applied to a wide variety of problems that do not seem to be such on the surface. The trick is to pose the problem in a manner that allows the solution to the problem to be the answer to a related language question. One problem related to the traveling salesman problem is the hamiltonian path problem which seeks to determine whether a given graph has a hamiltonian path. This can be phrased as a language problem for a Turing machine in the following way: The input for the Turing machine is a string of 0s and 1s representing the adjacency matrix of a graph. If the graph has n vertices, this string will have length n^2. We define language H to consist of all strings on the alphabet $\{0, 1\}$ which represent adjacency matrices of graphs that have hamiltonian paths. This is a recursive language, because a Turing machine can be designed to decide the language. Any graph G has a hamiltonian path if and only if its adjacency matrix belongs to language H.

Turing Machines and Functions

Turing machines are also used to represent numerical functions. In that case we are mainly interested in the output. The language accepted by the machine plays a secondary role in the sense that it is the domain of the function. For numerical functions it is traditional to deal only with nonnegative numbers, although a Turing machine can represent any number that can be represented by a computer.

Nonnegative numbers can be represented in a "unary" fashion by using a sequence of $n + 1$ ones to represent the number n. If an ordered pair of numbers is needed, the representation of the first number is followed by a blank, and that is followed by the representation of the second number. In Example 8.5 two nonnegative numbers are added.

EXAMPLE 8.5

The following Turing machine A adds two numbers. (This time we do not include comments, but instead ask you to supply the comments as an exercise.)

States(A) $= \{s_0, s_1, s_2, s_3, s_4, h\}$
Symbols$(A) = \{1, \#\}$
Instructions(A):

$d(s_0, 1)\ = (s_0, 1, R)$
$d(s_0, \#) = (s_1, 1, R)$
$d(s_1, 1)\ = (s_1, 1, R))$
$d(s_1, \#) = (s_2, \#, L)$
$d(s_2, 1)\ = (s_3, \#, L)$
$d(s_3, 1)\ = (s_4, \#, L)$
$d(s_4, 1)\ = (s_4, 1, L)$
$d(s_4, \#) = (h, \#, R)$

This machine inserts a 1 between the two strings and then removes two 1s from the right-hand end of the "merged" string. If the integers are m and n, this produces a string of length $m + n + 1$. ∎

Problem Set 2

345

8.2 TURING
MACHINES AND
LANGUAGE
RECOGNITION

1. Explain why a string of $n + 1$ ones is used instead of a string of n ones to represent the number n in unary notation.

2. Show how to represent the following data as numerical input for a Turing machine:
   (a) 4   (b) 0   (c) (2, 3)   (d) (3, 1, 1)

3. Supply comments to document machine A in Example 8.5.

4. How might a Turing machine represent negative numbers and still use unary notation? How about rational numbers?

5. Consider the following Turing machine:

   States(B) $= \{s_0, h\}$
   Symbols(B) $= \{\#, a, b\}$
   Instructions(B):
      $d(s_0, a) = (h, a, H)$
      $d(s_0, b) = (s_0, b, R)$

   Indicate how B processes the strings in (a) to (d) and whether the strings are accepted.
   (a) *abbbb*   (b) *bbbbabb*   (c) *bbbb*   (d) Λ
   (e) What language is defined by $L(B)$? Does B decide $L(B)$?

6. Consider the following machine:

   States(C) $= \{s_0, s_1, h\}$
   Symbols(C) $= \{\#, a, b\}$
   Instructions(C):
      $d(s_0, a)$ $= (s_1, a, R)$
      $d(s_1, b)$ $= (s_2, \#, R)$
      $d(s_1, \#)$ $= (h, \#, L)$
      $d(s_2, \#)$ $= (s_2, \#, R)$

   In (a) to (d) indicate how C processes the strings and whether they are accepted.
   (a) *aab*   (b) *bbb*   (c) *abb*   (d) *a*
   (e) What language is defined by $L(C)$? Does C decide $L(C)$?

7. Show how machine M in Example 8.4 processes the following strings, and indicate whether they are accepted.
   (a) ()()   (b) (())   (c) (()(()))   (d) (()))(()

8. Show how Turing machine A in Example 8.5 would perform the additions in (a) to (c).
   (a) $2 + 3$ (b) $0 + 1$ (c) $3 + 5$
   (d) What would happen to machine A if the tape had only the representation of a single number on it?

*9. Suggest a way of using a digraph to represent a Turing machine.

10. Design a Turing machine to take a nonnegative integer n as input and to provide $n + 2$ as output.

*11. Design a Turing machine which will count the number of a's that occur in a string from $\{a, b)*$.

12. Design a Turing machine to add 2 to the binary representation of a number.

*13. Design a Turing machine which accepts strings of the form $a^n b^n$. (The exponent indicates the number of consecutive occurrences of a substring; a^n indicates n consecutive a's.)

14. Design a Turing machine to accept the language $\{\Lambda\}$.

15. Design a Turing machine to accept ob and provide a translation from ob to English for each accepted string.

*16. Design a Turing machine to accept strings from pig Latin and to provide a proposed translation to English. (We say "proposed translation" because of the problems created by words which start with vowels in English, such as "English.")

8.3 CHURCH'S THESIS

Algorithms and Turing Machines

We have spent a great deal of effort describing processes through algorithms, and now we have introduced the idea of processing strings and numbers with a Turing machine. It is time to discuss the connection between these two ideas. *Church's thesis* (also known as the Church-Turing thesis or Turing's thesis) says that the description of a process by means of an algorithm and the use of a Turing machine to accomplish the process are actually equivalent. In this section we try to make that seem like a reasonable idea, and we consider a formal statement of Church's thesis.

In Example 8.2, comments were written to the right of each of the instructions of machine M. Taken together, these comments resemble the description of an algorithm. The instructions with first entry s_0 are intended to make sure we did not start with a blank tape. They function as a statement of the form "if the tape is blank, then quit." The instructions with first entry s_1 tell the machine to move the head left until the trailing blank is found. The s_1 instructions provide the loop "while current symbol is not blank, get next symbol." When this loop is finished, the machine goes into state s_2 and erases the symbol just before the trailing blank. These instructions function as the statement "erase preceding symbol." Then the machine is placed in s_3, and the instructions tell the machine to move to the left through the tape and stop when the leading blank is found. This is another loop, as in "while current symbol is not blank, move head to the left." We have actually set up a machine which implements the following algorithm:

ALGORITHM ERASELEFT

{The algorithm prints the string that would be the output from the Turing machine if the input were accepted. If the input is rejected, the algorithm prints the string '*error*'.}
begin
if current_symbol = '#' **then** print('error');
                                stop **fi**;
while current_symbol <> '#' **do** move right **od**;
move left;
erase current_symbol;
move left;
while current_symbol <> '#' **do** move left **od**;
move right;
print(string)
end.

Example 8.2 shows us that, by using appropriate changes of state, we can create instructions in a Turing machine which produce the same effect as sequencing, repetition, or selection structures do in an algorithm. Sequencing, selection, and repetition are the three fundamental constructs that we have used in our algorithm language. Based on these facts, it seems reasonable that a Turing machine could do anything that our algorithm language can do. This is reinforced by the fact that in Section 8.2 we discussed the construction of Turing machines to perform arithmetic computations. The comments we supplied with Examples 8.3 and 8.4 also help to show how the process works. It is also possible to combine Turing machines in a fashion quite similar to the way we combined finite automatons, so it is possible to construct Turing machines to accomplish complex tasks from simple parts. Some authors use that idea to provide a description language for Turing machines which is very close to our algorithm language.[1] Description languages can be used to design Turing machines in much the same way as computer programs are written.

Alonzo Church and Alan Turing, working independently in the late 1930s, formulated a principle which has come to be known as Church's thesis. Church's thesis essentially says that the collection of functions (or partial functions) which can be computed by an algorithm is the same as the collection of functions which can be computed by a Turing machine. (Church's computational device was different from the Turing machine, but it can be proved to be mathematically equivalent to a Turing machine.) Computer programs

---

[1] See Walter Savatich, *Abstract Machines and Grammars* (Little, Brown, Boston, 1982), for an example of such description languages.

look like algorithms, so Church's thesis can be interpreted to mean that Turing machines compute exactly the same things that computers compute.

To verify that Church's thesis is true, we have to show two things: (1) every Turing machine can be described as an algorithmic process, and (2) every algorithmic process can be computed by using a Turing machine.

Our example of describing the ERASELEFT machine in the form of an algorithm demonstrates how Turing machines can be described as algorithms. And, in fact, that part of Church's thesis can be demonstrated precisely.

The other half of Church's thesis—that every algorithm can be described as a Turing machine—is not so easy to verify and cannot be proved in any formal sense, because we can't offer a precise definition of an algorithmic process. In fact, depending on changes in the state of the art of computing equipment, we might even change our idea of an algorithm. There is, however, good reason to believe that for any given algorithm we can find a Turing machine. For one thing, if algorithms were more powerful than Turing machines, there would have to exist an algorithm which would compute a function that could not be computed by a Turing machine. It seems reasonable that there would be some machine that could execute this algorithm, and thus that machine would be more powerful than a Turing machine. However, every attempt to construct a machine more powerful than a Turing machine has met with failure. It can be shown that multitape Turing machines, Turing machines with two-dimensional or three-dimensional tapes, and other extensions of the Turing machine do *not* compute any functions that could not already be computed by a Turing machine. This situation is much like that with programming languages—although the languages appear different, the computational powers of the languages are all approximately the same. The main difference is that some languages make it much more convenient to implement certain kinds of operations than others. In a similar way, the "exotic" Turing machines do not have greater capabilities, but they do have more operations, which can make it easier to describe the computation of a function.

A Statement of Church's Thesis

The following is one of several equivalent statements of Church's thesis. It says that any algorithm can be reduced to one written in a very simple language. The language in question contains only statements that can be accomplished with a Turing machine.

CHURCH'S THESIS

Every problem which admits an algorithmic solution can be solved with algorithms which use only the following kinds of instructions. (Here A represents a variable, and all variables are assumed to be initialized to 0.)
(a) Add 1 to A, and go to the next instruction.
(b) If $A \neq 0$, then subtract 1 from A; otherwise, go to statement n.
(c) Halt and display the value of A.

Clearly a Turing machine has the capability of performing operations (a)

Clearly a Turing machine has the capability of performing operations (a) to (c), and thus this version of Church's thesis implies the equivalence of Turing machines and algorithms.

Church's thesis is remarkable because it was proposed when the modern electronic computer had not been developed and was, in fact, decades away from general availability. It is another example of a theoretical analysis of a problem that occurred long before its practical application was even remotely possible.

Church's thesis is *not* a theorem, but rather a statement of general principle, since it is a statement about the relationship between two ideas, one of which is well defined [the set of algorithms which can be described by instructions (a) to (c)] and one of which is *not* well defined (the set of all processes which can be described by algorithms). However, as we mentioned before, unless some drastic changes occur in our conception of computer programs and algorithms, most likely Church's thesis will continue to be accepted. Church's thesis is occasionally invoked in proofs as a kind of shorthand to avoid writing out a rather tedious and uninteresting proof.

One more point should be made. Turing machines themselves might best be thought of as algorithms, but it is actually possible to create a kind of Turing machine, known as a *universal Turing machine*, which accepts as an input tape a description of a Turing machine and its input (coded in some suitable form) and then produces the same results as would be produced by using the given data on the described machine. A universal Turing machine is really the theoretical equivalent of a programmable computer. In Chapter 9 we explore some consequences of universal Turing machines.

In Appendix 2, we include a Pascal program which emulates the operation of the Turing machine in Example 8.2. You may wish to study this as an illustration of the general process by which a Turing machine can be emulated by a computer program.

The program we present in Appendix 2 is not the most general possible— in fact, programs have been written to allow the user to input a description of the Turing machine and then have the output of the program emulate the Turing machine in question. Our program is simpler in that the code itself emulates the description of the Turing machine. However, this illustrates more clearly the close relationship between the description of a Turing machine and the code in a computer program.

Problem Set 3

1. Show that each of the three operations described by Church's thesis can be performed on a Turing machine (consider the value of A as a number with unary representation on the machine's tape, and show that a sequence of Turing machine operations can accomplish the appropriate action with A).

2. Show that our algorithm language can also perform the three operations described by Church's thesis.

Describe each of the following Turing machines by use of algorithms:

3. Turing machine M from Example 8.4.

4. Turing machine A from Example 8.5.

5. Turing machine O from Problem 7, Section 8.1.

6. Turing machine Q from Problem 8, Section 8.1.

7. Turing machine B from Problem 5, Section 8.2.

8. Turing machine C from Problem 6, Section 8.2.

9. (a) Use an algorithm to prove that if L_1 and L_2 are recursive languages, then $L_1 \cup L_2$ is also a recursive language.
(b) Describe how to construct a Turing machine to do the process described by the algorithm in (a).

10. Repeat Problem 9 for the intersection of two recursive languages.

11. What changes (if any) need to be made in Problems 9 and 10 if L_1 and L_2 are not recursive, but instead only recursively enumerable?

12. Suppose that L is language defined on Σ.
(a) Prove by constructing an algorithm that if L is recursive, then $\Sigma^* - L$ is also recursive.
(b) Does your argument work if L is recursively enumerable?

*13. A recursive function is one that can be computed by a Turing machine. Use Church's thesis to show that if f and g are recursive, then so are the functions $f + g, f - g, f * g$, and f/g.

*14. A function f is partially recursive if its domain is not the set all positive integers but for all values in its domain there is a Turing machine M which computes the values of the function. Machine M may or may not halt if the input is an integer not in the domain of f. If f and g are partially recursive, determine whether $f + g, f - g, f * g$, and f/g are partially recursive functions.

**15. The "Busy Beaver" n-problem is described in the following way: Using only the instructions described in Church's thesis, produce the largest possible value for A using an algorithm with n instructions. Clearly the solution to the 1-problem is 0, and the solution to the 2-problem is 1. Try to find a solution to the 8-problem.

P16. Write a Pascal program to emulate the action of Turing machine M of Example 8.4. Your output should be similar to the description of the states shown in the example. (*Hint:* Use the program in Appendix 2 as a model.)

P17. Write a Pascal program to emulate Turing machine A in Example 8.5.

P18. Write a Pascal program to emulate Turing machine B in Problem 5, Section 8.2.

The Efficiency or Complexity of a Turing Machine

We are now ready to consider some interesting questions about problems which can be solved by Turing machines (or algorithms). Not all functions can be computed by Turing machines, but that fact is not obvious at this point. (We discuss some of the impossibility theorems in Chapter 9.)

If an algorithm can solve a problem, we often feel as though that is enough, but we should recall our discussions of the efficiency of algorithms. Some algorithms are terribly inefficient. We can discuss the issue of algorithm efficiency conveniently in terms of Turing machines.

DEFINITIONS

Let L be a recursive language, T a function whose domain is the set of nonnegative integers, and M a Turing machine. Then:

M decides L in time T if and only if for any string w on the same alphabet as L, M decides whether or not $w \in L$ in $T(\lambda(w))$ or fewer steps.

M computes the function f in time T if M computes the function f and for every input string w, M either computes $f(w)$ or halts in a rejecting state within $T(\lambda(w))$ steps.

M has complexity T if and only if M decides a language L or computes a function f in time T. T is also called the **efficiency of M** or the **time complexity of M**.

The ideas of complexity of a Turing machine can be translated to similar ideas for algorithms. In discussing algorithms we noted that a traveling salesman problem with n vertices could be decided in some multiple of $(n - 1)!$ steps. And $(n - 1)!$ grows at an extremely rapid rate. Using the exhaustive algorithm with 10 cities requires some multiple of 9! steps, and $9! = 362,880$. This is a rather large number, but with modern computing equipment it would not be unreasonable to expect that a computer could solve the problem and halt in a reasonable time. However, if the exhaustive algorithm were applied to a traveling salesman problem with 30 cities, a multiple of 29! steps would be required, and 29! is so large that even if a computer could perform a *billion* steps per second, the completion of 29! steps would require 8.4×10^{15} *years*. This is what people mean when they say that a solution is theoretically possible, but practically infeasible. Even though the algorithm would solve the problem, we don't have the time to wait for the solution.

One measure of the solvability of a problem is the *complexity of the problem*, which can be defined as the "smallest" complexity of a Turing machine (or the algorithm) which solves the problem. This seems like a good theoretical definition, but not until we clarify what we mean by "smallest" function.

EXAMPLE 8.6

In Example 8.2 we presented a Turing machine M which erased the rightmost symbol of a string of 0s and 1s. We now find the complexity of M.

In this case, the easiest thing to do is to count the number of times each step is used. For M this is especially easy, because once the machine leaves a state, it does not return:

s_0 is used only once. (1)

s_1 is used one time for each symbol to the right of the first symbol and one additional time for the trailing blank—when we "run out of data," we move into s_2. (n)

s_2 is used only once. (1)

s_3 is used once for each symbol left in the string after the last entry has been erased and once for the leading blank. (n)

h is used only once. (1)

The numbers in parentheses indicate the number of times each state is used. We add those numbers to get $2n + 3$. In Example 8.2 we demonstrated M with an input string of length 3, and we found 9 steps; so this analysis seems to track for computations which reach the accepting state. If the string is rejected because of a blank tape, the machine will need only one step to halt. ■

The counting techniques discussed in Chapter 2 are often particularly important in determining the complexity of a Turing machine or algorithm.

Comparing Complexity Functions

Even after the complexity of a pair of algorithms is determined, direct comparison of their complexities is sometimes not entirely as straightforward as we might hope.

EXAMPLE 8.7

Suppose that machines M and N have complexity functions $T_1(n)$ and $T_2(n)$, respectively, where $T_1(n) = 4n$ and $T_2(n) = n^2$. Compare the complexities of machines M and N.

When n is 1, 2, or 3, we know $T_1(n)$ is larger than $T_2(n)$. So for those values of n, N is the more efficient machine; but when n is 5 or larger, $T_2(n)$ is larger, so M is the more efficient. On the surface, it would appear that the two complexity functions cannot be compared. However, T_1 is a more desirable function for the efficiency of an algorithm which solves a problem requiring large amounts of data. T_2 has the property that every time n is doubled, the value of the function is quadrupled; but for T_1, doubling n just doubles the value of the function. Our comparison of efficiency functions should take into consideration the rate at which the efficiency functions grow with n, and not just what happens to the efficiency functions for particular values of n. We should conclude that M is the more efficient machine for dealing with large inputs. ∎

EXAMPLE 8.8

Another way of describing Turing machines is to allow the machine to either write or move for any instruction, but not both. If we were asked to compute the efficiency of an algorithm to erase the rightmost symbol in a string of 0s and 1s, as in Example 8.2, then a Turing machine of this type would require one more step than we found in Example 8.6, because the step in which the last character was erased required both a write operation and a head move. The efficiency of this machine would be $2n + 4$ instead of $2n + 3$. The difference between these two should not be considered, because even though the efficiency function for the "one-operation" Turing machine is always larger, the algorithms are really the same—and the difference between the efficiency functions is relatively small in comparison to the value of the efficiency functions. We should say that the *growth rates* of the two functions are the same. ∎

EXAMPLE 8.9

Suppose we designed a Turing machine to completely erase all input, and again we compared the two types of Turing machines. Then we would find that for a Turing machine as we defined it, the efficiency function would be $T_1(n) = n + 1$, while for the one-operation Turing machine the efficiency function would be $T_2(n) = 2n + 1$. Here also, the same algorithm gives different complexity functions when it is executed on slightly different machines. This is unreliable in this case, because it seems reasonable that the "new" kind of Turing machine might execute each step faster because each operation involves fewer manipulations. The rate of growth of the two can be viewed as being about the same, because for either function as n is doubled (or tripled), the value of the efficiency function is also doubled (or tripled). ∎

DEFINITIONS

Suppose that T_1 and T_2 are functions whose domains are the set of non-negative integers.

T_1 **asymptotically dominates** T_2 if and only if there is a constant c and an integer N such that for all $n > N$, $T_1(n) \geq T_2(n)$.

T_1 **and** T_2 **are asymptotically equivalent** if and only if each dominates the other.

If T_1 asymptotically dominates T_2, we often write that T_2 is $O(T_1)$, which is read as "T_2 is big oh of T_1." Often $O(f)$ is used to also denote the set of all functions asymptotically dominated by f.

The relation of being asymptotically dominated can serve as an ordering on functions. Only the highest-order terms come into play with this relation, and coefficients are not significant, but the functions which grow the fastest are the ones considered dominant.

THEOREM 8.2

If R is the relation defined on functions f and g as fRg if and only if f is asymptotically dominated by g, then R is reflexive and transitive.

PROOF

The fact that R is reflexive is obvious—after all, for all values of n, $f(n) \leq f(n)$, so letting $N = 1$ and $c = 1$ in the definition of asymptotically dominated will do the job.

Transitive is slightly more difficult. Suppose that fRg and gRh. There is a constant c_1 and an integer N_1 such that if $n > N_1$ then $f(n) \leq c_1 g(n)$. There is a constant c_2 and an integer N_2 such that if $n > N_2$, then $g(n) \leq c_2 h(n)$. If we let $c_3 = c_1 2 c_2$ and let N_3 be the larger of N_1 and N_2, it follows that if $n > N_3$, we have $f(n) \leq c_1 g(n)$ and $g(n) \leq c_2 h(n)$. Thus $f(n) \leq c_1 [c_2 h(n)]$, as we needed to establish that fRh. ∎

If f is $O(g)$, we say that g *grows faster than* f. This is our basis for comparing the efficiency of algorithms (or Turing machines). If f and g are asymptotically equivalent, we say that f *and* g *grow at the same rate.*

THEOREM 8.3

The relation of asymptotic equivalence is an equivalence relation.

PROOF

This is left as an exercise. ∎

EXAMPLE 8.10

Show that the function $T_1(n) = 16n^2 + 5n + 6$ is asymptotically equivalent to $T_2(n) = n^2$ and is asymptotically dominated by $T_3(n) = 2n^3$.

SOLUTION

To show that functions T_1 and T_2 are asymptotically equivalent, we must show that they dominate each other. If $n > 1$, we know that $n < n^2$ and $1 < n^2$. Hence $16n^2 + 5n + 6 < 16n^2 + 5n^2 + 6n^2 = 27n^2$. In the definition of asymptotically dominant, we can let $N = 1$ and $c = 27$. This shows that T_1 is $O(T_2)$. Showing dominance the other way is easier, because for all $n \geq 1$, $T_2(n) \leq T_1(n)$, and thus T_2 is $O(T_1)$.

To show that T_3 dominates T_1, we could again find N and c which satisfy the definition, or we could use the theorems which we have already proved. Now T_2 dominates T_1 (actually they are equivalent), and we are aware that for all $n \geq 1$, $n^2 \leq n^3$, which means that n^3 dominates T_2. Since T_3 clearly dominates n^3, by the transitive property T_3 dominates T_1. ∎

Example 8.10 illustrates two interesting facts about asymptotic dominance. First, two polynomials of the same degree are asymptotically equivalent. Second, a polynomial of higher degree dominates a polynomial of lower degree regardless of the coefficients. This is a reasonable way to compare efficiencies of algorithms. If the complexity of one algorithm (Turing machine) is a polynomial of lower degree than that of another, it may well be that, for small amounts of data, the algorithm with the complexity given by the higher-degree polynomial may run faster. If the amount of information to be processed increases, the polynomial of larger degree will grow at a faster rate, and at some point the polynomial of smaller degree will take on significantly smaller values.

THEOREM 8.4

If $p(n)$ is any polynomial function, the exponential function 2^n asymptotically dominates $p(n)$.

PROOF
The proof is not difficult, but is rather tedious. All we really need to show is that for any power d, 2^n dominates n^d. We only need to worry about that because every polynomial of degree d is asymptotically equivalent to n^d. The hard part is to choose N and c to make things work as they are supposed to. Without going into any of the details of the proof, we note that N may be chosen as the first integer greater than $(r^{1/d} - 1)^{-1}$, and c can be chosen as $N^d/2^N$. The basic idea is that as soon as $n > N$, $(n + 1)^d/n^d$ will be less than 2, which means that going from n^d to $(n + 1)^d$ is accomplished by a multiplication of less than 2. The constant c is chosen to ensure that N^d is less than $c2^N$. ∎

There is nothing special about the 2 which we used as the base in the exponential function above. Any exponential function with base greater than 1 will asymptotically dominate any polynomial.

Hierarchy of Complexity Functions

There is a hierarchy of functions in terms of asymptotic dominance. We can order some commonly occurring functions, listed from smallest to largest, which occur as complexities of algorithms:

$$\log_2 n \quad n \quad n \log_2 n \quad n^2 \ldots n^d \ldots \quad 2^n \quad n!$$

Many algorithms have complexities which are asymptotically equivalent to one of the functions in the above list, and a common verbal shorthand is to say that an algorithm is an $n \log n$ algorithm when we mean that the complexity function of the algorithm is asymptotically equivalent to the function $n \log_2 n$. Similarly, when we say that an algorithm is an n^d algorithm, we mean that its complexity function is asymptotically equivalent to n^d.

Actually the smallest complexity that a Turing machine can have is n, since it requires at least one operation for each symbol on the tape just to scan the tape.

Algorithms versus Turing Machines

If we consider algorithms without reference to Turing machines, we can construct algorithms whose complexity we would compute at less than n, so we don't have an exact correspondence between complexities of Turing machines and algorithms. The main reason for this is that some computer operations which we consider to be a single operation are more complex when they are translated to a Turing machine. Multiplication is regarded as a single operation when we are doing operation counts for an algorithm, but to get a Turing machine to do a multiplication actually requires at least n^2 operations of the Turing machine.

THEOREM 8.5

Addition on a Turing machine can be accomplished with efficiency $O(n)$, and multiplication cannot be done in $O(n)$.

PROOF

In Example 8.4 we constructed a Turing machine to add two nonnegative integers. In that Turing machine we needed to scan both integers and their separating blank twice—once as we moved to the right as we concatenated the strings and erased the extra 1s and once as we moved to the left to find the beginning of the string. We needed two extra steps—one for the blank that marks the end of the input and one for the blank that marks the beginning of the input. The total number in steps is thus $2n + 2$, which is $O(n)$.

Suppose we wish to square the number n. We will have an input string of length $n + 1$. The algorithm to square n must place $n^2 + 1$ ones on the tape. It will require $n + 2$ steps to find the end of the input. Without even considering the head movement which may occur between the $n^2 + 1$ ones in the answer and the input string, we have accounted for at least $n^2 + n + 2$ operations of this Turing machine. A similar analysis can be applied to all but the most trivial multiplication problems. A multiplication process cannot be $O(n)$. ∎

The difficulty in using the Turing machine definition for complexity is that the results do not always correspond exactly to those we would like to have in dealing with algorithms. One problem is the one noted above, that different basic operations—arithmetic operations—have different complexities. However, the complexities of all the basic operations (addition, subtraction, comparison of two values, multiplication, division) are at worst polynomials. An algorithm which requires 5 multiplications and 4 additions in order to evaluate a function will have complexity of order $5n^2 + 4n$, which is still $O(n^2)$. If multiplication and addition could both be done in $O(n)$ time, the algorithm itself would be $O(n)$ as measured on a Turing machine. In any case, the functions are polynomials. Because of these kinds of considerations, it is useful to determine whether there is a Turing machine which will solve the problem in *polynomial* time. If there is, we can be sure that any computer program written to solve the problem will be able to run in polynomial time as well.

Polynomial versus Nonpolynomial Complexity

DEFINITIONS

An algorithm is a **polynomial time algorithm** if it can be implemented on a Turing machine whose complexity function is a polynomial.

An algorithm is a **nonpolynomial time algorithm** if its complexity dominates every polynomial function.

The question of polynomial complexity has very practical considerations. If M is a polynomial time algorithm, its growth rate is polynomial, and so the number of steps required for its completion will remain reasonable even for large values of n. Obviously the smaller the degree of a polynomial, the better, but any polynomial of any degree will be dominated by any exponential function. Exponential functions are thus nonpolynomial, and they get large at an alarming rate. So for an exponential time algorithm we may face a situation in which a problem may be solved in theory by an algorithm, but that algorithm may not terminate within a reasonable time. For comparison purposes, we can note the growth rates of n^2 and 2^n.

n	n^2	2^n
1	1	2
2	4	4
3	9	16
4	16	32
5	25	64
10	100	1,024
20	400	1,048,576
30	900	1,073,741,827
40	1,600	1.09×10^{12}
50	2,500	1.13×10^{15}

If a machine could perform a billion operations per second, a problem of size 50 with an n^2 algorithm would finish in 2.5 microseconds, while the 2^n

algorithm would require 13 *days*. By comparing the general trends of the values, we see that the situation deteriorates even more seriously as the size of the problem increases. In fact, for an input of size 100, the exponential algorithm requires 4×10^{11} *centuries*. Exponential algorithms are acceptable for small jobs, but for large jobs they are of little value.

Computer scientists and mathematicians have become quite interested in trying to determine which problems can be solved with polynomial time algorithms and which problems can be solved only with exponential time (or worse) algorithms.

Problem Set 4

1. Show that 2^n is $O(n!)$.

2. Show that 2^n is asymptotically equivalent to 2^{n+1}.

3. Let f be a function defined in the following way:

$$f(n) = \begin{cases} n & \text{if } n \text{ is even} \\ n^3 & \text{if } n \text{ is odd} \end{cases}$$

Determine whether each of the following is true, and explain your answers:

*(a) f is $O(n)$.   (b) f is $O(n^3)$.
  (c) n is $O(f)$.   *(d) n^3 is $O(f)$.

4. Determine which of the following functions asymptotically dominates the others. Present your answer in the form of a labeled digraph.

$$f_1(n) = 17 \qquad\qquad f_2(n) = 5n^3 + 2n + 3$$

$$f_3(n) = 0.5n + \frac{1}{n} \qquad f_4(n) = \log_2 n$$

$$f_5(n) = n + 5 \qquad\qquad f_6(n) = \log_2 n^2$$

$$f_7(n) = 2^n + n \qquad\qquad f_8(n) = 0.0005n^8$$

5. In Example 8.2, machine M erased the leftmost symbol of a string of 0s and 1s. Design a Turing machine N which either moves or writes on the tape in any operation, but never does both. Determine the complexity of machine M, and compare this to the complexity of machine N.

6. Design a Turing machine K which erases all input, but is subject to the same restraints as machine N in Problem 5. Determine the complexity of machine K, and compare its complexity to that of a standard Turing machine which does the same task.

By counting steps, determine the complexity of the following Turing machines. Explain any difficulties you encounter.

7. Machine M of Example 8.4.

8. Machine A of Example 8.5.

9. Machine O from Problem 7 of Section 8.1.

10. Machine C from Problem 6 of Section 8.2.

11. Prove that if L is a regular language, then the complexity of L is $O(n)$.

12. Prove that if L_1 has complexity f and L_2 has complexity g, then language $L_1 \cup L_2$ and language $L_1 \cap L_2$ both have complexity $O(f + g)$. (*Hint:* Use an algorithm.)

*13. Prove Theorem 8.3.

*14. Prove the following fact which we used in the text repeatedly: If f is a polynomial of higher degree than g, then f dominates g.

15. A language is of polynomial complexity if it can be decided by a Turing machine with polynomial complexity. Prove that if L_1 has polynomial complexity and L_2 has polynomial complexity, then language $L_1 \cup L_2$ and language $L_1 \cap L_2$ both have polynomial complexity. (*Hint:* Use Problem 12.)

16. A recursive function f has complexity T if it can be computed by a Turing machine with complexity T. Prove that if $f(n)$ is recursive with complexity function T_1 and $g(n)$ is recursive with complexity function T_2, then $f(n) + g(n)$ is recursive with complexity $O(T_1 + T_2)$.

17. A recursive function f has polynomial complexity if it can be computed by a Turing machine with polynomial complexity. Prove that if f and g have polynomial complexity, then so do $f(n) + g(n)$ and $f(n)g(n)$.

*18. Use the results of Problem 17 to prove that if an operation count for f as measured by an algorithm which uses only the operations of addition, subtraction, multiplication, and division is a polynomial, then f has polynomial complexity. (This verifies our earlier claim that algorithms which are polynomial in terms of operation count will also have polynomial complexity.)

*19. Use the results of Problem 17 to show that if f and g have polynomial complexity, so do $f \times g$ and f/g.

20. Prove that polynomial functions have polynomial complexity. (*Hint:* Use the result of Problem 19.)

Suppose that we wish to construct a machine to accept a string of a's and b's in which either there is an even number of a's or an even number of b's. This problem can be solved easily with a nondeterministic finite automaton. To use a nondeterministic finite automaton, we just connect an automaton which accepts strings with an even number of a's in parallel with one that accepts strings with an even number of b's. The same thing can be done with Turing machines, in order to create a *nondeterministic Turing machine*. The only change needed in the definition of a Turing machine, in order to provide a definition for a nondeterministic Turing machine, is to allow the instructions of the machine to be a relation—instead of a function—whose domain is a subset of $S \times \Sigma$ and range a subset of $S \times \Sigma \times \{L, R, H\}$. This means that every (deterministic) Turing machine is also a nondeterministic Turing machine. The nondeterministic machine operates in the same manner as nondeterministic finite automatons whenever there is more than one possible result for a state-symbol pair; we presume the machine will do all instructions and proceed as though it can process all active processes simultaneously. A string is accepted if any of the processes ends with the string being accepted.

EXAMPLE 8.11

Construct a nondeterministic Turing machine to test whether a string has either an even number of a's or an even number of b's.

SOLUTION

The trick is to do exactly as we said, namely, to put two Turing machines in parallel. From the start state then, we must allow a transition to the start states of each of two machines, one that tests for a's and one that tests for b's.

States$(M) = \{s_0, q_0, q_1, p_0, p_1\}$
Symbols$(M) = \{a, b, \#\}$
Instructions(M): $\{((s_0, a), (q_0, a, H)), ((s_0, a), (p_0, a, H)),$
$((s_0, b), (q_0, b, H)), ((s_0, b), (p_0, b, H)), ((q_0, a), (q_1, \#, R))$
$((q_0, b), (q_0, \#, R)), ((q_1, a), (q_0, \#, R)), ((q_1, b), (q_1, \#, R))$
$((p_0, a), (p_0, \#, R)), ((p_0, b), (p_1, \#, R)), ((p_1, a), (p_1, \#, R))$
$((p_1, b), (p_0, \#, R)), ((p_0, \#), (h, \#, H)), ((q_0, \#), (h, \#, H))\}$

The p states are used to check whether the number of b's is even or odd, by using p_1 to represent an odd number of b's and p_0 to represent an even number of b's. The q states do the same thing with the a's. The machine erases all its input and functions exactly as the corresponding finite automaton would. The notation for the instructions is easier to read if we write the instructions as though they were functions and use set notation to indicate the results. We can write $d(s_0, a) = \{(q_0, a, H), (p_0, a, H)\}$ instead of the formal relation notation. ∎

A tree can be used to represent the process of calculation in a nondeterministic Turing machine. In Example 8.11, the machine starts in s_0 and then is broken into two processes. Each process continues to completion. In the tree of Fig. 8.2, we show how the entire machine M is applied to the string *ababb*. This tree has two branches; the right branch ends in a rejecting state, and the left branch ends in an accepting state. The string is accepted because the process represented by the left branch accepts the string.

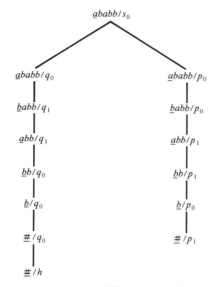

Figure 8.2 Tree representation of Turing machine in Example 8.10.

Each time a combination of state and current symbol is reached in which there is more than one possible instruction, a node with outdegree equal to the number of possible instructions is created, and the machine proceeds with all the processes. One can look at this tree in two ways: in a breadth-first fashion, by evaluating the condition of all vertices at depth n before proceeding to depth $n + 1$, or in a depth-first fashion, by taking one branch of the tree as deep as possible, perhaps to a leaf, before exploring the other possible choices. The main problem with this latter choice is that some paths through the tree may represent nonterminating computations, but if we use a breadth-first expansion of the tree, we have described a deterministic way of evaluating all the computations of a nondeterministic Turing machine. This leads us to the following conclusion:

THEOREM 8.6

Every language which is accepted by a nondeterministic Turing machine is also accepted by a deterministic Turing machine. ■

Actually for the problem in Example 8.11, there is a much easier way to accept the language in question with a deterministic Turing machine. We design a machine so that it first checks for an even number of a's. If the number of a's is even, the machine halts and accepts the string. If the number of a's is odd, the tape is "rewound" and tested for an even number of b's. If the number of b's is even, the string is accepted; otherwise, it is rejected. In the problems, you are asked to construct such a machine.

Deterministic and Nondeterministic Complexity

The questions of computational complexity can be studied in terms of nondeterministic Turing machines as well. For a string which is accepted, one of the leaves of the tree ends with acceptance of that string. The time of the calculation can be regarded as the smallest depth of a leaf which accepts the string. This represents the run time required if there were but a single processor, but the processor inevitably made the right choice among instructions to be executed. This number is called the *nondeterministic complexity* of the computation. By analogy with the deterministic case, we can define the *nondeterministic complexity* of a language which is decided by a nondeterministic Turing machine. (A little care is needed here, but there are no major problems.) The complexity, measured deterministically, will always dominate the nondeterministic complexity.

The nondeterministic Turing machine acts as a computer with infinitely many processors available. So whenever more than one process is needed, the Turing machine can replicate its current data and start processors. The number of processors required at any point is equal to the number of vertices in the tree at that depth.

EXAMPLE 8.12

Consider the following nondeterministic Turing machine:

States(M) = $\{s_0, s_1, s_2, s_3, s_4, h\}$
Symbols(M) = $\{\#, a, b\}$
Instructions(M):

$d(s_0, a)$ = $\{(s_0, \#, R), (s_1, \#, R)\}$
$d(s_1, b)$ = $\{(s_2, \#, R)\}$
$d(s_2, b)$ = $\{(s_2, \#, R), (s_3, \#, R)\}$
$d(s_3, a)$ = $\{(s_4, \#, R)\}$
$d(s_4, a)$ = $\{(s_4, \#, R)\}$
$d(s_4, \#)$ = $\{(h, \#, H)\}$

In Fig. 8.3 we illustrate the action of this machine on the string *aabbbaa*. The total number of leaves in the tree is 5. Thus to implement this Turing machine as described would require that five processors be used. The deepest

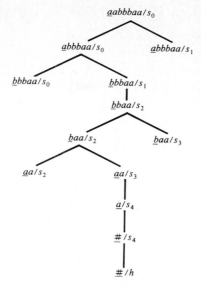

Figure 8.3 Action of M on *aabbbaa*.

leaf ends in an accepting state, so the string being processed should be accepted by this machine. We leave it as an exercise for the reader to determine exactly what properties a string of *a*'s and *b*'s must have to be accepted by this Turing machine. ∎

A formal proof of Theorem 8.6 involves a breadth-first expansion of the tree related to the processing of a string by the nondeterministic Turing machine by considering the status of the machine at each vertex at depth *n*, then continuing to depth $n + 1$, etc. (This way of implementing a multiprocessor activity on a single processor is how computers actually deal with that situation. Some computer languages allow for the implementation of a "fork" which splits a process into parallel computations, and computers with only one processor can accomplish this in a fashion analogous to the one described above.)

THEOREM 8.7

If a problem has complexity function $f(n)$ for a nondeterministic Turing machine, and if no symbol-state pair produces more than k instructions, then the complexity of the problem is $O(k^{f(n)})$.

PROOF

The worst case is that every instruction executed by the machine will require k branches. Then the tree representation of the computation will be a full k-ary tree. The total number of vertices in a full k-ary tree is $O(k^d)$, where d is the depth of the tree, and this is the number of vertices to be expanded in the worst case of converting the nondeterministic automaton into a deterministic automaton. ∎

Classes P and NP

Even though the use of nondeterminism does not change the kinds of problems that can be solved by Turing machines, it seems to have a major effect on the complexity of problems. The hamiltonian path problem can be solved quite easily in polynomial time by the exhaustive algorithm. Theorem 8.7 tells us that the problem itself can be solved in exponential time, but that is a huge difference!

DEFINITIONS

The **class P** of languages consists of those which can be recognized in polynomial time by a deterministic automaton.

The **class NP** of languages consists of those which can be recognized in polynomial time by a nondeterministic Turing machine.

One of the major unsolved problems in mathematics today is whether P equals NP. Most mathematicians are convinced that this is *not* true, but no one has been able to prove it.

NP-Complete Languages

In attacking the question of whether P equals NP, mathematicians have discovered a surprising class of languages known as the *NP-complete languages*. Every NP-complete language L has the following properties:

1. L is NP.
2. If L' is NP, then there is a polynomial time transformation from L' to L.

Part 2 means that there is a function T which can be computed in polynomial time that translates strings in the alphabet of L' to strings in the alphabet of L in such a way that string $A \in L'$ if and only if $T(A) \in L$. The upshot of all this is that if any NP-complete language can be found to actually be in P, then P = NP.

DEFINITION

A problem is **NP-complete** if it is equivalent to the decision problem for an NP-complete language.

Some Well-Known NP-Complete Problems	
(a) The Hamiltonian Path	Find a path which includes each vertex once.
(b) The Traveling Salesman	Find the shortest hamiltonian path
(c) The Chromatic Number	Find the minimum number of colors needed to color the graph.

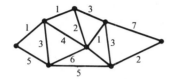

Figure 8.4 NP-complete graph theory problems.

It is most remarkable that some well-known problems turn out to be NP-complete. The traveling salesman problem, the hamiltonian path problem, and the determination of the chromatic number of a graph can all be expressed in form of a decision problem for NP-complete languages (see Fig. 8.4). Since most mathematicians feel that P and NP are not equal, this should tell us that the NP-complete problems do not have practical, exact algorithmic solutions, so they must be approached in some other fashion. This is why many people will use the term "hard" as a synonym for NP-complete.

Upon the discovery that a problem is NP-complete, it becomes useful to find a heuristic algorithm which will provide a "good" estimate to the actual solution to the problem. The Welch-Powell algorithm as used to find the chromatic number of a graph is a good example. Although the algorithm does not find the exact chromatic number, it does run in polynomial time, and it provides an upper bound on the chromatic number of the graph.

Problem Set 5

1. Show the effect of the nondeterministic Turing machine M in Example 8.11 on the following strings:
   (a) *aabbbab*   (b) *aaaaa*   (c) *bb*   (d) *abababab*   (e) Λ

2. Show the effect of the nondeterministic Turing machine M from Example 8.12 on the following strings:
   (a) *ab*   (b) *aaaaa*   (c) *bbbb*   (d) *bababa*   (e) *abbababb*

3. Determine the properties required of a string for machine M of Example 8.12 to accept it.

4. Construct a nondeterministic Turing machine to connect machine O from Problem 7, Section 8.1, in parallel with machine Q of Problem 8, Section 8.1.

5. Repeat Problem 4 with machines B and C from Problems 5 and 6 of Section 8.2.

6. Construct a nondeterministic Turing machine to connect the machines of Problem 4 in series (O first, then Q).

*7. Construct a deterministic Turing machine to accept either an even number of a's or an even number of b's in a string from $\{a, b\}^*$, as described in the remarks *following* Theorem 8.6.

*8. Use the technique discussed *preceding* Theorem 8.6 to construct a Turing machine to accept a string from $\{a, b\}^*$ which has either an even number of a's or an even number of b's.

9. What modification must be made in our algorithm language to make it function as a nondeterministic automaton?

10. Prove that the hamiltonian path problem is NP by describing how to construct a nondeterministic Turing machine to accept it. (*Hint:* Try to implement an exhaustive process as a nondeterministic Turing machine, and show that the depth of any leaf in the resulting tree can be no more than n.)

11. Conceptually describe a nondeterministic algorithm to find the chromatic number of a graph.

12. Conceptually describe a nondeterministic algorithm to find a solution to the traveling salesman problem.

*13. Prove that if some NP-complete language can be decided in polynomial time, then P = NP. (*Hint:* Remember the properties of NP-complete languages.)

*14. Prove the following theorem by constructing an appropriate nondeterministic Turing machine: Let L be a language on Σ. L is recursive if and only if both L and $\Sigma^* - L$ are recursively enumerable.

CHAPTER SUMMARY

In this chapter we continued our discussion of mathematical models of computation. In Section 8.1 we briefly discussed the pushdown automaton, and then we took up the main topic of this chapter, the Turing machine. Turing machines are the model of computation which has come to be accepted as the most useful model of a computer executing a computer program.

The Turing machine provides a level of complexity and sophistication considerably beyond that of the finite automaton, and we discussed the Turing machine as a device for both recognizing languages and computing the values of functions.

In Section 8.3, we introduced Church's thesis, which essentially states that any problem which can be solved with an algorithm can be solved with a Turing machine. The fact that most people are willing to accept Church's thesis means that the Turing machine provides us with a mathematical model for studying a number of properties of algorithms, including the efficiency of algorithms. We discussed the issue of computational complexity (or efficiency) in the last two sections.

In Section 8.5 we also introduced the idea of a nondeterministic automaton, and the discussion of this device finally enabled us to describe the concept of the NP-complete problem in terms of the efficiencies of deterministic and nondeterministic Turing machines.

KEY TERMS

The following terms were introduced in this chapter:

asymptotic dominance (8.4)
asymptotic equivalence (8.4)
big "oh" (8.4)
Church's thesis (8.3)
class P (8.5)
class NP (8.5)
complexity (8.4)
context-free language (8.1)
nondeterministic complexity (8.5)

nondeterministic Turing machine (8.5)
nonpolynomial time (8.4)
NP-complete language (8.5)
polynomial time (8.4)
pushdown automaton (pda) (8.1)
recursive language (8.2)
recursively enumerable language (8.2)
Turing machine (8.1)

9
Formal Languages and Grammars

9.1 A SUBSET OF SPANISH

In previous chapters, we introduced the idea of a language on Σ as being any subset of Σ^*, the set of all strings on Σ. Our first approach to languages has been to study them in terms of the machines that can be used to recognize them. A regular language is one which can be decided by a finite automaton, a pushdown language is one which can be accepted by a pushdown automaton, a recursive language is one which can be *decided* by a Turing machine, and a recursive enumerable language is one which can be *accepted* by a Turing machine. All regular languages are pushdown languages, all pushdown languages are recursive and all recursive languages are recursive enumerable. There are a number of interesting questions about this classification of languages: Are all those categories necessary? Are there languages which are recursively enumerable but not recursive? Are there languages which are recursive and are not pushdown languages? Are there pushdown languages which are not regular?

Descriptions of Languages

We need to consider other ways of describing languages. Languages are, of course, sets of strings, and one reasonable way to describe them is by set-theoretic notation. If $\Sigma = \{a, b\}$ and $L = \{s \in \Sigma^* \mid s = a^n b^n\}$, it would be reasonable to try to classify the language according to the method discussed above. In discussing a language, or set of strings, a string followed by an exponent indicates that that string is to be repeated that number of times, so L consists of strings which contain a certain number of a's followed by the same number of b's.

One of the most useful ways of constructing a language is by providing a grammar, or set of rules, for producing strings. This method is usually followed in the definition of computer programming languages. The idea of a grammar has its roots in linguistics, because linguists attempt to describe a spoken or written language in terms of the rules for constructing statements in that language. In this section we present an example of how that process might be used to describe a very small subset of Spanish. With this example, we begin to see some of the nuances involved in the process of providing a mathematical description of a language. We give a formal definition of a phrase structure grammar in the next section, but studying this example will enable us to introduce definitions of some other important terms associated with the description of a language by means of a grammar.

Our Subset of Spanish

Our subset of Spanish will have a rather limited structure, with each sentence consisting of a subject phrase followed by a predicate phrase. We symbolize that in the following manner:

$$\langle \text{Sentence} \rangle \rightarrow \langle \text{subject phrase} \rangle \langle \text{predicate phrase} \rangle$$

This notation means that to make a sentence, we can concatenate a subject phrase and a predicate phrase. The brackets indicate that these terms need to be further broken down.

In our language, subject phrases will consist of either a subject or a subject followed by a possessive phrase, and predicate phrases will consist of the word 'es' followed by an adjective. Symbolically these rules become

$$\langle \text{subject phrase} \rangle \quad \rightarrow \langle \text{subject} \rangle$$

$$\langle \text{subject phrase} \rangle \quad \rightarrow \langle \text{subject} \rangle \langle \text{possessive phrase} \rangle$$

$$\langle \text{predicate phrase} \rangle \rightarrow \text{'es'} \langle \text{adjective} \rangle$$

At this point the word 'es' is the only example of what we call a *terminal symbol* since it is a string which *can* occur in a sentence in our language. All the symbols enclosed in angle brackets are *nonterminal symbols* because they must be eliminated before we have a valid sentence.

The subject consists of an article followed by one of two nouns, while the possessive phrase consists of the word 'de' followed by a possessive and a noun chosen from a different collection. The articles are the words 'el' and 'un', and the nouns for the subject are the words 'gato' and 'perro'. The possessives are 'mi' and 'su'. The choices for the second noun are 'tia' and 'madre'. Finally the permissible adjectives are 'gris' and 'blanco'. Finishing the symbolic description of the language we have the following:

⟨subject⟩ → ⟨article⟩ ⟨noun1⟩

⟨possessive phrase⟩ → 'de' ⟨possessive⟩ ⟨noun2⟩

⟨article⟩ → 'un'

⟨article⟩ → 'el'

⟨noun1⟩ → 'gato'

⟨noun1⟩ → 'perro'

⟨possessive⟩ → 'mi'

⟨possessive⟩ → 'su'

⟨noun2⟩ → 'madre'

⟨noun2⟩ → 'tia'

⟨adjective⟩ → 'blanco'

⟨adjective⟩ → 'gris'

Determining Legal Sentences

We can list some legal sentences in our subset of Spanish:

1. El gato de su madre es gris.
2. El perro es blanco.
3. Un perro de mi tia es blanco.

We get a sentence in the language by following the rules of the grammar one step at a time and replacing the nonterminal symbols with any of their replacements as listed in the rules. The rules are called *production rules*, and the use of one is called a *production*. A sentence is in the language if and only if a person can start with ⟨sentence⟩ and use productions to create the sentence in question. The process of starting with ⟨sentence⟩ and performing productions until we reach a string with all terminal symbols is called a *derivation*. In this language, we place spaces between successive symbols. In a more formal description, we would actually have to specify that in the grammar. A derivation of sentence 1 goes something like this:

⟨sentence⟩ →

⟨subject phrase⟩ ⟨predicate phrase⟩ →

⟨subject⟩ ⟨possessive phrase⟩ ⟨predicate phrase⟩ →

⟨article⟩ ⟨noun1⟩ ⟨possessive phrase⟩ ⟨predicate phrase⟩ →

el ⟨noun1⟩ ⟨possessive phrase⟩ ⟨predicate phrase⟩ →

el gato ⟨possessive phrase⟩ ⟨predicate phrase⟩ →

el gato de ⟨possessive⟩ ⟨noun2⟩ ⟨predicate phrase⟩ →

el gato de su ⟨noun2⟩ ⟨predicate phrase⟩ →

el gato de su madre ⟨predicate phrase⟩ →

el gato de su madre es ⟨adjective⟩ →

el gato de su madre es gris

There are a number of derivations of this same sentence. We have used what might be called a "leftmost" derivation because we always replaced the non-terminal symbol farthest left in each of our productions.

Parse Trees

It is possible to describe this process by means of a tree. In constructing the tree, the root is labeled with the starting symbol ⟨sentence⟩. When a production is used to replace a nonterminal symbol, an edge is drawn from the vertex labeled by that nonterminal symbol for each symbol used to replace it. The new vertices are labeled with the symbols that replace the nonterminal symbol. The leftmost vertex is given the leftmost symbol used to replace the nonterminal symbol, etc. At any stage in the development of the tree, the current string can be read by listing the leaves of the tree from left to right. When the derivation is finished, the leaves will consist of only terminal symbols. A tree constructed in this fashion is called a *parse* or *derivation, tree*. For sentence 1, every derivation produces the same tree. We say that sentence 1 is *unambiguous*.

> DEFINITIONS
>
> In a language described by a grammar, a **sentence is unambiguous** if and only if every derivation of the sentence produces the same parse tree.
>
> A **language is unambiguous** if and only if every sentence in the language is unambiguous.

Figure 9.1 shows the parse tree for 'El gato de su madre es gris'.

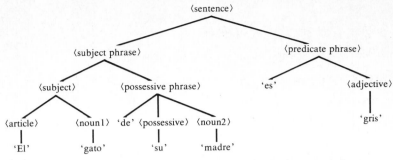

Figure 9.1 Parse tree for 'El gato de du madre es gris'.

> **DEFINITION**
>
> **Parsing** is the process of recovering a parse tree from a string in a given language.

In computer languages, parsing is an important part of translating from the language in which a program was written to machine language. The parsing makes sure that the sentence is a legal sentence, and the parse tree provides an interpretation of the intent of the statement. It is particularly important in those circumstances that the entire language be unambiguous, that is, that all the strings in the language be unambiguous.

Expanding Our Language

We can expand our language slightly by changing the rules for predicate phrases to allow verb-adverb phrasing in the predicate. These new rules can be listed:

$\langle$predicate phrase$\rangle \rightarrow \langle$verb$\rangle$ $\langle$adverb$\rangle$

$\langle$verb$\rangle \rightarrow$ 'habla'

$\langle$verb$\rangle \rightarrow$ 'canta'

$\langle$adverb$\rangle \rightarrow$ 'despacio'

$\langle$adverb$\rangle \rightarrow$ 'rapido'

These new rules add to the number of possible sentences in our language.
    The following two sentences belong to the language:

4. Un gato de mi madre habla rapido.

5. El perro de su tia canta despacio.

In Fig. 9.2 we indicate a parse tree for sentence 4.

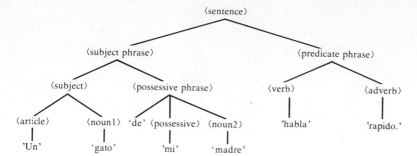

Figure 9.2 Parse tree for 'Un gato de mi madre habla rapido'.

Syntax and Semantics Revisited

Up to this point, our study of this language has been entirely concerned with determining whether a particular string belongs to the language. This study is called the study of the *syntax* of the language. *Formal language theory* is the study of languages solely in reference to their syntax. We would be remiss if we did not mention that the meaning of sentences in a language is obviously important. For a natural language, or for a computer language, just being able to construct legal statements in the language or being able to recognize that sentences belong to the language is hardly enough to make use of the language. The issue of meaning is important. The meaning of a language is referred to as its *semantics*. In our mathematical study of languages, we deal almost exclusively with the syntax of languages, but often we must also have some concern for the semantics.

To describe the semantics of our subset of Spanish, we need only provide a dictionary for translating into a language whose semantics we know— namely English. In this particular case translation is quite easy, but the process of translation does mirror the process of compiling a computer program. First the syntax is checked. If there is an error in the syntax of the program, the error is reported. If the syntax is all right, then the computer functions as a transducer and turns the string into a language known by the computer, namely, machine language.

In our example, we can check whether a sentence belongs to the language. If it does not, we can stop and say so. If it does belong to the language, all that is necessary is to apply the following dictionary to the terminal symbols of the language:

el = the	tia = aunt
un = a	es = is
gato = cat	blanco = white
perro = dog	gris = gray

de = of	habla = talks
mi = my	canta = sings
su = your	rapido = rapidly
madre = mother	despacio = slowly

From this dictionary, we can see that some of the sentences that we can construct in this language are slightly bizarre. Sentence 4 translates into English as "A cat of my mother speaks rapidly." The semantics of this particular statement might be questioned, but this is not unlike dealing with computer programs. Some statements are syntactically proper in a programming language but are semantically absurd. The BASIC statement "10 GO TO 10" is syntactically correct, but the semantics produces nothing useful.

The Question of Context

Another note about our language is that it is "context-free" in the sense that replacements for nonterminal symbols do not depend in any sense on the surrounding text. A full-grown version of Spanish would not be context-free. If we had added a "feminine" noun such as "casa" (house) to our listing of ⟨noun1⟩ nouns, then we would have needed to require that the adjective "blanco" be changed to "blanca" when the feminine noun was used, and we would also have had to change the possible articles from "el" and "un" to "la" and "una." Making these changes in the grammar turns it into a "context-sensitive" grammar because the legality of some words in sentences seems to depend on the context in which the words are used. Syntax checking for context-sensitive grammars is much more involved than is syntax checking for context-free grammars. One issue which we address in this chapter is the question of what kind of machine is necessary to check the syntax of what kinds of grammars. Our model of a subset of Spanish is quite simple, but it does illustrate some of the issues involved in the theory of languages.

Problem Set 1

1. Find a derivation of the sentence 'el gato de su madre es gris' which is different from the one found in the text. Find the parse tree for your derivation, and compare it to the parse tree in Fig. 9.1.

2. Find a derivation and a parse tree for sentence 2.

3. Find a derivation and a parse tree for sentence 3.

4. Find a derivation and a parse tree for sentence 5.

5. Determine whether the string 'el gato de mi madre es rapido' belongs to our language. If it does, find a parse tree. If it does not, explain why not.

6. Determine whether the string 'el perro de mi tia canta rapido' belongs to our language. If it does, find a parse tree. If it does not, explain why not.

*7. Show that there are a finite number of sentences in the language of Section 9.1, and determine the number of sentences.

Consider the following grammar:

⟨sentence⟩ → ⟨subject⟩ ⟨predicate⟩

⟨subject⟩ → ⟨article⟩ ⟨noun⟩

⟨subject⟩ → ⟨pronoun⟩

⟨predicate⟩ → ⟨verb⟩ ⟨adverb⟩

⟨predicate⟩ → ⟨verb⟩

⟨article⟩ → 'a'

⟨article⟩ → 'the'

⟨noun⟩ → 'boy'

⟨noun⟩ → 'girl'

⟨verb⟩ → 'eats'

⟨verb⟩ → 'runs'

⟨verb⟩ → 'sleeps'

⟨adverb⟩ → 'noisily'

⟨adverb⟩ → 'quickly'

⟨pronoun⟩ → 'he'

⟨pronoun⟩ → 'she'

8. Determine derivations for the following sentences:
(a) 'A girl eats noisily'
(b) 'The boy sleeps quickly'

9. Find parse trees for the following sentences:
(a) 'He runs noisily'
(b) 'A girl sleeps'

10. Comment on the semantics of the above language.

*11. Determine whether the grammar discussed above is unambiguous.

*12. Show that the language determined by the grammar above is finite, and determine the number of sentences in the language.

*13. Show that any language which is finite is a regular language by describing how to construct a finite automaton to accept it.

9.2 PHRASE STRUCTURE GRAMMARS

A phrase structure grammar, or simply a grammar, is a set of syntax rules which allow for the creation of a language. A grammar provides the means of describing many languages, including all languages which are used as computer languages.

DEFINITIONS

A **phrase structure grammar** is an ordered 4-tuple (V, T, S, P), where V is the set of **nonterminal symbols,** T is the set of **terminal symbols,** S is an element of V called the **start symbol,** and P is a relation on $(V \cup T)^*$ in which each first element of an ordered pair includes at least one element of V.

The language described in Section 9.1 is an example of a language described by a phrase structure grammar. Phrase structure grammars can be used to describe languages on T by making use of the production rules as we did in Section 9.1. We will make that more precise momentarily, but first let us describe the grammar from Section 9.1:

$V = \{\langle$sentence$\rangle$, $\langle$subject phrase$\rangle$, $\langle$predicate phrase$\rangle$, $\langle$possessive phrase$\rangle$, $\langle$subject$\rangle$, $\langle$article$\rangle$, $\langle$noun1$\rangle$, $\langle$noun2$\rangle$, $\langle$possessive$\rangle$, $\langle$verb$\rangle$, $\langle$adverb$\rangle$, $\langle$adjective$\rangle\}$

$T = \{$'de', 'un', 'el', 'gato', 'perro', 'mi', 'su', 'madre', 'tia', 'blanco', 'gris', 'habla', 'canta', 'rapido', 'despacio'$\}$

$S = \langle$sentence$\rangle$

$P = \{(\langle$sentence$\rangle$, $\langle$subject phrase$\rangle$ $\langle$predicate phrase$\rangle$), $(\langle$subject phrase$\rangle$, $\langle$subject$\rangle$), $\ldots$, $(\langle$adverb$\rangle$, 'despacio'$)\}$

(We left out a number of elements from P, but you should get the general idea.)

With a grammar, we can construct strings in T^* by use of the relation P. Any occurrence of the first entry in an element of P as a substring of the current string can be replaced by the second entry of the same ordered pair. We may do this with any (or none) of the occurrences of that string. And if the same string occurs as the first entry in two or more ordered pairs of P, we can use whichever entries of P we would like to use. Thus if the symbol $\langle$adverb$\rangle$ occurs in a string, we are developing, we may replace it by 'rapido' or by 'despacio', and having chosen that once does *not* restrict us in any way the next time the same possibility occurs.

DEFINITION

A **production** is any use of the elements of set P as described above.

In the same spirit as for our terminology for machines, if G is a grammar, we will indicate T as Terminals(G), and V will be Nonterminals(G), S will be written as Start(G), and P will be referred to as Productions(G). The set of all strings which can be produced for a grammar is called the *language* of the grammar and is denoted by $L(G)$.

Grammars and "Real" Languages

Every spoken or written language has some collection of rules which describe it, often in a form not unlike productions in our grammars. All computer languages are described in terms of a grammar. Mathematicians and computer scientists are interested in grammars and the languages produced by them for their own sake. And an understanding of the grammatical structure of "natural languages" may provide insight into the development of computers which would be able to process natural language commands instead of relying on current programming languages. If we wanted to compute $x + 2$ and have the result printed, instead of a statement like "writeln($x + 2$)" as in Pascal or "100 PRINT X + 2" as in BASIC, it would be nice to be able to just ask the computer, "What is $x + 2$?" and get an answer. One way to accomplish natural language processing may lie in a complete understanding of the grammar of natural languages and the process of translating from natural languages to computer languages.

Pascal, as we discussed earlier, is often described in terms of syntax diagrams, but it is also possible to use a formal grammar. Part of the process can be described by the following productions:

⟨program⟩ → ⟨program heading⟩ ⟨main block⟩

⟨main block⟩ → ⟨declaration section⟩ ⟨executable section⟩

⟨executable section⟩ → ⟨compound statement⟩'.'

⋮

⟨assignment statement⟩ → ⟨identifier⟩ ':=' ⟨expression⟩

⋮

Several Examples

The next four examples examine successively more complex grammars.

EXAMPLE 9.1

Let G_1 be defined in the following way:

Nonterminals $(G_1) = \{A, B, C, D\}$
Terminals$(G_1) = \{a, b\}$
Start$(G_1) = A$
Productions$(A) = \{(A, aB), (B, bB), (B, aC), (B, D), (C, aC), (C, D), (D, b)\}$

Usually we replace the notation of ordered pairs by the $\rightarrow$ symbol, and it is much more convenient if we condense the notation. So whenever several productions share the same initial symbols (same left-hand side), we write all possible replacements for that left side separated by a vertical bar ($|$). This means that the productions above can be written as

$$A \rightarrow aB$$
$$B \rightarrow bB \mid aC \mid D$$
$$C \rightarrow aC \mid D$$
$$D \rightarrow b$$

The following strings all belong to $L(G_1)$: '*aaaaab*', '*abbbbb*', and '*ab*'. In many cases, a parse tree such as we discussed in Section 9.1 offers an easy way to demonstrate how a string is produced. See Fig. 9.3.

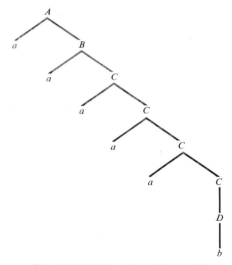

Figure 9.3 Parse tree for *aaaaab*.

The following strings do not belong to $L(G_1)$:

  '*baaa*'   '*aaa*'   '*abbba*'   '*abababbb*'

String '*baaa*' does not belong to $L(G_1)$ because A is the start symbol, and the only production using A places an '*a*' at the far left of the string produced. All the remaining productions place the nonterminal symbols at the far right of the string, and thus each string in the language must start with an '*a*'. We leave as exercises the proof that the remaining strings in our list do not belong to the language. ∎

We mentioned the use of exponents in describing strings, but let us review the idea. In any string the appearance of an exponent means to repeat the substring referenced by the exponent as many times as the exponent indicates. The exponent refers to a single character if no parentheses appear; otherwise, the exponent refers to all the characters enclosed by the parentheses immediately preceding it. Thus 'a^2b^2' is the string '$aabb$', and '$(ab)^2$' is the string '$abab$'. Another notation uses a superscript R to indicate that a substring is to appear in reverse order. Thus '$(ab)^R$' means 'ba'. In describing a language we also use an asterisk as an exponent to indicate that the substring is to be repeated any number of times (including zero) and a plus as an exponent to indicate that the string is to appear at least once and then may be repeated any number of times. Thus $\{s \mid s = a^+b*\}$ would be a language consisting of at least one 'a' followed by some number of additional a's and then a string (possibly empty) of b's. That language would include strings like '$aaabb$' and 'aaa' but not 'bbb' or Λ. We leave it as an exercise for the reader to describe the language generated by G_1.

EXAMPLE 9.2

In this example, we present a grammar which may be considered more complicated than G_1.

Nonterminals$(G_2) = \{S, A, B\}$
Terminals$(G_2) = \{a, b\}$
Productions(G_2):
$\quad S \rightarrow aA$
$\quad A \rightarrow aAB \mid a$
$\quad B \rightarrow b$
Start$(G_2) = S$

The following strings belong to $L(G_2)$:

$\quad$ 'a^2' $\quad$ 'a^4b^2' $\quad$ 'a^3b'

The following strings do not belong to $L(G_2)$:

$\quad$ 'a^2b' $\quad$ 'ab^2' $\quad$ 'ab^{17}'

Grammar G_2 is a little trickier to work with than G_1 because we now have a production of the form $A \rightarrow aAB$, which places some restrictions on the relationship that can exist between the number of a's and the number of b's. This grammar can still be dealt with by using parse trees because there is but a single nonterminal symbol on the left side of each of the production rules. ■

EXAMPLE 9.3

We can create a more complicated grammar by allowing more than one symbol on the left-hand side of a production. The resulting grammar cannot be dealt with through the use of a parse tree.

Nonterminals(G_3) = {S, A, B}
Terminals(G_3) = {a, b}
Productions(G_3):
  $S \rightarrow aAB$
  $AB \rightarrow bB$
  $B \rightarrow b$
  $A \rightarrow aBA \mid aB$
Start(G_3) = S

The fact that the productions of G_3 include one which has more than one symbol on the left side means that the production of this grammar cannot be used in the same straightforward way, as are the productions in G_1 and G_2. The construction of the tree is made possible in those grammars because only one nonterminal symbol is replaced at a time, and that replacement can be represented by a vertex in a tree. We would actually need a directed graph with cycles to represent the construction of a string in *this* language. ■

EXAMPLE 9.4

A final, even more complicated grammar is the following:

Nonterminals(G_4) = {S, A, B}
Terminals(G_4) = {a, b}
Productions(G_4):
  $S \rightarrow aAB$
  $AB \rightarrow c \mid bB$
  $B \rightarrow AB \mid b$
  $A \rightarrow b$
Start(G_4) = S

This grammar is more complicated because some productions cause the string being produced to become shorter. This makes the process of testing a string to see whether it belongs to the grammar even more complicated. With the preceding grammar, at least we could tell when we had passed the point of being able to construct a target string. (As soon as our string got too long, we could stop.) ■

EXAMPLE 9.5

Produce a grammar to generate the language $L = \{x \in \{a, b\}^* \mid x$ is of the form $a^*b^*\}$.

We can describe a grammar which first allows us to construct as many a's as we would like, and then we can create productions which allow us to produce as many b's as we like. We also need to allow the empty string as a possibility in our language. We obviously need to begin with the start symbol S and to allow a production which takes us to the empty string. Otherwise, we need productions to allow us to start making a's or to start making b's. We use the symbol A to produce our a's and the symbol B to produce our b's. The following productions should seem reasonable at this point:

$$S \rightarrow \Lambda \mid A \mid B$$

The A symbol should allow us to produce as many a's as we like. One way of accomplishing this is to include a production which replaces the 'A' with 'aA' or to quit producing a's by replacing the 'A' with 'B'. The productions needed are thus:

$$A \rightarrow aA \mid B$$

We can accomplish the same thing with B, allowing termination by replacing B with an empty string

$$B \rightarrow bB \mid \Lambda$$

This collection of productions, and the appropriate descriptions of the other parts of a grammar, allows for the production of the language which we seek. For various technical reasons, there are some undesirable features in this grammar, which you are asked to fix in the problems at the end of the section.

■

We have introduced the concept of a phrase structure grammar and provided some examples. We return to the examples of this section later and discuss them in relation to questions of classifying grammars and producing automatons for testing membership in languages generated by grammars.

Problem Set 2

1. Produce the derivation and parse trees to show that 'ab' and '$abbbbb$' belong to $L(G_1)$.

2. Show that the strings 'aaa', '$abbba$', and '$ababababb$' do not belong to $L(G_1)$.

3. Describe the language generated by G_1 as a set.

4. Produce derivations and parse trees for the following members of $L(G_2)$: 'a^2', 'a^4b^2', and 'a^3b'.

5. Show that the strings 'a^2b', 'ab^2', and 'ab^{17}' do not belong to $L(G_2)$.

6. Describe the language generated by G_2 as a set.

7. Determine whether each of the following strings belongs to $L(G_3)$. For those that belong, provide a derivation. For those that don't, explain why not.

$$`ab^2{'} \qquad `a^2bab^3{'} \qquad `ba^2bab^3{'}$$

8. Determine whether each of the following strings belongs to $L(G_4)$. For those that belong, provide a derivation. For those that don't, explain why not.

$$`abc{'} \qquad `ab^3c{'} \qquad `ab^4{'}$$

9. Develop a grammar for the language $\{s \mid s = a^*b^*c^*\}$.

10. Describe the language of balanced-parentheses strings by means of a grammar. Show the derivation of the string `(())()' in your grammar.

11. A *palindrome* is a string which reads the same from left to right as it does from right to left. Find a grammar to describe the set of all palindromes on $\{0, 1\}$, and show the derivation of the string 0010110100.

12. Define G_4 by

$$S \rightarrow xST \mid yS \mid z$$

$$T \rightarrow Tw \mid aST \mid z$$

(The capital letters represent nonterminal symbols.) Determine whether each of the following belongs to $L(G_4)$:
(a) *xyzzw*
(b) *wyaxzyx*
(c) *yxyzwxy*
(d) *xyzazzw*

*13. The grammar in Example 9.5 has the undesirable feature that it has an "erasing" production which allows for the string being produced to become shorter. (The production in question is $B \rightarrow \Lambda$.) Show that in this case there is a grammar which produces the same language but which does not contain an erasing production. We need two things:

a. A way to stop producing a's and produce no b's

b. A way to stop producing b's

Modify the productions of the grammar from Example 9.5 so as to accomplish the two tasks described above, and show that your new grammar produces the same language as the original grammar.

In making use of a language, we must consider both the syntax and the semantics. The main focus of our study has been, and will continue to be, the syntax of languages, but at this point it is appropriate to spend a little time discussing how syntax and semantics together affect the operation of computer programming languages.

Errors can be detected in computer programs at two different points in the process of compiling and running a program, depending in part on the type of error that occurs. The most obvious kinds of errors to recognize are those detected by the computer as the computer begins translating the "source code" (programming language instructions) to machine language. The second kind of error occurs when the computer is running the translated program. Some of these errors are not detected at all by the computer; they are found only when the results of the program are discovered to be incorrect. The first kind of error is generally a syntax error, and the second is a semantic error.

Syntax errors occur in programs when we write statements like the following statements in Pascal:

```
X : X + 2;
for i = 1 to 3 do t := t + 2;
```

In each of the above statements, the strings listed fail to meet the specifications of the Pascal grammar. The symbol used in an assignment statement in Pascal is :=, and to leave out either part of the symbol causes the computer to report an error.

In BASIC, the situation is much the same. The following statements will cause error messages to appear before the program runs. (The exact message depends on the particular machine being used.)

```
10 WET X = 3
20 LET X = (X + 3/2
```

Very often, especially with microcomputers the error message may be simply "syntax error" or even " ?SN error." The message means that an error in the syntax of the BASIC language has occurred, and the translating program cannot parse the given statement.

Semantic errors are usually harder to detect than syntax errors. Often programs are syntactically correct, but do not have the effect originally intended. In Pascal, for example, there are two different division operations for integers, one denoted by DIV and the other by MOD. DIV takes the quotient for the division of two integers; thus 5 DIV 2 is 2, because in dividing the integer 5 by the integer 2, we have a result of 2 with a remainder of 1. MOD produces the remainder of the division as a result. Both

$$w := x \text{ MOD } y \qquad \text{and} \qquad w := x \text{ DIV } y$$

are syntactically correct, but to use the wrong one is a semantic error. Semantic errors can be very hard to detect in a program, unless they cause problems which make the program fail.

In BASIC, the following program is syntactically correct, but contains a semantic error:

```
10 DIM X(3)
20 FOR I = 1 TO 4
30 INPUT X(I)
40 NEXT I
50 END.
```

The computer will detect this error, although a similar error in some other languages may go undetected, and produce really bizarre results. When that happens, it is very difficult to discern the nature of the problem. Other common semantic errors include things like failing to zero variables being used to accumulate totals, use of parentheses which produces unintended groupings of terms, and in general all errors which are not detected by the translating program.

Our main concern here is the study of *formal languages*, and as such, we shall not deal further with the semantics of languages. But readers must be aware of the importance of the semantics of any language that they are using.

Problem Set 3

1. What kind of errors in English might be regarded as syntax errors? Give an example.

2. What kind of errors in English might be regarded as semantic errors? Give an example.

Each of the following statements in Pascal contains either a syntax error or a semantic error. Classify the errors.

3. **for** $i := 1.5$ **to** 10 **do** $x := x + 1$;

4. **for** $i := 6$ **to** 5 **do** $x := x + 1$;

5. $x(3) := 3$;

6. **if** $3 < x < 4$ **then** writeln('x is in the interval');

7. **if** $x < 0$ **then** writeln(tiny);

8. **if** $x = (y = 10)$ **then** writeln('both are 10')

Each of the following BASIC statements contains either a syntax error or a semantic error. Classify the errors.

9. 10 X = Y = 2

10. 10 PRINT (X + Y = X + Y)

11. 10 PRINT ("X + Y = ";X)

12. 10 IF X = 3, Y = 4

13. 10 GOSUB 10

14. 10 IF X < 3 THEN WRITE ('SMALL')

15. The following BASIC program contains both semantic and syntactic errors. Find the errors and determine which kind of error each is. (The program is supposed to print the roots of the quadratic equation $Ax^2 + Bx + C = 0$.)

```
10 INPUT A, B
20 LED = B∧2 − 4∗A∗C
30 LER1 = SQRT(D) − B/2A
40 LET R2 = (B − SQR(D))/(2∗A)
50 PRONT " THE ROOTS ARE ";R1:R2
60 END
```

P16. Correct the errors in the above program so that it will run and produce the appropriate output.

17. The following Pascal program contains both semantic and syntactic errors. Find the errors and determine which kind of error each is. (The program is supposed to print the roots of the quadratic equation $Ax^2 + Bx + C = 0$.)

```
PROGRAM QUAD (INPUT, OUTP)

var A, B, C: integer;
    R!, R2 : real;
begin
writeln('Type in the coefficients);
readln(A < B, C);
D := B∗B − 4∗A∗C;
R1 := sqrt(D) − B/2∗A;
R2 := b − sqrt(D)/2∗A;
writeln('the roots are ',R!, R2);
end.
```

P18. Correct the errors in the program above so that the program will run and produce the correct output.

9.4 CLASSIFICATION OF GRAMMARS AND LANGUAGES

In Section 9.2, we presented four examples of grammars. Let us summarize them before we continue.

G_1 is defined in the following way:

Nonterminals$(G_1) = \{A, B, C, D\}$
Terminals$(G_1) = \{a, b\}$
Start$(G_1) = A$
Productions(A):
   $A \rightarrow aB$
   $B \rightarrow bB \mid aC \mid D$
   $C \rightarrow aC \mid D$
   $D \rightarrow b$

G_2 is defined as follows:

Nonterminals$(G_2) = \{S, A, B\}$
Terminals$(G_2) = \{a, b\}$
Start$(G_2) = S$
Productions(G_2):
   $S \rightarrow aA$
   $A \rightarrow aAB \mid a$
   $B \rightarrow b$

This is a description of G_3:

Nonterminals$(G_3) = \{S, A, B\}$
Terminals$(G_3) = \{a, b\}$
Start$(G_3) = S$
Productions(G_3):
   $S \rightarrow aAB$
   $AB \rightarrow bB$
   $B \rightarrow b$
   $A \rightarrow aBA \mid aB$

Finally, here is G_4:

Nonterminals$(G_4) = \{S, A, B\}$
Terminals$(G_4) = \{a, b\}$
Start$(G_4) = S$
Productions(G_4):
   $S \rightarrow aAB$
   $AB \rightarrow c \mid bB$
   $B \rightarrow AB \mid b$
   $A \rightarrow b$

The Chomsky Hierarchy

The *Chomsky Hierarchy* classifies grammars and the languages they generate according to the complexity of the productions. The grammars are classified according to a numbering system with type 0 being the least restrictive (and thus allowing the most complex productions) and type 3 being the most restrictive (and thus having the simplest productions). Every grammar belongs to at least one of the four types. The types are set up so that if a grammar satisfies the rules for belonging to type n, it also satisfies the rules for belonging to all the classifications with numbers less than n.

> **DEFINITION**
>
> Any phrase structure grammar is a **type 0 grammar.**

There are no restrictions at all on type 0 grammars.

DEFINITION

A **type 1 grammar** is a phrase structure grammar which satisfies the requirement that, except for the possible production $\text{Start}(G) \to \Lambda$, the number of symbols on the left-hand side of each production is less than the number of symbols on the right-hand side. $\text{Start}(G) \to \Lambda$ is allowed in a type 1 grammar if $\text{Start}(G)$ never occurs on the *right* side of any production.

Grammars of type 1 are often called *context-sensitive grammars*. This term comes from the fact that the right-hand side of the production may involve several symbols, and the manner in which those symbols are replaced depends on the information surrounding (the context of) the nonterminal symbols in the production. A production in which the length of the left is greater than the length of the right is called an *erasing* production.

DEFINITION

A grammar is a **type 2 grammar** if the left-hand side of each production consists of only a *single* nonterminal symbol and the grammar is non-erasing [i.e., there are no productions of the form $A \to \Lambda$ unless A is $\text{Start}(G)$ and $\text{Start}(G)$ never occurs on the right side of any production].

Clearly every type 2 grammar is also type 1. Type 2 grammars are often called *context-free grammars* because the replacement of nonterminal symbols by strings does not require that the nonterminal symbols occur in any particular context. The term "context-sensitive" is sometimes used to indicate that a grammar is *not* context-free.

DEFINITIONS

A **type 3 grammar** is a type 2 grammar that satisfies the additional requirement that the right-hand side of any production include no more than one nonterminal symbol, which must be the last (rightmost) symbol on the right side of the production.

A language is a **type n language** if it is generated by a type n grammar.

Of the grammars in Section 9.2, all are type 0; G_1, G_2, and G_3 are type 1; G_1 and G_2 are type 2; and only G_1 is type 3. Strings in languages of either of the two context-free types can be parsed conveniently by means of a tree. The two context-sensitive types would require much more complicated graphs in order to check their membership.

When we say that a language is type n, it may also be of some higher type. For example, if G is type 2, it is quite possible that it may be type 3 as well. The grammar that we described in Example 9.5 was only a type 0 grammar. But by following the suggestions in the problems at the end of Section 9.2, we actually showed that the same language can be produced by a type 2 grammar. Saying that the language in question is type 0 is not incorrect, but saying that it is type 2 conveys more information.

Backus-Naur Form

A standard way of describing a context-free grammar is by a notation called *Backus-Naur form* (BNF). In the BNF notation, all nonterminal symbols are placed within angle brackets. The production symbol $\rightarrow$ is replaced by the symbol $::=$; and if more than one production has the same left-hand side, we list the left-hand side once and then on the same line list all the right-hand sides separated by vertical bars. In BNF, the start symbol is always denoted $\langle Start \rangle$. From a listing of the BNF productions one can determine the start symbol, the nonterminal symbols, the terminal symbols, and the production rules. BNF allows the specification of a grammar through a listing of the productions only.

Context-free grammars are often called BNF grammars, and large parts of computer languages can be described by using BNF. Most computer languages are largely context-free. It is necessary to use the context of a Pascal program to know if an identifier has been declared, and the validity of a Pascal **goto** statement depends both on a label being declared and that label actually being present in the program. But by and large, the definition of Pascal can be made in terms of a context-free grammar. Similarly in BASIC, the validity of a FOR statement depends on the existence of the matching NEXT, and GOTO statements must reference existing statements in the program. But, again, much of the language can be described by BNF (often for BASIC, an equivalent form known as a *CBL grammar* is used to describe the language). It is much easier to write a translator for a context-free language than for a less structured one.

EXAMPLE 9.6

The grammar G_1 can be described in BNF notation as follows:

$\langle Start \rangle ::= a \langle B \rangle$
$\langle B \rangle ::= b \langle B \rangle \mid a \langle C \rangle \mid \langle D \rangle$
$\langle C \rangle ::= a \langle C \rangle \mid \langle D \rangle$
$\langle D \rangle ::= b$ ∎

The subset of Spanish that we developed in Section 9.1 is also context-free and can be denoted with the BNF notation. We conclude this section with a theorem which is obvious from the way that the types were defined.

THEOREM 9.1

Every type n language is also type k for $k < n$.

PROOF
This is left as an exercise. ∎

Classify each of the following grammars. In each case,

$$\text{Terminals}(G_i) = \{x, y\}$$

$$\text{Nonterminals}(G_i) = \{S, X, Y\}$$

$$\text{Start}(G_i) = S$$

1. Productions(G_1):

$S \rightarrow xX$

$X \rightarrow yXY \mid x$

$Y \rightarrow y$

2. Productions(G_2):

$S \rightarrow xXY$

$XY \rightarrow yY$

$Y \rightarrow y$

$X \rightarrow xY$

3. Productions(G_3);

$S \rightarrow xXY$

$XY \rightarrow y$

$X \rightarrow y$

$Y \rightarrow xXY$

4. Productions(G_4):

$S \rightarrow xY$

$Y \rightarrow yX \mid x \mid y$

$X \rightarrow xY \mid x$

5. Productions(G_5):

$S \rightarrow xXY$

$X \rightarrow Yyx$

$Y \rightarrow yY \mid y$

6. Productions(G_6):

$S \rightarrow xY$

$X \rightarrow xY$

$Y \rightarrow yX \mid y$

7. Productions(G_7):

$S \rightarrow xSX \mid yS \mid x$

$X \rightarrow Xy \mid xSX \mid y$

8. Productions(G_8):

$S \rightarrow X \mid y$

$X \rightarrow x \mid xX$

9. Productions(G_9):

$S \rightarrow \Lambda \mid X \mid Y$

$X \rightarrow xY \mid x$

$Y \rightarrow yX \mid y$

10. Find a type 3 grammar to generate the language in $\{a, b\}$ which consists of all strings with an *odd* number of b's.

11. Find a type 3 grammar to generate the language in $\{a, b\}$ which consists of all strings with an *even* number of a's.

12. Find a context-free grammar which will generate the language $L = \{s \mid s = a^n b^n\}$.

13. Find a type 3 grammar which will generate the language $L = \{s \mid s = a^n b^m c^p\}$.

*14. Find a context-free grammar which will generate the language $L = \{s \mid s = xx^R$, where x is any string in $\{a, b\}\}$.

15. Find a context-free grammar to find all binary representations of odd numbers.

16. Find a context-free grammar to find all unary representations of odd numbers.

17. Use BNF to describe grammar G_2 of Section 9.2.

18. Use BNF to describe the subset of Spanish from Section 9.1.

19. Use BNF to describe the context-free languages in Problems 1 to 7.

20. Prove Theorem 9.1.

9.5 GRAMMARS AND MACHINES

We have seen two different ways of generating and classifying languages; one used abstract machines to test membership of strings in a language and the other used different kinds of grammars to generate the language.

Regular languages are accepted by finite automatons, pushdown languages are accepted by pushdown automatons, recursive languages are *decided* by Turing machines, and recursively enumerable languages are *accepted* by Turing machines. The four classifications of languages by means of grammars are described in the previous section. Figure 9.4 illustrates the two classifications. There is a remarkably close relationship between the two classifications.

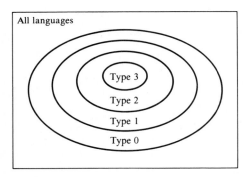
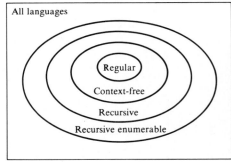

(a) By grammar   *(b)* By machine

Figure 9.4 Two ways of classifying languages.

Type 3 Grammars and Finite Automatons

We recall grammar G_1 from Example 9.1:

$$\langle \text{Start} \rangle :: = a\langle B \rangle$$
$$\langle B \rangle :: = b\langle B \rangle \,|\, a\langle C \rangle \,|\, \langle D \rangle$$
$$\langle C \rangle :: = a\langle C \rangle \,|\, \langle D \rangle$$
$$\langle D \rangle :: = b$$

We can construct a finite automaton to accept the language generated by this grammar because in the process of producing a string with the grammar, new terminal symbols are always added only on the left, and we can use a finite automaton to make sure that symbols were added to the string in an order consistent with the grammar. Each of the nonterminal symbols acts very much as a state in a finite automaton, because it controls what symbols can occur in a string at a given time. Several productions are possible from some of the nonterminal symbols. In some cases the production may have the same *terminal* symbols occurring on the left side of the production, and thus it is convenient to make this automaton nondeterministic. We need a state for each nonterminal symbol, a rejecting (trapping) state, and an accepting state.

When the process starts, we must be in the $\langle \text{Start} \rangle$ state. And since the only production here requires us to produce '$a\langle B \rangle$', if the first symbol in the string is not 'a', we must reject the string by entering the rejecting state. The automaton looks like Fig. 9.5 at this stage.

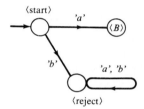

Figure 9.5 Step 1 in construction of automaton for G.

From $\langle B \rangle$ there are three productions: $\langle B \rangle \rightarrow b \langle B \rangle$, $\langle B \rangle \rightarrow a \langle C \rangle$, and $\langle B \rangle \rightarrow \langle D \rangle$. The first two are easy to deal with. If the next symbol in the string is a 'b', we can remain in state $\langle B \rangle$. If the next symbol is an 'a', we can move to state $\langle C \rangle$. The transition to just $\langle D \rangle$ with no terminal symbols preceding it is a little bit more annoying. This does give us an opportunity to use the fact that any language which is accepted by a nondeterministic automaton can also be accepted by a deterministic automaton. The transition to $\langle D \rangle$ can be accomplished by noting that any transition which led to $\langle B \rangle$ could just as well have led to $\langle D \rangle$. We will allow any input which causes a transition to $\langle B \rangle$ to also produce a transition to $\langle D \rangle$. This produces a nondeterministic automaton. Graphically the finite automaton now looks like Fig. 9.6.

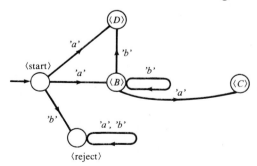

Figure 9.6 Step 2.

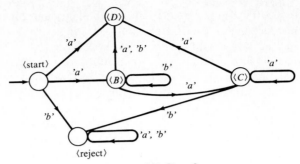

Figure 9.7 Step 3.

The productions involving $\langle C \rangle$ are $\langle C \rangle \rightarrow a\langle C \rangle$ and $\langle C \rangle \rightarrow \langle D \rangle$, so, as before, any transition which leads to $\langle C \rangle$ should also be allowed to lead to $\langle D \rangle$. (See Fig. 9.7.)

Finally, the nonterminal $\langle D \rangle$ allows only one production $\langle D \rangle \rightarrow b$. That production has no nonterminal symbols, so the transition from $\langle D \rangle$ must be to an accepting state. If any more input occurs from the accepting state, we do not have the string parsed properly, and we should go to the rejecting state. (See Fig. 9.8.) It should be clear (although it really requires proof) from our construction that a string is generated by G_1 if and only if it is accepted by this finite automaton. A similar process can be followed for any *type 3* grammar.

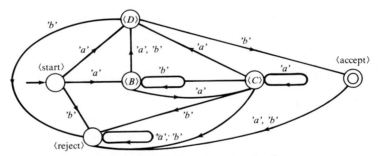

Figure 9.8 Finite automaton for G_1.

We can reverse the process to construct a type 3 grammar for any finite automaton. We illustrate that process with the following:

States$(M) = \{s_0, s_1, s_2\}$
Symbols$(M) = \{a, b\}$
Instructions(M):
  $d(s_0, a) = s_0$
  $d(s_0, b) = s_1$
  $d(s_1, a) = s_0$
  $d(s_1, b) = s_2$
  $d(s_2, a) = s_2$
  $d(s_2, b) = s_2$
Start$\langle M \rangle = s_0$
Accepting$(M) = \{s_0, s_1\}$

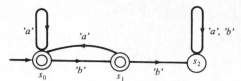

Figure 9.9 The automaton M.

We will use the graph representation of M (Fig. 9.9) as motivation in our construction of G. Clearly, the ⟨Start⟩ of the grammar corresponds to s_0. We need to set up our grammar in such a way that any path through the graph which ends in an accepting state can be realized by a derivation in the grammar which leads to a string with no nonterminal symbols in it. We will provide nonterminal symbols for each of the states. State s_0 has outdegree 2 and thus should have two productions which leave us with nonterminal symbols. Also s_0 is an accepting state, so a derivation should be allowed to end in s_0 without producing any more symbols. We have the following:

⟨Start⟩ :: = a⟨Start⟩ | a⟨S⟩ | Λ

From s_1, which we have rendered here as ⟨S⟩, an a takes us back to s_0 and b takes us to s_2. Here again s_1 is an accepting state. The productions we need are

⟨S⟩ :: = a⟨Start⟩ | b⟨T⟩ | Λ

Finally we can see that in s_2, which has become ⟨T⟩, the appropriate productions are

⟨T⟩ :: = a⟨T⟩ | b⟨T⟩

This is not quite a legitimate type 3 or regular grammar because there are erasing productions from ⟨S⟩ and ⟨Start⟩. (⟨Start⟩ is an issue because it occurs on the right side of a production as well as on the left side.) In the problems we will clean that up to produce a type 3 grammar which generates exactly the same language. Also note that the nonterminal ⟨T⟩ is not really needed because once a ⟨T⟩ is produced, it is not possible to construct a string with no nonterminal symbols in it. With those few caveats, clearly the same process can be applied to any finite automaton to produce a type 3 grammar that will generate the same set of strings as is accepted by the automaton. Thus we state the following theorem without proof.

THEOREM 9.2

Every regular language is generated by a type 3 grammar, and every type 3 language is regular. ∎

Type 2 Grammars and Pushdown Automatons

Similar constructions can be made for type 2 grammars using pushdown automatons.

THEOREM 9.3

Every pushdown language is generated by a type 2 grammar, and every type 2 grammar is accepted by a pushdown automaton.　■

We need a pushdown automaton (pda) to recognize context-free languages because type 2 grammars may produce more than one nonterminal symbol in the same production, and it becomes necessary to store the states corresponding to the nonterminal symbols that are not being considered so they can be processed later. We do not prove this theorem, but it should seem reasonable. Because of Theorem 9.3, we usually see the term "context-free" used instead of "pushdown" in discussing the languages accepted by a pda. The use of nondeterminism is necessary because there may be multiple productions with the same left side. In view of other results, it is surprising that there are context-free languages which cannot be accepted by a deterministic pda. For this reason, context-free languages accepted by deterministic pda's are called *deterministic context-free languages*.

Type 0 and 1 Grammars and Turing Machines

We complete the discussion of the relationship between the two language classification hierarchies with the following theorem.

THEOREM 9.4

The class of recursively enumerable languages is exactly the same as the class of type 0 languages; furthermore, every type 1 language is recursive.　■

The proof of this theorem is not particularly difficult to follow, but does involve considerable use of notation, and the principles involved in the proof are essentially the same as we dealt with in Theorem 9.2. The fact that type 1 languages are recursive follows from the fact that we can keep track of the number of symbols on the tape at any time, (as soon as we have used too many in an attempted parsing of a string, we can stop and reject the string).

The following table summarizes the relationship between the classification schemes from languages.

Machine Classification	Chomsky Hierarchy	Relationship
Recursively enumerable	Type 0	Identical
Recursive	Type 1	Type 1 $\rightarrow$ recursive
Context-free	Type 2	Identical
Regular	Type 3	Identical

Find finite automatons for the grammars in Problems 1 to 4.

1. $\langle Start \rangle :: = x\langle A \rangle | y\langle B \rangle$
$\langle A \rangle :: = z\langle B \rangle | y\langle C \rangle | z$
$\langle B \rangle :: = y\langle A \rangle$
$\langle C \rangle :: = x\langle B \rangle | x$

2. $\langle Start :: = a\langle B \rangle | b\langle C \rangle | b | \Lambda$
$\langle B \rangle :: = b\langle A \rangle | a\langle C \rangle | a$
$\langle C \rangle :: = b\langle C \rangle | a\langle B \rangle | b$

3. $\langle Start \rangle :: = a\langle O \rangle | b\langle E \rangle | a$
$\langle O \rangle :: = a\langle E \rangle | b\langle O \rangle | a$
$\langle E \rangle :: = a\langle O \rangle | b\langle E \rangle | b$

4. $\langle Start \rangle :: = a\langle A \rangle | b\langle B \rangle | a$
$\langle A \rangle :: = a\langle C \rangle | b\langle A \rangle | b$
$\langle B \rangle :: = a\langle B \rangle | b\langle C \rangle$
$\langle C \rangle :: = a\langle A \rangle | b\langle B \rangle | a$

Find type 3 grammars for the finite automatons in Problems 5 to 8.

5. States$(M) = \{q_0, q_1, q_2, q_3\}$
Symbols$(M) = \{a, b\}$
Instructions(M):

$d(q_0, a) = q_2$ $d(q_2, a) = q_2$
$d(q_0, b) = q_1$ $d(q_2, b) = q_1$
$d(q_1, a) = q_2$ $d(q_3, a) = q_3$
$d(q_1, b) = q_3$ $d(q_3, b) = q_0$
Accepting$(M) = \{q_1, q_2\}$

6. States$(N) = \{s_0, s_1, s_2\}$
Symbols$(N) = \{0, 1\}$
Instructions(N):

$d(s_0, 0) = s_0$ $d(s_1, 1) = s_2$
$d(s_0, 1) = s_1$ $d(s_2, 0) = s_2$
$d(s_1, 0) = s_1$ $d(s_2, 1) = s_0$
Accepting$(N) = \{s_0\}$

7. States$(P) = \{q_0, q_1, q_2, q_3\}$
Symbols$(P) = \{a, b\}$
Instructions(P):

$d(q_0, a) = q_1$ $d(q_2, a) = q_1$
$d(q_0, b) = q_0$ $d(q_2, b) = q_3$
$d(q_1, a) = q_2$ $d(q_3, a) = q_3$
$d(q_1, b) = q_0$ $d(q_3, b) = q_3$
Accepting$(P) = \{q_0, q_1, q_2\}$

8. States$(Q) = \{s_0, s_1, s_2\}$
Symbols$(Q) = \{x, y\}$
Instructions(Q):

$d(s_0, x) = s_1$ $d(s_1, y) = s_1$
$d(s_0, y) = s_2$ $d(s_2, x) = s_0$
$d(s_1, x) = s_2$ $d(s_2, y) = s_1$
Accepting$(Q) = \{s_2\}$

*9. "Clean up" the grammar developed for the automaton in Fig. 9.9, so that the grammar is actually a type 3 grammar. *Hint:* (1) Create a new state $\langle Start2 \rangle$ so that any production which has $\langle Start \rangle$ on the right side has $\langle Start2 \rangle$ on the right side instead, and then make the productions for $\langle Start2 \rangle$ identical to those for $\langle Start \rangle$ *except* for $\langle Start \rangle \to \Lambda$. (2) For the erasing productions, such as $\langle S \rangle \to \Lambda$, provide for each production of the form $\langle A \rangle \to X\langle S \rangle$ the additional production $\langle A \rangle \to X$, and remove the erasing productions. (3) Prove that the new grammar produces exactly the same language as the old.)

*10. Our discussion of the relationship between grammars of type 3 and finite automatons assumed that all productions in the grammar were of the form $\langle B \rangle :: = x \langle C \rangle$ where x was a single character. Actually this need not be the case, and we want grammars where that is not the case. Construct a finite automaton to accept the language generated by the following grammar:

$$\langle \text{Start} \rangle :: = b\langle A \rangle \,|\, a\langle B \rangle \,|\, \Lambda$$
$$\langle A \rangle :: = aba\langle C \rangle \,|\, aba$$
$$\langle B \rangle :: = bab\langle C \rangle \,|\, bab$$
$$\langle C \rangle :: = b\langle A \rangle \,|\, a\langle B \rangle$$

(*Hint:* Provide intermediate states so that the entire string *aba* must occur in order to get from the state corresponding to $\langle A \rangle$ to the state corresponding to $\langle C \rangle$ and from $\langle A \rangle$ to the accepting state. Do the same kind of thing for *bab*.)

9.6 THE PUMPING LEMMA AND SOME IMPOSSIBILITIES

The Pumping Lemma

Up to this point, we have studiously avoided the issue in our classification schemes of whether there were languages which belonged to one class and not to another. Except for a discussion in which we made it seem reasonable that a finite automaton could not accept a language which required balanced parentheses, the relevant questions in this area have been avoided. The main reason for this has been a lack of mathematical tools for considering such problems.

In this section we develop some tools for attacking these kinds of problems, and then we apply those tools to produce some results, which illustrate some potential limitations of the computer. The pumping lemma for a class of languages is a theorem which states that if strings of a certain kind belong to a language, then we can produce other members of the language. This can be used to show that a language is infinite, or it can be used to show that some "undesirable" strings also get into a language when some "desirable" strings are accepted. We consider the pumping lemmas for regular and for context-free languages.

Balanced Parentheses and Regular Languages

We introduce the techniques used in these proofs by proving that no regular language can require balanced parentheses. This will verify the statement which we made earlier to that effect. It will also show us that there are nonregular languages—in particular, since most programming languages require that algebraic expressions have balanced parentheses, those languages are *not* regular. The strategy we use in this proof is to exploit the fact that there are only finitely many states in a finite automaton.

THEOREM 9.5

No regular language can require that all its strings have balanced parentheses.

PROOF

Let L be a regular language, and let M be the finite automaton which accepts L. Let $k = |\text{States}(M)|$. If L requires balanced parentheses, then a string of the form $y = (^k x)^k$ must belong to L. If we consider the operation of M on string y, the machine begins in the state $\text{Start}(M)$ and changes to a new state after each left parenthesis is read. After the last left parenthesis has been read, the machine will have been in $k + 1$ states (the first state plus a new state after each left parenthesis). Since M has only k states, some state, say s^*, must have occurred twice in the processing of y.

If we consider the digraph representation of M, there is a cycle in the digraph which occurs between the first time (i) that the machine enters s^* and the second (j), and $j - i > 0$. Since y is accepted by M, there is a path from s^* to an accepting state, and that path requires only $k - j$ parentheses. Thus the string $(^i(^{k-j}x)^k$ must be accepted by M. In this string there are $i + k - j = k - (j - i)$ right parentheses, and this number is less than k. The parentheses are not balanced in this string, but the string is a member of L, and thus L cannot require balanced parentheses. ∎

The technique used here is very similar to the technique that we use in proving the pumping lemma for regular languages. We find a path in the state diagram (a graph, of course) which includes enough vertices that some vertex must occur more than once in the path. This produces a circuit and will thus allow the string represented by this cycle to be included as a substring as many (or as few) times as we would like in a string which is accepted by the automaton.

The Pumping Lemma for Regular Languages

THEOREM 9.6

If L is a regular language, then there is a number k, depending on L, such that if $A \in L$ and $A = xyz$ with $\lambda(y) \geq k$, then y can be written as $y = uvw$ with $\lambda(v) > 0$ and for all n $xuv^n wz \in L$.

PROOF

Let M be a finite automaton which accepts L, and let $k = |\text{States}(M)|$. As M processes A, the string y is eventually entered. Since the length of y is at least

k, the machine will enter k more states as y is processed. These states, plus the state that the machine was in when y was entered, guarantee that at least $k + 1$ states will be used in processing y. Thus some state, say $s*$, must be entered twice. We can write $y = uvw$, where $s*$ is entered following the completion of u and $s*$ is reentered following the processing of v. Strings u and w may be empty, but v is nonempty. The labels in v describe a loop on $s*$. In processing a string, starting at $\text{Start}(M)$ and processing xu bring us to $s*$, and processing wz when starting in $s*$ will lead us to an accepting state. We may include as many copies of the string v as we like, for using a copy of v and starting in $s*$ cause the machine M to return to $s*$. Thus any string of the form $xuv^n wz$ will be accepted by M. ∎

Applications of the Pumping Lemma

The pumping lemma allows us to conclude that if the regular language L has strings of length at least k, then those strings can be both "pumped up" and "pumped down" to produce as many strings in L as we like. This capability turns out to be quite useful, and the theorem also gives us some choice in the matter of determining what portion of the string is to be "pumped."

EXAMPLE 9.7

Use the pumping lemma for regular languages to produce the same result as Theorem 9.5.

SOLUTION
Let k be as in Theorem 9.6. Consider the string $A = {}^{\cdot}(^k C)^k{}^{\cdot}$. Let $x = \Lambda$, $y = (^k$, and $z = C)^k$. By the pumping lemma, y can be written as uvw with v not empty, so for all n, $xuv^n wz \in L$. Let $b = \lambda(v)$. The string $xuwz \in L$. Since y was chosen in such a way as to include only left parentheses, the string $xuwz$ has $k - b$ left parentheses. Since b is not zero, the string $A = {}^{\cdot}(^{k-b} C)^k{}^{\cdot}$ is in L because of the pumping lemma, but is not balanced because $k - b < k$. ∎

The pumping lemma also provides us with a simple test to determine whether a regular language has finitely many strings, infinitely many strings, or no strings at all. We determine the number k and then generate all strings in Σ^* of length less than $2k$. If any are in L, then obviously L is not empty. If any of length k or greater are members of L, then L is infinite. If no strings are found in L, then L is empty; and if the only strings found have length less than k, then L is finite. Much of this is obvious. It is not so obvious that an infinite language must have strings of length between k and $2k$. If we find any string of length greater than $2k$, we can "pump it down" until such time as we find a string of length less than $2k$. The pumping lemma allows the removal of certain substrings of length less than or equal to k. So even if we start with a very long string, by removing enough of it, we will find an element of L of the appropriate length. You are asked to make this precise in the problems at the end of the section.

We should remember that regular languages and type 3 languages are the same. Our algorithm for determining whether a type 3 language is empty, finite, or infinite can use the number of states in a finite automaton for the language to determine the sizes of strings which need to be tested.

The Pumping Lemma for Context-Free Languages

There is also a pumping lemma for context-free languages. It is interesting to note that the proof of the pumping lemma for regular languages made use of the "machine" description of regular languages, while our proof of the pumping lemma for context-free (type 2, pushdown) languages will rely on the grammatical description of such languages. This pumping lemma is quite similar to the one for regular languages, except that we end up pumping a different part of the string.

THEOREM 9.7

Let L be a context-free language. There is a number k, depending on L, such that if $A \in L$ and $|A| > k$, then $A = uvwxy$ with at least one of v and x non-empty and $uv^n wx^n y \in L$ for all n.

PROOF

We need to show that in a derivation for A, the following construction occurs. $\langle Start \rangle \to \cdots u\langle S \rangle y \to \cdots uv\langle S \rangle xy \to \cdots uvwxy$. If this can be shown to occur, then the derivation which begins with $\langle S \rangle$ and ends with $v\langle S \rangle x$ can be repeated as many times as we would like or simply left out (since there is a derivation which begins with $\langle S \rangle$ and ends with w).

We consider a parse tree for A. The parse tree must have exactly $\lambda(A)$ leaves. Let p be the maximum number of productions associated with any nonterminal symbol in the grammar for L. Let t be the number of nonterminal symbols in the grammar for L. We choose k to be p^t. Recall that a full p-ary tree of depth d has p^d leaves. Suppose that $\lambda(A) > k$. The parse tree for A has $\lambda(A)$ leaves. Since this number is greater than k, the depth d of the parse tree for a must be greater than t. If we follow a path in the parse tree from $\langle Start \rangle$ to a leaf at maximum depth, it must pass through $d + 1$ vertices and d of them must be labeled with nonterminal symbols. Since $d > t$, there must be at least two vertices in the path labeled with the same nonterminal symbols.

Figure 9.10 shows what the situation must be like. The path in the parse tree from $\langle Start \rangle$ to $\langle S \rangle$ indicates that a derivation of the form $\langle Start \rangle \to \cdots u\langle S \rangle y$ is possible. If we expand all nonterminal symbols on both sides of $\langle S \rangle$ fully, then u and y can be regarded as terminal strings. The path from $\langle S \rangle$ to $\langle S \rangle$ shows that a derivation of the form $\langle S \rangle \to \cdots v\langle S \rangle x$ is possible, and the final path from $\langle S \rangle$ to a leaf shows that a derivation of the form

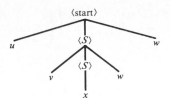

Figure 9.10 Parse tree for *A*.

$\langle S \rangle \rightarrow \cdots w$ is possible. This is exactly the situation we need to prove our theorem. It is necessary to verify that either v or x is nonempty, but that is relatively easy. The details are left to the problems. ∎

If we are more careful in our choice of $\langle S \rangle$ and the occurrences of $\langle S \rangle$ that we select in the parse tree, we can sharpen the result stated above to add that $\lambda(vwx) = k$. We leave the details of these selections as an exercise.

By using the fact that $\lambda(vwx)$ can be made less than or equal to k, an algorithm almost identical to the one described for regular languages can be constructed to determine whether a context-free language is empty, finite, or infinite. The pumping up or down is guaranteed to be done with a string of length less than or equal to k, just as in using the pumping lemma for regular languages.

It is fascinating that the effort involved in constructing this proof actually used almost all the mathematical resources that we have developed in this book.

1. Mathematical logic was needed to develop the ideas involved in the proof.

2. Counting techniques are used in proving the existence of a repeated label in the parse tree.

3. Graph theory techniques were needed to analyze the parse tree.

4. The proof requires an understanding of the concepts of grammars, languages, and the relationship between languages and machines.

Some Impossibilities

At this point, we have developed a very rich background in the theory of formal languages from either a grammatical or a machine point of view. We are now ready to conclude our studies by using that background to prove that some tasks are impossible for a machine to do. In doing this, we use the pumping lemmas and other theorems to show that each category of languages includes languages which do not belong to any smaller category. We also prove that there are languages which are not even type 0 languages, that is, languages which are not recursively enumerable. We begin with a simple illustration.

EXAMPLE 9.8

The language generated by the following grammar is context-free, but not regular:

$\langle\text{Start}\rangle::= \Lambda \mid (\langle B \rangle) \mid BB \mid ()$
$\langle B \rangle::= () \mid (B) \mid BB$

This grammar is certainly a type 2 grammar, and thus the language which it generates is context-free. It is also clear that the language generated by this grammar requires (in fact, it consists of) balanced parentheses. In Theorem 9.5 we proved that no regular language could require balanced parentheses; thus this language is context-free, but not regular.   ∎

Recursive, but Not Context-Free

Our next task is to produce a language which is recursive and not context-free. In doing so we also show that the intersection of context-free languages is not context-free. We let $L = \{s \in \{a, b, c\}^* \mid s = a^n b^n c^n\}$. The languages $M = \{s \in \{a, b, c\}^* \mid s = a^n b^n c^k\}$ and $N = \{s \in \{a, b, c\}^* \mid s = a^j b^n c^n\}$ are context-free, and their intersection is L.

EXAMPLE 9.9

Show that L is recursive but not context-free.

SOLUTION
We describe the operation of a Turing machine which tests a string for membership in L:

1. Test the string to determine whether it is of the form $a^k b^l c^m$ without changing the string on the tape.

2. If the string has the proper format, rewind the tape. If it does not, halt in a rejecting state.

3. For each 'a' found on the tape as the head scans to the right, replace it with a dirty blank; locate one 'b' and one 'c', and replace them with dirty blanks. Repeat this process for each 'a' on the tape. If not enough b's or c's are found, halt in a rejecting state.

4. When no more a's can be found, test the tape to see whether it consists entirely of dirty blanks. If it does, accept the string. If the tape is not empty, halt in a rejecting state.

We next show that L is not context-free by using the pumping lemma. If L is context-free, then there is an integer k as determined by the pumping lemma for L. The string $s = a^k b^k c^k$ must belong to L and has length greater than k. By the pumping lemma, s can be written in the form $s = uvwxy$ where for all n the string $uv^n wx^n y$ is also in L. If v is nonempty (the argument is identical if x is nonempty), then v must consist of only one kind of symbol, since otherwise a pumped string would not be of the form $a^n b^n c^n$. If v consists only of a's and $\lambda(v) = q$, then the string uwy will have only $k - q$ a's in it. The string x can consist also of only one type of symbol, so the string uwy must have either k b's or k c's in it. This string does not have the same number of a's as it does b's and c's, so it cannot belong to L. The only solution is to conclude that L is not context-free. ∎

We have thus shown that it is *impossible* to define a context-free grammar which produces only strings of the form required for L. Any context-free grammar which produces all the strings of L must also produce other strings which are not contained in L.

Our Final Task

Our final task is to show there are languages which are not recursive and not recursively enumerable. We do this by showing that it is impossible to construct machines to do certain tasks. If we accept Church's thesis, then no algorithm can do these tasks.

To do this, we need to make a couple of observations about recursive and recursively enumerable languages.

1. If L is a recursive language on Σ, then $\Sigma^* - L$ is also recursive.

2. If L is a recursively enumerable language on Σ and $\Sigma^* - L$ is recursively enumerable, then L is recursive.

The second fact can be verified in the following way: Construct Turing machines for both languages, and connect them in parallel. For any string, either the machine which accepts L or the machine which accepts $\Sigma^* - L$ must halt. If the string in question belongs to L, it will be accepted by L's machine; otherwise, it will be accepted by the other machine, which we can treat as a rejection. In fact, we can prove (see Problems 14 and 15) that L is recursive if and only if both L and $\Sigma^* - L$ are recursively enumerable.

Thus if we can find a language which is recursively enumerable and whose complement is not, we will have found *both* a language which is recursively enumerable and not recursive and a language (the complement) which is not recursively enumerable.

In Chapter 8 we noted that it was possible to produce a string to represent a directed graph and to define a language in terms of all strings representing directed graphs with certain properties. In a very similar fashion, it is possible to produce a way of coding Turing machines as strings. This can be done in a number of ways. One way is to use an algorithm language for describing Turing machines. In that case, the alphabet of the language is the set of all characters allowed in the language. If we do this, we can also express the input symbols of the machine in the same alphabet. Another way develops a coding scheme for the states and instructions of the machine and its input symbols in such a way as to reduce the coding of the machine to a string of 0s and 1s. The method used is actually not relevant except to note that there is an alphabet Σ such that any Turing machine and its input can be expressed as a string on alphabet Σ. In more advanced textbooks, the process we have described is made precise, but here we simply accept that it can be done. We call String(M) the coding of machine M into a string.

EXAMPLE 9.10

There is a language L_1 which is recursively enumerable and whose complement is not.

We start by defining the language L as $L = \{s \in \Sigma^* \mid s = \text{String}(M)x$ and M accepts string $x\}$. We mentioned in Chapter 8 that it is possible to prove that there is a Turing machine U, called a *universal Turing machine*, which can take as its input a description of a Turing machine in string form concatenated with an input string for the described machine, and given this input, U simulates the action of the described machine on the input. We can use a universal Turing machine to accept the language L, and thus L is recursively enumerable.

The language L will be our tool for constructing L_1. To do this, we flirt with Russell's paradox by attempting to make a machine refuse to accept itself. As with set theory, this will ultimately lead to a contradiction. If L were a recursive language, the language $L_1 = \{s \in \Sigma^* \mid s = \text{String}(M)$ and M accepts String(M)$\}$ would also be a recursive language. If L were recursive, then we could find a machine T such that, given a string A, T would either accept or reject A as a member of L in finite time. To test s for membership in L_1, we must produce the string ss and determine if s represents a machine M that accepts the string s.

T must decide this issue. If L_1 is recursive, then so is $L_2 = \Sigma^* - L_1$, and this is what finally causes the contradiction. For a string to be in L_2, either it is not the coding of a Turing machine or else it is the coding of a Turing machine M which does not accept String(M). Suppose that L_2 is decided by the machine M_2. We then attempt to determine whether $x = \text{String}(M_2) \in L_2$. If x is in L_2, then it follows that M_2 accepts x. Since x is the string representation of a Turing machine, the only way that x can be in L_2 is if x is String(M) and M does not accept x. Thus if x is in L_2, then x is not in L_2. However, if x is not in L_2, then M_2 must reject x; but M_2 rejecting x means that x is in L_2. The only resolution to this contradiction is to conclude that L_2 is not recursive. If L_2 is not recursive, then neither is L_1. L is recursively enumerable, and hence so is L_1. L_2 cannot be recursively enumerable, for if it were, both L_1 and its complement would be recursively enumerable, which would mean that L_1 would be recursive. ∎

At this point, we have achieved our goal—we have found a language which is recursively enumerable, but not recursive, and we have found a language which is not recursively enumerable. This actually completes our task of showing that our classification scheme of languages is not redundant and that for each category there are languages in the larger category which do not belong to the smaller category. This fact is true for *both* classification schemes considered in this book. The Venn diagrams of the language classifications as seen in Fig. 9.11 are correct, and none of the regions in the diagrams represent empty sets.

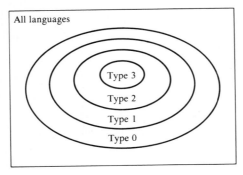

(*a*) By grammar

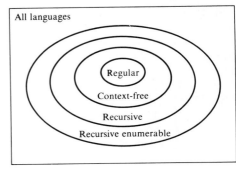

(*b*) By machine

Figure 9.11 The classification of languages.

Turing's Halting Theorem

Actually we can make one more observation before we close. We are now in a position to prove one of the most famous theorems of computer science, the famous *halting theorem*, which is credited to Turing.

To do this, we apply methods similar to the ones used above and show that language $K = \{\text{String}(M)x \mid M \text{ halts on } x\}$ is recursively enumerable, but not recursive. This statement is actually equivalent to the halting theorem, which says that it is not possible to design an algorithm to test whether a given algorithm will halt with given input.

THEOREM 9.8

The language $K = \{\text{String}(M)x \mid M \text{ halts on } x\}$ is recursively enumerable but not recursive.

PROOF

Let $K_1 = \{\text{String}(M) \mid M \text{ halts on String}(M)\}$. As in the previous discussion, it is clear that K_1 is recursively enumerable and is recursive if K is recursive. We let K_2 be the complement of K_1. Language K_1 is recursive if and only if K_2 is recursive. Let M_2 be a Turing machine which decides K_2. We modify M_2 slightly to define M_3. We define M_3 as the Turing machine which will halt and accept a string if it belongs to K_2 but will loop infinitely if the string does not belong to K_2. We then ask if $s = \text{String}(M_3)$ is in K_2. If it is, then M_2 will halt on $\text{String}(M_3)$, and as a consequence so will M_3. This means that M_3 halts on $\text{String}(M_3)$, but then $s = \text{String}(M_3)$ is not in K_2. (Why?) Thus if s is in K_2, it is not in K_2. However, if s is not in K_2, then M_2 rejects s and hence M_3 will not halt on s. This means that s is in K_2. ■

There are some tasks which are not possible for computers to perform. In particular, there are languages which cannot be checked by computer, and it is not possible to determine with an algorithm whether a program will halt. There certainly are some checks which can be made—an algorithm can look for things like the BASIC statement "10 GOTO 10" but there is no computer program which can determine for every program whether it will stop.

In conclusion, we note that computers and algorithms are valuable tools for solving many problems, but not all problems can be solved by computer. It may be impossible to write a program to do what we want. Or, as we discovered in Chapter 8, it may be possible to write the program, but totally impractical to use it.

Problem Set 6

1. Prove that if k is the constant for language L in the pumping lemma for regular languages, then L is infinite if and only if L has a string of length l such that $k \leq l < 2k$.

2. Prove that if k is the constant for language L in the pumping lemma for regular languages, then L is finite if and only if the only strings that belong to L are of length less than k.

3. *Formally* describe the algorithm for determining whether a regular language is finite, infinite, or empty.

*4. Complete the proof of the pumping lemma for context-free languages by showing that at least one of v and x must be nonempty. (*Hint:* If both v and x are empty, what must all productions in the derivation $\langle S \rangle \to \cdots v \langle S \rangle x$ be? "Prune" all such productions from the tree, and show that there still must be a derivation of this form.)

*5. We stated, following the proof of Theorem 9.7, that we can force $\lambda(vwx) \leq k$. Do this by showing that $\langle S \rangle$ can be chosen so that both occurrences of $\langle S \rangle$ are in the last $p + 2$ vertices in the path of length d.

6. Give an algorithm to determine whether a context-free language is finite, infinite, or empty. Prove that your algorithm works.

7. Use the pumping lemma for regular languages to show that $\{s \mid s = a^n b^n\}$ is not a regular language.

8. Use the pumping lemma for regular languages to show that $\{s \mid s = xx^R$ (x is a string on $\{a, b\}$)$\}$ is not a regular language.

9. Use the pumping lemma for regular languages to show that $\{s \mid s = xx^2$ (x is a string on $\{a, b\}$)$\}$ is not a regular language.

10. Use the pumping lemma for context-free languages to show that $\{s \mid s = a^p$ (p is prime number)$\}$ is not a context-free language.

*11. Use the pumping lemma for context-free languages to show that $\{s \mid s = x^3$ (x is a string on $\{a, b\}$)$\}$ is not a context-free language.

*12. Prove formally that Pascal (or BASIC, if you prefer) is not a regular language. (*Hint*: Use the pumping lemma.)

13. Design the Turing machine described in Example 9.9.

14. Let L be a language on Σ. Prove that L is recursive if and only if $\Sigma^* - L$ is recursive.

*15. Let L be a language on Σ. Prove that L is recursive if and only if both L and $\Sigma^* - L$ are recursively enumerable.

*16. Prove that it is impossible for an algorithm to be designed to test all algorithms for correctness (*Hint*: Try the techniques used at the end of the section.)

CHAPTER SUMMARY

In Chapter 9, we completed our discussion of languages by introducing the idea of a language generated by a grammar. We compared the idea of using a grammar to determine a language with the concept of using a mathematical machine to decide whether a given string belongs to a particular language.

We began with a simplified subset of Spanish to illustrate how a grammar can be used to define a language. From this idea we went to the formal definition of a grammar and the use of a grammar to classify language. This classification of language provides us with two different ways of classifying languages, and one of the major tasks of the remainder of the chapter was to show the relationship between the two classification methods.

Finally, given our background in dealing with abstract machines and languages and the rest of our mathematical techniques available to us, we were able to prove two very important facts: (1) Our language classifications are not redundant (i.e., for each class there are languages in the larger class which are not present in the smaller one), and (2) we demonstrated the famous Turing halting theorem, which says that no computer program can check all programs for infinite loops.

Having completed the material in this book, the reader is in a very good position to continue with the study of mathematics and of the mathematical foundations of computer science.

Terms introduced in this chapter include the following:

BNF (9.4)

Chomsky's Hierarchy (9.4)

derivation (9.1)

derivation tree (9.1)

formal language (9.1)

nonterminal symbols (9.1)

phrase structure grammar (9.2)

production (9.2)

pumping lemma for context-free languages (9.6)

pumping lemma for regular languages (9.6)

start symbol (9.2)

terminal symbol (9.1)

Turing's halting theorem (9.6)

unambiguous grammar (9.1)

1

The Algorithm Language

In this book, we used an algorithm language to describe the solution of problems which can be solved by means of algorithms. The language we use is loosely modeled on Pascal, and in a number of cases it is easy to translate algorithms from our algorithm language directly into Pascal code.

Our language is of considerably freer form than Pascal (and most other programming languages), and in some cases we even take liberty with our "advertised" structure if that is useful. In particular, we have allowed ourselves to use statements which are themselves a free-form description of other algorithms. In some such cases (particularly in some of the early graph theory examples) it may not be clear exactly how to implement the subalgorithms described in the language. It is sufficient to note that these " subalgorithms " *can* be made precisely correct if necessary, but doing so could cost us some of the understanding which goes with the simplified description of the algorithm.

Our language does not require declaration of the types of variables, and in many cases we have avoided issues relating to the passing of parameters and results between a calling procedure and the called procedure. Our main goal with this language is to convey the ideas involved in the algorithm. Most of the necessary information related to data types and the passing of parameters is described by the use of comments in the algorithm.

When an algorithm is to return a value, we often use the convention of assigning a value to a variable with the same name as the algorithm. This device is often used in function calls in certain programming languages. Thus if the algorithm VIRGINIA were to return the value 2, our last line before the end of the program would say " VIRGINIA := 2;".

Heading

Each algorithm begins with a header line of the form algorithm NAME(list of parameters). The NAME is the name of the algorithm, and the list of parameters is a list of the names of variables which need to be imported from elsewhere for the algorithm to work. Types of parameters are *not* displayed; the type (if it is not obious from context) is described in the comments.

Comments

Statements used to explain the algorithm, but which are not actually part of the algorithm, are written as comments and enclosed in braces. The comments generally include whatever general explanatory material is appropriate for the algorithm at hand.

Beginning and End of the Algorithm

A **begin** statement is used to mark the beginning of the algorithm. Unlike in Pascal, there is usually only one **begin** for a given algorithm, and that **begin** is matched with **end.**, which marks the end of the algorithm. The period following **end** should be the last symbol in the algorithm.

Form of Statements

Every statement (except for **begin**, **end**, and comments) ends with a semicolon.

ASSIGNMENT

The assignment operation is indicated with the symbol $:=$. Thus the statement "$x := 3$," assigns the value 3 to x. The left side of an assignment must be a variable name—the right-hand side can be any arithmetic expression. The notation for arithmetic expressions is the same as in Pascal, with the addition of an exponentiation operator $\wedge$, which is given the highest precedence.

SELECTION

Selection is effected by two statements: the **if** $\cdots$ **then** $\cdots$ **fi** statement and the **if** $\cdots$ **then** $\cdots$ **else** $\cdots$ **fi** statement.

The **if** $\cdots$ **then** $\cdots$ **fi** statement is of the form

if (condition) **then** (list of statements) **fi**;

Here (condition) is any logical expression which can be evaluated as either true or false. If the condition is true, the list of statements which follows the **then** is done. The **fi** indicates the end of the list of statements. If the condition is false, then the algorithm proceeds to the next statement.

The **if** $\cdots$ **then** $\cdots$ **else** $\cdots$ **fi** statement is similar, except that the **else** indicates the end of the list of statements to be executed when the condition is true. Also there is a list of statements between the **else** and the **fi** which are to be executed when the condition is false.

In either of the two above statements, the **fi** indicates the end of the structure. Some people do not like the use of "if" spelled backward to indicate the end of an **if** construct. But **fi** is used in this language because it is a little easier to see what is being ended with **fi** than with the structure used in Pascal. The proliferation of **begin**'s and **end**'s that can occur in Pascal constructs is somewhat avoided by this scheme. When **fi** is encountered, we know that we just ended an **if**. In Pascal, when **end**; is found, it is not clear what just ended.

REPETITION

Three types of loop structures are provided. The first is the **for** loop. The form of the **for** loop is

for index := start **to** stop **do** (list of statements) **od**

In the **for** loop, the index variable (an integer) is set to start and is incremented by 1 each time the last statement in the list is executed. If the new value of index is greater than stop, the loop is finished and the statement following the **do** is the next to be done. And **od** marks the end of the list of statements. A variation of this is the **for downto** statement. Again the variable begins at start, and the process continues until it reaches stop, but now the program *subtracts* 1 from the index until the value of index is less than stop.

The second of the looping statements is the **while** statement. This statement has the form

while (condition) **do** (list) **od**

In the case of the **while** statement, the statements in the list are executed, and after each pass through the list, condition is checked. If (condition) becomes false, then the loop is finished.

The final loop control statement has the form

repeat (list) **until** (condition)

In the **repeat** ··· **until** statement, the statements in the list are done and the condition is checked, and this process continues until condition becomes true. This means that the list is done at least one time.

Note that any structure can be nested inside any other, and each of the statements above is regarded as being a single statement. (This last comment is a technical one which indicates that in any place where a single statement is appropriate, a structured statement may be placed.)

2

A Pascal Program
to Emulate a Turing Machine

The following Pascal program provides output which is similar in form to the way in which we have shown the status of the tape of a Turing machine in Chapter 8. The only real difference is that the underscore character used to indicate the current position of the tape is printed *following* the current symbol on the tape. The program we list emulates the Turing machine in Example 8.2. It is relatively easy to modify this program to emulate other Turing machines. The main changes needed are in the procedure instruction, in the declaration of the type status, and in the procedure which writes out the name of the current state.

It is relatively easy to write a universal Turing machine program for which the input provides a description of the Turing machine being simulated as well as the string being processed. We have chosen not to follow that route in order to emphasize the fact that there is a quite direct translation from a Turing machine to a computer program.

```
PROGRAM  SIMULATE(INPUT,OUTPUT);

const bounds = 70;
      start = 25;
```

```pascal
type string = array [1..bounds] of char;
     status = (S0, S1, S2, S3, H);
var I, pointer: integer;
    X: string;
    state: status;
```

PROCEDURE INITIALIZE(Var X: String; Var Pointer: Integer; Var State: Status);

```pascal
var I: integer;
begin
     for I := 1 to bounds do X[I] := ' # ';
     pointer := start;
     state := S0;
     writeln('Input initial string one character at a time');
     writeln('Terminate input with the symbol "$"');
     I := start - 1;
     while X[I] <> '$' do
     begin
          write(' ? ');
          I := I + 1;
          readln(X[I])
     end;
     X[I] := ' # '
end;
```

PROCEDURE INSTRUCTION(Var X: string; Var Pointer: Integer; Var State: Status);

```pascal
begin
     case state of
     S0; begin
          pointer := pointer + 1;
          state := S1
          end;
     S1:
          case X[pointer] of
             '0','1': pointer := pointer + 1;
           ' # ': begin
                      pointer := pointer - 1;
                      state := S2
                   end
             end;
     S2: begin
          X[pointer] := ' # ';
          pointer := pointer - 1;
          state := S3
          end;
```

```
        S3: case X[pointer] of
                '0','1': pointer := pointer - 1;
                '#': begin
                        pointer := pointer + 1;
                        state := H
                     end
             end
        end
end;


PROCEDURE OUT(Var X: String; Pointer: Integer; State: Status var I:
integer);

begin
        for I := 1 to bounds do
        begin
            write X[I];
            if I = pointer then write('_');
        end;
        write('/');
        case state of
                S0: write('S0');
                S1: write('S1');
                S2: write('S2');
                S3: write('S3');
                H: write('H')
        end;
        writeln;
end;
begin {Main line code begins here.}
initialize(X,pointer,state);
OUT (X,pointer,state);
while not (state = H) do
    begin
    INSTRUCTION(X,pointer,state);
    OUT(X,pointer,state)
    end
end.
```

Solutions to Selected Problems

CHAPTER 1

Problem Set 1

No solutions given for this problem set.

Problem Set 2

1. (a) Proposition; (b) proposition; (c) not a proposition; (d) not a proposition

5. George Washington was the first president of the United States. True.

7. George Washington was not the first president of the United States. False.

9. Neil Armstrong walked on the moon or IBM makes computers. True.

11. Neil Armstrong did not walk on the moon. False.

13. The capital of Nebraska is Omaha or North Dakota borders on Canada. True.

15. The capital of Nebraska is not Omaha. True.

17. Either Neil Armstrong walked on the moon or IBM makes computers. False.

19.

p	g	if p then q
T	T	T
T	F	F
F	T	T
F	F	T

Problem Set 3

1. A byte has 7 bits and a word is 2 bytes. False.

3. A word is not 2 bytes. False.

5. A byte does not have 7 bits or a word is not 2 bytes. True.

7. $p \wedge q$ **9.** $p \wedge r$ **11.** $\sim(p \vee q)$

13.

p	q	p XOR q
T	T	F
T	F	T
F	T	T
F	F	F

17.

p	q	p NAND q
T	T	F
T	F	T
F	T	T
F	F	T

Problem Set 4

(Our solutions for truth tables list the final results, where the truth table has been set up in the order described in the text.)

1. T F T T   **3.** F T F F T F T F   **5.** Not a tautology.

7. p: The Cubs win the pennant.
q: The Padres win the World Series.
$p \vee \sim q$

9. p: The company needs a mainframe.
q: The company can buy a new mainframe.
r: The company can lease a used computer.
$p \wedge (q \vee r)$

11. p: The train leaves from Seattle.
q: The train stops in Ellensburg.
r: The train stops in Yakima.
$(p \wedge q) \vee (p \wedge r)$

Problem Set 5

1. Construct truth table for each. The final result is T T T F in each case.

3. The final result of each truth table is T T T F F F F F.

5. If p is T, q is F, and r is T, then the first statement will have a truth value of F, and the second statement will have a truth value of T.

7. Each statement has a truth table with final result F F F T.

9. The first statement is $(p \lor q) \land r$; the second is $(p \land r) \lor (q \land r)$.

11. The first statement is $\sim(p \land q)$; the·second is $\sim p \lor \sim q$.

13. The first statement is $p \lor \sim(q \land r)$; the second is $(p \lor \sim q) \lor \sim r$. (This is the pair of statements shown equivalent in Problem 8.)

15. $((x <> 0) \text{ AND } (n <> 7)) \text{ OR } ((x <> 0) \text{ AND } (a > 5))$

17. $((x <> 0) \text{ OR } (n <> 7)) \text{ AND } (a > 5)$

19. $((n = 7) \text{ OR } (a > 5)) \text{ or } (x <> 0)$

Problem Set 6

1. Construct truth tables for each. In each case the final result is T F T T.

3. (a) For a series to be absolutely convergent, a necessary condition is that it be convergent. (b) A sufficient condition for a series to be convergent is that it be absolutely convergent. (c) A series is absolutely convergent only if it is convergent.

5. Hypothesis: Being elected president
Conclusion: Being a politician

7. Hypothesis: Having rich parents
Conclusion: Becoming rich

9. Hypothesis: Being a good student
Conclusion: Getting all A's

11. Hypothesis: The program runs.
Conclusion: There are no typing errors.

13. T F F T F T T F   15. The statement is a tautology.

17. Inverse: If the weather is not cold, then it will not snow.
Converse: If it will snow, then the weather is cold.
Contrapositive: If it will not snow, then the weather is not cold.

19. Inverse: If you practice, then you will learn how to play your horn.
Converse: If you don't learn how to play your horn, then you did not practice.
Contrapositive: If you learn how to play your horn, then you practiced.

21. The statement $[(p \land q) \lor r] \leftrightarrow (p \lor r) \land (q \lor r)$ is a tautology.

23. The statement $(p \leftrightarrow q) \leftrightarrow [(p \land q) \lor (\sim p \land \sim q)]$ is a tautology.

Problem Set 7

1. The statement $[p \land (p \rightarrow q)] \rightarrow q$ is a tautology.

3. The statement $p \rightarrow (p \lor q)$ is a tautology.

5. $(p \land q) \rightarrow (p \land q)$ is obviously a tautology.

7. The statement $q \rightarrow (p \rightarrow q)$ is a tautology.

9. The statement $(p \leftrightarrow q) \rightarrow [(p \rightarrow q) \land (q \rightarrow p)]$ is a tautology.

19. The argument is of the form $p \rightarrow \sim q, q \vdash p$, which is valid.

21. The argument is of the form $p \rightarrow \sim q, q \vdash p$, which is valid.

23. The argument is of the form $(p \wedge q) \to r, r \vdash p$, which is not valid.

25. The argument is of the form $(p \vee \sim q), (\sim q \to r), p \vdash \sim r$, which is not valid.

Problem Set 8

1. $|3.01 - 3| < 0.02$ **3.** $5^2 = 9$ **5.** If $|x - 3| < d$, then $|x^2 - 9| < d$.

7. Mike is taller than Sam, and Sam weights more than 200 pounds.

9. If y is taller than x, then x is not more than 200 pounds.

11. For all x and y, if $|x - y| < 0.01$, then $|x - 3| < 3$.

13. For all x, $|x - 3| < 0.01$ and $|y - x| < 0.01$.

15. For some x, both $x^2 = 9$ and $|x - 3| < y$ for all y.

17. $P(x): x$ pays no taxes. The universe of x is all people who make more than $100,000. The statement given is $\exists x P(x)$.

19. $P(x, y): x$ is faster than y. The universe of x is the collection of all cars, and the universe of y is the collection of all horses. The statement given is $\forall x\ \exists y P(x, y)$

21. $\exists x\ \exists y(S(x, y) \wedge \sim P(x, 3))$ **23.** $\exists x(\sim P(x, 0.01) \vee S(y, x))$

25. $\forall x(\sim R(x) \vee \exists y \sim P(x, y))$

27. All people in the $100,000 income class pay some taxes.

29. Some cars are slower than all horses.

31. The problem has solutions, and for any x there is a y such that y is a solution and x is less than y.

Problem Set 9

1. The form of the argument is $\forall x(C(x) \to I(x))$, $C(a) \vdash I(a)$. This argument is valid.

3. The form of the argument is $\exists x(C(x) \wedge X(x))$, $\forall x(X(x) \to S(x)) \vdash \exists x(C(x) \wedge S(x))$. This argument is valid.

5. The form of the argument is $I(2)$, $\sim R(2^{0.5}) \vdash (I(x) \wedge \sim R(x^{0.5}))$. This argument is valid.

CHAPTER 2

Problem Set 2

(There is no problem set associated with Section 2.1.)

1. $\{x \mid x$ is a letter of the alphabet and x is a vowel$\}$.

3. $\{x \mid x$ is an odd positive integer$\}$

5. $\{x \mid x$ was a president of the United States$\}$

7. $\{3, 6, 9, 12\}$ **9.** $\{\ldots, -15, -10, -5, 0, 5, 10, 15, \ldots\}$

11. $\{6, 12\}$ **13.** $\{-3, 2, 3, 4, 6, 8, 10, \ldots\}$

15. $\{b, c, d, f, g, h, j, k, l, m, n, p, q, r, s, t, u, w, x, y, z\}$

17. $|A| = 5$, C is infinite, $|E| = 39$, $|G| = 4$, I is infinite, $|K| = 2$, M is infinite, $|O| = 21$

19. $\{1, 2, 3, 4\}$ **21.** 0

23. $A \subset B$, $B \subset A$, $D \subset C$, $D \subset E$, $E \subset D$, all sets are subsets of U.

25. A and D are disjoint. A and E are disjoint.

Problem Set 3

1. (a) $\{a, b, c, d, e, f, g\}$ (b) $\{d, e\}$ (c) $\{a, c, e, f, g, h, i\}$ (d) $\{e, g, i\}$
    (e) $\{a, b, d, e, f, h, i\}$ (f) 0   (g) $\{a, b, c, d, e, f, g, h, i\}$ (h) $\{e\}$

3. (a) $\{d, e\}$ (b) $\{a, b, c, h, i\}$ (c) $\{a, b, c\}$
    (d) $\{f, g\}$ (e) $\{c, g\}$ (f) $\{c, g\}$

5. (a) $\{x \mid (-11 < x \leq 0) \lor (3 \leq x < 11)\}$
    (b) $\{x \mid (-11 < x \leq 2) \lor (6 \leq x < 11)\}$
    (c) $\{x \mid 0 < x \leq 2\}$
    (d) $\{x \mid 3 \leq x < 6\}$
    (e) $\{x \mid (-11 < x \leq 0) \lor (10 \leq x < 11)\}$
    (f) Same answer as (e)

7. (a) The set of all numbers which are multiples of 12;   (b) 12
    (c) $\{4, 6, 8, 12, 16, 18, 20, 24, 28, 30, 32, \ldots\}$
    (d) $\{1, 2, 3, 5, 7, 9, 10, 11, 13, 14, 15, 17, 19, 21, 22, 23, \ldots\}$
    (e) $\{6, 18, 30, 42, \ldots\}$

Problem Set 4

21. Region 2 is $A \cap B' \cap C'$ Region 6 is $A \cap B \cap C$
    Region 3 is $A \cap B \cap C'$ Region 7 is $A' \cap B \cap C$
    Region 4 is $A' \cap B \cap C'$ Region 8 is $A' \cap B' \cap C$
    Region 5 is $A \cap B' \cap C$

Problem Set 5

1. (a) $\{1, a), (1, b), (1, c), (2, a), (2, b), (2, c)\}$
    (b)

3. (a) $((1, 3), 5)$; (b) $\{\{1, \{1, 3\}\}, \{\{1, \{1, 3\}\}, 5\}\}$

7. Reflexive, symmetric, transitive, equivalence relation

9. Reflexive, symmetric   **11.** Symmetric

13. Reflexive, transitive   **15.** Symmetric

17. Reflexive, symmetric, transitive, equivalence relation

21. None of the relations are functions.

23. (a) Asymmetric, antisymmetric, transitive; (b) none; (c) symmetric; (d) symmetric

25. The equivalence classes are $\{1, 2, 3\}$ and $\{4\}$.

27. $R = \{(1, 1), (2, 2), (1, 2), (2, 1), (3, 3), (4, 4), (5, 5), (4, 5), (5, 4)\}$

Problem Set 6

1. (a) 0; (b) 1; (c) 39   **3.** (a) 175; (b) 115   **5.** 19   **7.** 22 days

9. (a) 3; (b) 0; (c) 9; (d) 9; (e) 16   **11.** (a) 5; (b) 10; (c) 8; (d) 6

13. $|A| + |B| + |C| + |D| - |A \cap B| - |A \cap C| - |A \cap D| - |B \cap C| - |B \cap D| - |C \cap D| + A \cap B \cap C| + |A \cap B \cap D| + |A \cap C \cap D| + |B \cap C \cap D| - |A \cap B \cap C \cap D|$

15. (a) 49; (b) 33; (c) 19; (d) 27   **17.** 23

Problem Set 7

1. 15   **3.** (a) 260,000; (b) 131,040; (c) 25,200; (d) 234,800
5. 120   **7.** (a) 55,440; (b) 35,280; (c) 18,144; (d) 12,432
9. 60; 18   **11.** 24

Problem Set 8

1. $\{\{a, b, c\}, \{a, b\}, \{a, c\}, \{b, c\}, \{a\}, \{b\}, \{c\}, 0\}$ **3.** 4096

5. (a) 90; (b) 1,814,400; (c) 210; (d) 840   **7.** (a) 45; (b) 45; (c) 35; (d) 35

13. 1225   **15.** 280   **17.** 23,760; 380,160   **19.** 5040; 3600

21. (a) 5148; (b) 624; (c) 54,921; (d) 3744; (e) 123,552

23. (a) 4653; (b) 12; (c) 2868; (d) 125,484; (e) 5976; (f) 106,920

25. (a) 154; (b) 616

CHAPTER 3

Problem Set 1

7. $\overline{(x + y)}$ **9.** $\overline{(x + y)z}$

11. $(p \wedge \sim q \wedge r) \vee (p \wedge \sim q \wedge \sim r) \vee (p \wedge q \wedge r) \wedge (p \wedge q \wedge \sim r) \vee (\sim p \wedge q \wedge r)$
$\vee (\sim p \wedge q \wedge \sim r)$

Problem Set 2

1. (a) $(ab)(ac) + a + c'$ (b) $(a + b' + c')(a + b' + c)$
(c) $(ab + c)' + ab$ (d) $a + a'b$

3. (a) $a(a + b) = a$ (b) $a(b'a + b)' = 0$
(c) $(a + b')b = ab$ (d) $(a + 0) + (1a') = 1$

Problem Set 3

No answers given.

Problem Set 4

3.

x	y	z	$f(x, y, z)$
0	0	0	0
0	0	1	0
0	1	0	1
0	1	1	0
1	0	0	1
1	0	1	0
1	1	0	1
1	1	1	0

5.

x	y	z	$h(x, y, z)$
0	0	0	0
0	0	1	0
0	1	0	1
0	1	1	1
1	0	0	0
1	0	1	0
1	1	0	1
1	1	1	1

7.

x	y	z	$k(x, y, z)$
0	0	0	1
0	0	1	1
0	1	0	0
0	1	1	0
1	0	0	1
1	0	1	1
1	1	0	0
1	1	1	1

9. $x'yz' + xy'z' + xyz'$ **11.** $x'yz' + x'yz + xyz' + xyz$

13. $x'y'z' + x'y'z + xy'z' + xy'z + xyz$

Problem Set 5

1. $(p \wedge q \wedge r) \vee (p \wedge \sim q \wedge r) \vee (\sim p \wedge q \wedge r)$

3. $(p \wedge q \wedge r) \vee (p \wedge q \wedge \sim r) \vee (\sim p \wedge q \wedge r) \vee (\sim p \wedge q \wedge \sim r)$

5. $(A \cap B \cap C) \cup (A' \cap B \cap C)$

7. $(A \cap B \cap C') \cup (A \cap B \cap C) \cup (A \cap B' \cap C)$

9. The first circuit is $(a + b)c$. The second is $ac + bc$. They are equivalent by the distributive property.

11. $(((p \mid p) \mid (q \mid q)) \mid ((p \mid p) \mid (q \mid))) \mid ((r \mid r) \mid (r \mid r))$

13. $(((P \# Q) \# (P \# Q)) \# (R \# R)) \# (((P \# Q) \# (P \# Q)) \# (R \# R))$

17. $(((p$ NOR $q)$ NOR $(p$ NOR $q))$ NOR $(r$ NOR $r))$ NOR $(((p$ NOR $q)$ NOR $(p$ NOR $q))$ NOR $(r$ NOR $r))$

Problem Set 6

1. $xy + x'y'$

3. $xyz + xyz' + x'yz' + x'y'z' + xy'z + x'y'z$

9. xy' **11.** $xy + x'z'$ **13.** $xz + y'$ or $z + y'z'$

15. $w'z' + xzw + x'yw'$ **17.** $xz + zw + x'yw + x'y'z'w'$

CHAPTER 4

Problem Set 2

(There is no problem set associated with Section 4.1)

1. $(5 \quad -7 \quad -3)$ **3.** $(4 \quad -14 \quad 2)$ **5.** $(-1 \quad -7 \quad 5)$

7. $(-6 \quad -21 \quad 19)$ **9.** $(8 \quad 11 \quad -50)$ **11.** $\begin{bmatrix} 2 & -1 & 6 \\ 6 & 6 & -2 \end{bmatrix}$

13. Cannot be computed: dimensions do not match.

15. $\begin{bmatrix} -1 & -7 & -3 \\ 12 & -13 & 11 \end{bmatrix}$

17. $\begin{bmatrix} 2 \\ -7 \\ 1 \end{bmatrix}$ **19.** $\begin{bmatrix} 2 & 0 \\ 2 & 10 \\ 6 & -6 \end{bmatrix}$ **21.** $\begin{bmatrix} 2 & 5 \\ -1 & 8 \\ 4 & -3 \end{bmatrix}$

Problem Set 3

1. $\begin{bmatrix} 9 \\ 5 \end{bmatrix}$

3. Not possible   **5.** (8   0   8   −14)   **7.** 27; 12

9. (a) 190 input; 630 CPU; 910 output; (b) program 1 costs $460; program 2 costs $6020; (c) $7400.

Problem Set 4

1. $\begin{bmatrix} 1 & 1 & 3 \\ 5 & 3 & 3 \\ -1 & -1 & 2 \end{bmatrix}$ **3.** $\begin{bmatrix} 1 & 0 & 1 \\ 1 & 2 & 1 \\ 0 & -1 & 1 \end{bmatrix}$

5. $\begin{bmatrix} 3 & 1 & 0 & 2 \\ 7 & 3 & 2 & 4 \\ -2 & 1 & 0 & 1 \end{bmatrix}$ **7.** $\begin{bmatrix} 4 & -7 & 9 \\ 3 & 3 & 7 \\ 1 & 6 & 2 \end{bmatrix}$

9. $\begin{bmatrix} -0.5 & 1.5 & 2 \\ 4.5 & 0.5 & -2 \\ -2 & 1 & 3 \end{bmatrix}$ **11.** $\begin{bmatrix} 2 & 1 & 3 \\ 5 & 4 & 3 \\ -1 & -1 & 3 \end{bmatrix}$

13. $\begin{bmatrix} 2 & 1 & 3 \\ 5 & 4 & 3 \\ -1 & -1 & 3 \end{bmatrix}$ **15.** $\begin{bmatrix} 1 & 5 & -1 \\ 1 & 3 & -1 \\ 3 & 3 & 2 \end{bmatrix}$

CHAPTER 5

Problem Set 2

(There is no problem set associated with Section 5.1.)

1. (a) $G = (V, E)$ where $V = \{a, b, c, d\}$ and $E = \{\{a, b\}, \{a, c\}, \{b, c\}, \{b, e\}, \{c, d\}, \{c, e\}\}$
(b) $\deg(a) = 2$; $\deg(b) = 3$; $\deg(c) = 3$; $\deg(d) = 1$; $\deg(e) = 3$
(c) There are six edges.
(d) The sum of the degrees is $2 + 3 + 3 + 1 + 3 = 12$.

3. (a) $G = (V, E)$ where $V = \{a, b, c, d, e, f\}$ and $E = \{\{a, b\}, \{a, c\}, \{a, e\}, \{a, f\}, \{b, c\}, \{b, d\}, \{b, f\}, \{c, d\}, \{d, e\}, \{d, f\}\}$
(b) $\deg(a) = 4$; $\deg(b) = 4$; $\deg(c) = 3$; $\deg(d) = 4$ $\deg(e) = 2$; $\deg(f) = 3$
(c) There are 10 edges.
(d) The sum of the degrees of the vertices is $4 + 4 + 3 + 4 + 2 + 3 = 20$.

5. (a) *abcf*; (b) *adefbef*; (c) *abebbf*   **7.** (a) *abdf*; (b) none; (c) *abbacdf*

9. (a) *eadebce*; (b) *aeabde*; (c) *abea*   **11.** (a) *abdcefdba*; (b) *cdbacefdc*; (c) *abdca*

13. $V = \{A, B, C, F\}$; $E = \{\{A, B\}, \{B, C\}, \{B, F\}, \{C, F\}\}$

17. (a) The first and third represent graphs. (b) Only the first is a connected graph. (c) The components of the third are the subgraphs determined by $\{a, c, e\}$ and $\{b, d\}$.

Problem Set 3

1. $v = 12, e = 18, r = 8$; $v = 10, e = 13, r = 5$; $v = 4, e = 6, r = 4$

3. (First graph) $8 \leq (2/3)18$; (second graph) $5 \leq (2/3)13$; (third graph) $4 \leq (2/3)6$

7. The graphs cannot be isomorphic, because one of them has a vertex of degree 4 and the other has all of its vertices with degree 3 or less.

9. $a \leftrightarrow c \qquad e \leftrightarrow e \qquad d \leftrightarrow d \qquad c \leftrightarrow b \qquad b \leftrightarrow a$

11. Remove vertices f and c from the first graph and remove b and f from the second. The resulting graphs are isomorphic.

13. If a vertex of degree 2 is added to the edge $\{a, f\}$ in the second graph, the two graphs will clearly be isomorphic.

15. The graph is planar.

17. The gaph is not planar. The subgraph determined by the vertices a, c, d, e, f is K_5.

Problem Set 4

1. (a) 10^4; (b) 3; (c) 30,000   **3.** (a) 10; (b) 300,000   **5.** (a) 6; (b) 3; (c) 5; (d) 1; (e) 6

7. The edges of one spanning tree are $\{a, b\}, \{b, d\}, \{d, c\}, \{d, e\}, \{e, f\}, \{f, g\}, \{f, h\},$ $\{h, j\}, \{j, i\}, \{i, k\}$. (Other answers are possible.)

9. Using the tree from our answer to Problem 7:

$\{\{a, b\}, \{a, d\}\}$	$\{\{d, e\}, \{c, e\}\}$	$\{\{h, j\}, \{g, i\}\}$
$\{\{b, d\}, \{a, d\}\}$	$\{\{e, f\}\}$	$\{\{i, j\}, \{g, i\}\}$
$\{\{c, d\}, \{c, e\}\}$	$\{\{f, g\}, \{g, h\}, \{g, i\}\}$	$\{\{i, k\}\}$

11. Adding $\{a, d\}$ to the tree from Problem 7 produces the fundamental cycle $abda$. $\{a,d\}$ is an edge only in the fundamental cut sets $\{\{a, b\}, \{a, d\}\}$ and $\{\{b, d\}, \{a, d\}\}$, which are the cut sets determined by the edges $\{a, b\}$ and $\{b, d\}$.

Problem Set 5

1.

	1	2	3	4	5	6	7	8	9	10
a	1	0	0	0	0	0	0	0	0	0
b	1	1	1	0	0	0	0	0	0	0
c	0	1	0	1	1	0	0	0	0	0
d	0	0	1	1	0	1	1	0	0	0
e	0	0	0	0	1	1	0	1	1	0
f	0	0	0	0	0	0	1	1	0	1
g	0	0	0	0	0	0	0	0	1	1

3. $\deg(1) = 3; \deg(2) = 2; \deg(3) = 2; \deg(4) = 2; \deg(5) = 3$

5.
$$\begin{bmatrix} 0 & 1 & 0 & 0 & 0 & 0 & 0 \\ 1 & 0 & 1 & 1 & 0 & 0 & 0 \\ 0 & 1 & 0 & 1 & 1 & 0 & 0 \\ 0 & 1 & 1 & 0 & 1 & 1 & 0 \\ 0 & 0 & 1 & 1 & 0 & 1 & 1 \\ 0 & 0 & 0 & 1 & 1 & 0 & 1 \\ 0 & 0 & 0 & 0 & 1 & 1 & 0 \end{bmatrix}$$

7.
$$\begin{bmatrix} 0 & 0 & 1 & 1 & 1 \\ 0 & 0 & 0 & 1 & 1 \\ 1 & 0 & 0 & 0 & 1 \\ 1 & 1 & 0 & 0 & 0 \\ 1 & 1 & 1 & 0 & 0 \end{bmatrix}$$

9. (a) $\deg(1) = 2; \deg(2) = 2; \deg(3) = 1; \deg(4) = 4; \deg(5) = 6$
$\deg(6) = 3; \deg(7) = 1; \deg(8) = 2; \deg(9) = 1$

(c)
	a	b	c	d	e	f	g	h	i	j	k
1	1	1	0	0	0	0	0	0	0	0	0
2	0	0	1	1	0	0	0	0	0	0	0
3	0	0	0	0	1	0	0	0	0	0	0
4	0	1	0	0	0	1	1	1	0	0	0
5	1	0	1	0	0	1	0	0	1	1	1
6	0	0	0	1	1	0	0	0	1	0	0
7	0	0	0	0	0	0	1	0	0	0	0
8	0	0	0	0	0	0	0	1	0	1	0
9	0	0	0	0	0	0	0	0	0	0	1

11.
$$M = \begin{bmatrix} 0 & 1 & 0 & 0 & 0 \\ 1 & 0 & 1 & 0 & 0 \\ 0 & 1 & 0 & 0 & 0 \\ 0 & 0 & 0 & 0 & 1 \\ 0 & 0 & 0 & 1 & 0 \end{bmatrix}$$
$$M^2 = \begin{bmatrix} 1 & 0 & 1 & 0 & 0 \\ 0 & 2 & 0 & 0 & 0 \\ 1 & 0 & 1 & 0 & 0 \\ 0 & 0 & 0 & 1 & 0 \\ 0 & 0 & 0 & 0 & 1 \end{bmatrix}$$

$$M^3 = \begin{bmatrix} 0 & 2 & 0 & 0 & 0 \\ 2 & 0 & 2 & 0 & 0 \\ 0 & 2 & 0 & 0 & 0 \\ 0 & 0 & 0 & 0 & 1 \\ 0 & 0 & 0 & 1 & 0 \end{bmatrix}$$
$$M^4 = \begin{bmatrix} 2 & 0 & 2 & 0 & 0 \\ 0 & 4 & 0 & 0 & 0 \\ 2 & 0 & 2 & 0 & 0 \\ 0 & 0 & 0 & 1 & 0 \\ 0 & 0 & 0 & 0 & 1 \end{bmatrix}$$

Sum of
the matrices:
$$\begin{bmatrix} 3 & 3 & 3 & 0 & 0 \\ 3 & 6 & 3 & 0 & 0 \\ 3 & 3 & 3 & 0 & 0 \\ 0 & 0 & 0 & 2 & 2 \\ 0 & 0 & 0 & 2 & 2 \end{bmatrix}$$
Since row 1, column 5, of the sum is 0, no path exists between vertex 1 and vertex 5.

13. The sum of the first four powers of the adjacency matrix is

$$\begin{bmatrix} 8 & 5 & 5 & 6 & 6 \\ 5 & 8 & 6 & 5 & 6 \\ 5 & 6 & 8 & 6 & 5 \\ 6 & 5 & 6 & 8 & 5 \\ 6 & 6 & 5 & 5 & 8 \end{bmatrix}$$
The sum has no zeros, and thus the graph is connected.

SOLUTIONS TO
SELECTED PROBLEMS

Problem Set 6

1. No eulearan path. Vertices a, d, f, h, e, and i all have odd degree.

3. *mihjkhlikljm* is an eulearan path.

5. *adgihebcf* is a hamiltonian path.

7. *abcdhefg* is a hamiltonian path.

9. 3   **11.** 4   **13.** 3

21. If the vertices are ordered as b, d, g, h, c, f, i, j, a, e, the Welch–Powell algorithm yields a coloring with three colors. Vertices b, d, f, and i will have color 1, g, c, j, a, e will have color 2, and h will have color 3.

CHAPTER 6

Problem Set 1

3. Several answers are possible. The simplest is to go on A Street to 2nd Avenue, turn right, and then turn right onto D Street.

9. Yes. Going down from above floor 5 to floor 5 or below, go up the escalator to reach the 10th floor, and take the elevator down to 1. It is then possible to ride the escalator to the desired floor.

11. (a) Two paths exists, each taking 10 minutes. The first is to ride the escalator up to the 10th floor, walk to the elevator, ride the elevator down to the 1st floor, walk to the escalators, and ride the escalators up to the 5th floor. The second is to reach the 1st floor as above, but ride the elevator to the 5th floor and walk to the escalators. (b) Results are similar to (a), except that the escalator ride to the 10th floor is eliminated.

13. Yes. To get from below floor 4 to a floor above 4, one can go down to floor 1 and take the elevator to floor 5. From 5, any point in the building can clearly be reached.

Problem Set 2

1. (b) $\begin{bmatrix} 0 & 1 & 1 & 1 & 0 \\ 1 & 0 & 1 & 0 & 0 \\ 0 & 0 & 0 & 0 & 0 \\ 0 & 0 & 0 & 0 & 1 \\ 0 & 0 & 0 & 0 & 0 \end{bmatrix}$

(c) Indegrees $a : 1$, $b : 1$, $c : 2$, $d : 1$, $e : 1$.
   Outdegrees $a : 3$, $b : 2$, $c : 0$, $d : 1$, $e : 0$.
(d) *ade*

3. (b) $\begin{bmatrix} 0 & 0 & 1 & 0 & 0 & 0 & 0 \\ 0 & 0 & 0 & 1 & 0 & 0 & 0 \\ 0 & 0 & 0 & 0 & 1 & 0 & 0 \\ 0 & 0 & 0 & 0 & 0 & 1 & 0 \\ 0 & 0 & 0 & 1 & 0 & 0 & 0 \\ 0 & 0 & 0 & 0 & 0 & 0 & 0 \end{bmatrix}$

(c) Indegrees $1:0, 2:0, 3:1, 4:2, 5:1, 6:1, 7:0$.
   Outdegrees $1:1, 2:1, 3:1, 4:1, 5:0, 6:1, 7:0$.
(d) There is no path from 1 to 6.

5. $\begin{bmatrix} 0 & 0 & 0 & 0 & 0 \\ 1 & 0 & 0 & 0 & 1 \\ 0 & 1 & 0 & 0 & 1 \\ 1 & 0 & 0 & 0 & 0 \\ 0 & 0 & 0 & 0 & 0 \end{bmatrix}$ **7.** $\begin{bmatrix} 0 & 1 & 1 & 0 \\ 0 & 0 & 1 & 1 \\ 0 & 0 & 0 & 0 \\ 0 & 1 & 0 & 0 \end{bmatrix}$

9. The lowest cost is \$260, which is found by going from Boston to Chicago to Kansas City.

15. (a) Accepted; (b) not accepted; (c) accepted; (d) not accepted; (e) not accepted.

Problem Set 3

1. The shortest path is *ADGZ*, and its length is 7.

3. The shortest path is *ADFZ*, and its length is 7.

5. The shortest path is *ADGZ*, and its length is 5.

Problem Set 4

1. The sum of the first three powers of $M \vee M^T$ is

$\begin{bmatrix} 2 & 5 & 2 & 5 \\ 5 & 2 & 5 & 2 \\ 2 & 5 & 2 & 5 \\ 5 & 2 & 5 & 2 \end{bmatrix}$ Since this has no zeros, the graph is connected.

3. The sum of the first five powers of $M \vee M^T$ is

$\begin{bmatrix} 3 & 7 & 3 & 0 & 0 & 0 \\ 7 & 6 & 7 & 0 & 0 & 0 \\ 3 & 7 & 3 & 0 & 0 & 0 \\ 0 & 0 & 0 & 20 & 21 & 21 \\ 0 & 0 & 0 & 21 & 20 & 21 \\ 0 & 0 & 0 & 21 & 21 & 20 \end{bmatrix}$ Since this matrix has zeros in nondiagonal positions, the graph is not connected.

5. The final matrix from Warshall's algorithm is

$\begin{bmatrix} 1 & 1 & 1 & 0 & 0 & 0 \\ 0 & 1 & 1 & 0 & 0 & 0 \\ 0 & 0 & 1 & 0 & 0 & 0 \\ 0 & 0 & 0 & 1 & 1 & 1 \\ 0 & 0 & 0 & 0 & 1 & 1 \\ 0 & 0 & 0 & 0 & 0 & 1 \end{bmatrix}$ The final matrix includes zero entries; so the graph is not strongly connected. The 1s in the matrix indicate paths from vertex 1 to 2 and 3, from 2 to 3, from 4 to 5 and 6, and from 5 to 6.

7. The final matrix is

$$\begin{bmatrix} 1 & 1 & 1 & 0 \\ 0 & 1 & 1 & 0 \\ 0 & 0 & 1 & 0 \\ 1 & 1 & 1 & 1 \end{bmatrix}$$

All vertices can be reached from vertex 4, but vertex 4 cannot be reached from any other vertex. Vertex 3 can be reached from all other vertices, but no vertices can be reached from vertex 3. It is also impossible to get from vertex 2 to vertex 1.

9. The graph is not connected. The algorithm shows no path possible from vertex 1 to either 7 or 8.

11. The graph is not connected. The algorithm shows no path possible from vertex 1 to any of the vertices 2, 4, 6, 7, or 8.

15. Vertex i and j will be in the same component if and only if the final result has a 1 in row i, column j.

Problem Set 5

1. 6   **3.** 24   **5.** 9

9. $A:0$ (the root); $B:1$; $C:2$; $D:1.1$; $E:1.2$; $F:1.1.1$; $G:1.2.1$; $H:1.2.2$; $I:1.1.1.1$; $J:1.1.1.2$

17. The root

19. The left descendant of a vertex will be $3n-1$. The middle descendant will be $3n$. The right descendant will be $3n+1$.
$T[1]=a$; $T[2]=b$; $T[3]=c$; $T[4]=d$; $T[5]=e$; $T[6]=f$; $T[7]=g$; $T[11]=j$; $T[12]=k$; $T[14]=h$; $T[15]=i$; $T[35]=l$; $T[36]=n$; $T[37]=o$; $T[104]=m$; $T[110]=p$; $T[111]=q$

The remaining entries in the array will be @. The minimum size of such an array is 10, and the maximum is 88,573.

Problem Set 6

7. (a) The tree is not balanced. (b) (i) 2; (ii) 4; (iii) 5; (iv) 6; (v) 1

9. (a) The tree is balanced. (b) (i) 1; (ii) 3; (iii) 2; (iv) 4.

11. One possible arrangement is as follows:

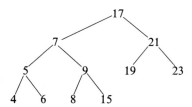

(a) This tree is balanced.
(b) (i) 3; (ii) 3; (iii) 3; (iv) 4

15. The almost full tree requires the least amount of storage.

Problem Set 7

5. Preorder: $ABDEFCG$
Postorder: $EFDBGCA$
Inorder: $EDFBAGC$

7. Preorder: $ABGIHJCDGEF$
Postorder: $IGJHBDEFGCA$
Inorder: $IGBHJADCEGF$

9. Result is 20.

11. Result is -0.04.

13. (a) $* + 2 \quad 3 \quad 4$ (b) $+ / - 2 \quad 4 + + 3 \quad 5 \quad 2$
    (c) $/ 1 - 2 * 3 + 4 \quad 5$ (d) $+ 1 * + 3 \quad -4 - 2 * 3 - 4 \quad 1$

17. The value is -3.

19. The value is -5.

Problem Set 8

5. Relations in Problems 1 and 3 describe functions. Relations in Problems 2 and 4 do not.

7. 1: domain is $\{1, 2, 3, 4\}$; range $\{3, 4, 5, 6\}$.   2: domain is A; range is A.
   3: domain is A; range is $\{a\}$.   4: domain is A; range is B.

10. 1:
$$\begin{bmatrix} 0 & 0 & 1 & 0 & 1 & 0 \\ 0 & 0 & 0 & 1 & 0 & 1 \\ 0 & 0 & 0 & 0 & 1 & 0 \\ 0 & 0 & 0 & 0 & 0 & 1 \\ 0 & 0 & 0 & 0 & 0 & 0 \end{bmatrix}$$
2:
$$\begin{bmatrix} 1 & 1 & 1 & 1 & 1 & 1 \\ 0 & 1 & 0 & 1 & 0 & 1 \\ 0 & 0 & 1 & 0 & 0 & 1 \\ 0 & 0 & 0 & 0 & 1 & 0 \\ 0 & 0 & 0 & 0 & 0 & 1 \end{bmatrix}$$

11. (We supply the adjacency matrices for the given digraphs; you should draw a representation as well.)

Reflexive closure
$$\begin{bmatrix} 1 & 0 & 1 \\ 0 & 1 & 1 \\ 0 & 1 & 1 \end{bmatrix}$$
Symmetric closure
$$\begin{bmatrix} 1 & 0 & 1 \\ 0 & 0 & 1 \\ 1 & 1 & 0 \end{bmatrix}$$
Transitive closure
$$\begin{bmatrix} 1 & 1 & 1 \\ 0 & 1 & 1 \\ 0 & 1 & 1 \end{bmatrix}$$

13. Symmetric closure
$$\begin{bmatrix} 0 & 1 & 0 & 0 & 0 \\ 1 & 0 & 0 & 0 & 0 \\ 0 & 0 & 0 & 1 & 0 \\ 0 & 0 & 1 & 0 & 1 \\ 0 & 0 & 0 & 1 & 0 \end{bmatrix}$$
Transitive closure
$$\begin{bmatrix} 1 & 1 & 0 & 0 & 0 \\ 1 & 1 & 0 & 0 & 0 \\ 0 & 0 & 0 & 1 & 1 \\ 0 & 0 & 0 & 0 & 1 \\ 0 & 0 & 0 & 0 & 0 \end{bmatrix}$$

15. The graph of an equivalence relation will have as components subgraphs that are complete digraphs with the vertices of each being the equivalence classes.

17. False. Let R have adjacency matrix

$$\begin{bmatrix} 0 & 1 & 0 \\ 0 & 0 & 0 \\ 0 & 1 & 0 \end{bmatrix}$$

19. False. Use the same relation as in Problem 17.

Problem Set 1

1. (a) 1; (b) 5; (c) 27; (d) 17; (e) 103

3. (a) 1001; (b) 100011; (c) 111110; (d) 1110001; (e) 1110110111

Problem Set 2

1. $AB = $ 'thisisatest'
$BA = $ 'testthisa'
$BC = $ 'testhisisa'
$BD = $ 'testhistory'
$DC = $ 'historyhisisa'

3. C is a substring of A.

5. $BC = $ 'testhisisa'
$(BC)D = $ 'testhisisahistory'

7. $\lambda(C) = 6$; $\lambda(D) = 7$; $\lambda(CD) = 13$

Problem Set 3

1. b, c, and d are accepted.   **3.** b and c are accepted.

5.

	0	1
s_0	s_0	s_1
s_1	s_0	s_2
s_2^*	s_2	s_2

7.

	0	1
s_0^*	s_0	s_1
s_1^*	s_0	s_2
s_2	s_2	s_2

Problem Set 4

1. 0*

3. a*bb*as where s is any string in $\{a, b\}$*.

7. All strings of a's; all strings which alternate a's and b's

9.

	0	1
s_0	s_0	s_1
s_1	s_0	s_2
s_2	s_3	s_1
s_3^*	s_3	s_4
s_4^*	s_3	s_5
s_5^*	s_4	s_3

(s_3 represents $\{s_0, s_1\}$ from the original machine, s_4 represents $\{s_1, s_3\}$, and s_5 represents $\{s_2, s_3\}$.)

11.

	a	b
s_0^*	s_{10}, s_{01}	s_{11}, s_{02}
s_{01}^*	s_{05}	s_{03}
s_{02}^*	s_{04}	s_{05}
s_{03}^*	s_{05}	s_{05}
s_{04}^*	s_{05}	s_{05}
s_{05}	s_{05}	s_{05}
s_{10}	s_{10}	s_{11}
s_{11}	s_{11}	s_{11}

13.

	a	b
s_0	s_{13}	s_4
s_{13}^*	s_{15}	s_{24}
s_{15}^*	s_{15}	s_{25}
s_{23}^*	s_{25}	s_{24}
s_{24}^*	s_{23}	s_{25}
s_{25}	s_{25}	s_{25}
s_3^*	s_5	s_4
s_4^*	s_3	s_5
s_5	s_5	s_5

Notation: s_{ij} represents $\{s_i, s_j\}$. Only those states which can be reached from s_0 are included in the table.

Problem Set 5

No answers given.

Problem Set 6

1. (a) s_0 $\quad s_0$ $\quad s_1$ $\quad s_2$ $\quad s_0$ $\quad s_0$
$\quad\quad$ J $\quad o$ $\quad b$ $\quad o$ $\quad b$ $\quad\quad\quad\quad\quad$ (accepted)

(b) s_0 $\quad s_1$ $\quad s_2$ $\quad s_0$ $\quad s_0$ $\quad s_1$ $\quad s_2$
$\quad\quad$ o $\quad b$ $\quad e$ $\quad x$ $\quad o$ $\quad b$ $\quad a$

$\quad\quad$ s_0 $\quad s_0$ $\quad s_0$ $\quad s_0$ $\quad s_1$ $\quad s_2$ $\quad s_0$
$\quad\quad$ m $\quad p$ $\quad l$ $\quad o$ $\quad b$ $\quad e$ $\quad\quad\quad$ (accepted)

(c) s_0 $\quad s_1$ $\quad s_2$ $\quad s_0$ $\quad s_0$ $\quad s_1$ $\quad s_2$
$\quad\quad$ o $\quad b$ $\quad a$ $\quad n$ $\quad o$ $\quad b$ $\quad o$

$\quad\quad$ s_0 $\quad s_0$ $\quad s_0$ $\quad s_1$ $\quad s_2$ $\quad s_0$ $\quad s_0$
$\quad\quad$ t $\quad h$ $\quad o$ $\quad b$ $\quad e$ $\quad r$ $\quad\quad\quad$ (accepted)

(d) s_0 $\quad s_0$ $\quad s_0$ $\quad s_0$ $\quad s_1$ $\quad s_2$ $\quad s_0$ $\quad s_0$ $\quad s_0$
$\quad\quad$ s $\quad t$ $\quad r$ $\quad o$ $\quad b$ $\quad i$ $\quad n$ $\quad g$ $\quad\quad$ (accepted)

3.

	o	b	Other Vowel	Other Consonant
s_0^*	s_1	s_0	s_3	s_0
s_1	s_3	s_2	s_3	s_3
s_2	s_0	s_3	s_0	s_3
s_3	s_3	s_3	s_3	s_3

5.

	o	b	Other Vowel	Other Consonant
s_0^*	s_1, Λ	s_0, b	s_3, Λ	s_0, same
s_1	s_3, Λ	Λ	s_3, Λ	Λ
s_2	s_0, o	s_3, Λ	s_0, same	s_3, Λ
s_3	s_3, Λ (for all possible inputs in this state)			

13. A state table is as follows:

	0	1
s_0^*	s_2, Λ	s_1, Λ
s_1	$s_0, 2$	$s_0, 3$
s_2	$s_0, 0$	$s_0, 1$

(a)
state	s_0	s_1	s_0	s_1	s_0
input	0	0	0	0	0
output		0		0	

Output = '00'

(b)
state	s_0	s_2	s_0	s_2	s_0
input	1	0	1	0	
output		2		2	

Output = '22'

(c)
state	s_0	s_2	s_0	s_1	s_0
input	1	1	0	1	
output		3		1	

Output = '31'

(d)
state	s_0	s_1	s_0	s_2	s_0
input	0	1	1	0	
output		1		2	

Output = '12'

(e)
state	s_0	s_1	s_0	s_2	s_0	s_2
input	1	0	1	1	1	
output		2		3		

No output because string is rejected

15. (a) Since 5 is 101 in binary and 7 is 111 in binary, the input string for 5 + 7 would be 11 01 11 00.

state	s_0	s_1	s_1	s_1	s_0
input	11	01	11	00	
output	0	0	1	1	

Output = '0011', which we interpret as the binary number 1100

(b) Since 8 is 1000 in binary and 9 is 1001 in binary, the input string for 8 + 9 would be 01, 00, 00, 11, 00.

state	s_0	s_0	s_0	s_0	s_1	s_0
input	01	00	00	11	00	
output	1	0	0	0	1	

Output = '10001', interpreted as the binary number 10001

CHAPTER 8

Problem Set 1

3. (a) Keystrokes 2 (enter) 3 + 4 *

Stack	2	2	3	5	4	20
			2		5	

(b) Keystrokes 1 (enter) 2 (enter) 3 (enter) 4 + * +

Stack	1	1	2	2	3	3	4	7	14	15
			1	1	2	2	3	2	1	
					1	1	2	1		

(c) Keystrokes 4 (enter) 3 (enter) 2 (enter) 1 − + −

Stack	4	4	3	3	2	2	1	1	4	0
			4	4	3	3	2	3	4	
					4	4	3	4		
							4			

(d) Keystrokes   6 (enter) 5 (enter) 2 + 11 − /
   Stack   6   6   5   5   2   7 11 −4 −0.6666667
             6   6   5   6   7   6
                     6   6

7. (a) $\underline{a}bbaab/s_0$
$\bar{a}bbaab/s_1$
$a\underline{b}baab/s_1$
$ab\bar{b}aab/s_1$
$abb\underline{a}ab/s_0$
$abba\underline{a}b/s_1$
$abbaa\bar{b}\,\#/h$

(b) $\underline{a}bbbabb/s_0$
$\bar{a}bbbabb/s_1$
$a\underline{b}bbabb/s_1$
$ab\underline{b}babb/s_1$
$abb\bar{b}abb/s_1$
$abbb\underline{a}bb/s_0$
$abbba\underline{b}b/s_0$
$abbbab\underline{b}/s_0$
$abbbabb\,\#/s_0$ halts rejecting

(c) There must be an odd number of a's in the string.
(d) Provides no output.

9. States$(M) = \{s_0, h\}$
Symbols$(M) = \{\#, a\}$
Instructions(M):
$d(s_0, a) = (s_0, \#, R)$
$d(s_0, \#) = (h, \#, H)$

Problem Set 2

3. $d(s_0, 1) = (s_0, 1, R)$ {Searches for right end of first summand.}
  $d(s_0, \#) = (s_1, 1, R)$ {Replaces blank between summands with a 1;
                    number is now $n + m + 2$.}

  $d(s_1, 1) = (s_1, 1, R)$ {Searches for right end of second summand.}
  $d(s_1, \#) = (s_2, \#, L)$ {Begins to move left.}
  $d(s_2, 1) = (s_3, \#, L)$ {Removes first extra 1.}
  $d(s_3, 1) = (s_4, \#, L)$ {Removes second extra 1.}
  $d(s_4, \#) = (h, \#, R)$ {Stops when finds leading blank.}
  $d(s_0, 1) = (s_0, 1, R)$

5. (a) $\underline{a}bbbb/s_0$
$\bar{a}bbbb/h$
(b) $\underline{b}bbbabb/s_0$
$b\underline{b}bbabb/s_0$
$bb\underline{b}babb/s_0$
$bbb\underline{b}abb/s_0$
$bbbb\underline{a}bb/s_0$
$bbbba\underline{b}b/h$

(c) $\underline{b}bbb/s_0$
$b\underline{b}bb/s_0$
$bb\underline{b}b/s_0$
$bbb\underline{b}/s_0$
$bbbb\,\#/s_0$ halts rejecting
(d) $\underline{\#}/s_0$ halts rejecting

7. (a) $\underline{(}\,)(\,)\,)/s_0$ $\&\&\&\&\&/s_1$
$(\underline{)}(\,)\,)/s_1$ $\&\&\&\&\&/s_1$
$\underline{(}\&(\,(\,)/s_1$ $\&\&\&\&\&/s_1$
$\&\underline{\&}(\,)\,)/s_0$ $\underline{\#}\&\&\&\&\&/s_1$ halts rejecting
$\&\&\underline{(}\,)\,)/s_0$
$\&\&(\underline{)}\,)/s_0$
$\&\&(\underline{)}\,)/s_0$
$\&\&\underline{(}\&)/s_1$
$\&\&\&\underline{\&})/s_0$
$\&\&\&\underline{\&})/s_0$
$\&\&\&\&\underline{\&}/s_1$

(b) $\underline{(}(\,)\,)/s_0$ $\&\&\&\&/s_0$
$(\underline{(}\,)\,)/s_0$ $\&\&\&\&/s_0$
$((\underline{)}\,)/s_0$ $\&\&\&\&/s_0$
$(\underline{(}\&)/s_1$ $\&\&\&\&\,\#/s_0$
$(\underline{\&}\&)/s_0$ $\&\&\&\&/s_2$
$(\&\underline{\&})/s_0$ $\&\&\&/s_2$
$(\&\&\underline{)}/s_0$ $\&\&/s_2$
$(\&\&\underline{\&}/s_1$ $\underline{\&}/s_2$
$(\&\&\underline{\&}/s_1$ $\underline{\#}/s_2$
$\underline{(}\&\&\&/s_1$ $\#/h$

(c) Details omitted, but string is accepted.

(d) String is rejecting halts with machine status being $\underline{\#\,\&\&\&\&\&(\,(\,)\,)}/s_1$.

15. A complete formal description of this machine is lengthy, but the basic idea is as follows: Read symbols from left to right, remaining in a "scan" state and leaving the scanned symbols unchanged until a vowel is found. If an '*o*' is found, replace it with a "dirty" blank, and examine the next symbol to see if it is a '*b*'. If it is, replace it with a dirty blank and examine the next symbol to see if it is a vowel. If none of these occurs, halt in a failed state. If all is correct, continue to scan in the same way. If the end of the string is reached, go back and "squeeze out" the dirty blanks by shifting symbols.

Problem Set 3

1. The operation of adding 1 is accomplished by putting the head at the leftmost symbol of the string, moving to the adjacent blank on the left, and writing a 1. The operation of subtracting 1 from a nonzero number is accomplished by finding the leftmost symbol of the string, erasing it, and moving right one character. The machine can halt if, after moving right, it finds a blank character. (It should also put the 1 back on the tape so as to indicate the number 0.) The unconditional halt is accomplished by finding the leftmost symbol in the string and halting.

3. Algorithm *M*

```
begin
while current_symbol < > '#' do
    while (current_symbol = '(') or (current_symbol = '&') do
            scan_right od;
    if (current_symbol = ')' then
            current_symbol := &;
            scan_left;
            while (current_symbol = '&') do scan_left od;
            if (current_symbol = '(' then current_symbol := '&'
                                    else print ('fails');
                                        halt fi
od;
scan_left;
while current_symbol = '&' do scan_left od;
if current_symbol = '#' then print('string accepted')
                    else print('string rejected') fi
end.
```

5. Algorithm *O*

```
begin
state := odd;
while current_symbol < > '#' do
    if (current_symbol = 'a') AND (state = even)
                then state := odd fi;
    if (current_symbol = 'a') AND (state = odd)
                then state := even fi;
    scan_right;
```

```
od;
if (state = odd) then print('accepted')
                 else print ('rejected')
fi
end.
```

7. Algorithm B

```
begin
state := no
while (current_symbol <> '#') do
    if (current_symbol = 'a') then state := yes; fi
od
if (state = yes) then print('accepted'); stop
                 else print ('rejected'); stop fi
end.
```

9. An algorithm to find if a string x is in the union of L_1 and L_2 will first test the string for membership in L_1, and if it is there will return a value of TRUE. If the string is not in L_1, we can then test if for membership in L_2. If the string is in L_2, then it is in the union. If not, since it is in neither set, we can return a result of FALSE.

11. The problem is that the tests for membership in L_1 and L_2 might not halt on all inputs. The string in question might fail to be in L_1, but the test might not halt. A parallel processing of tests for both L_1 and L_2 is needed so that, if the string is in either, the process will eventually stop. The machine will still not halt for all input, but it will halt in an accepting state for all strings belonging to the union of L_1 and L_2.

Problem Set 4

3. (b) True because for all $nf(n) \leq n^3$. (c) True because for all $nn \leq f(n)$. We leave (a) and (d) as puzzles, but they are false.

5. States$(M) = \{s_0, s_1, s_2, s_3, h\}$
   Symbols$(M) = \{0, 1, \#\}$
   Instructions(M):
   $d(s_0, 0) = (s_0, 0, R)$
   $d(s_0, 1) = (s_0, 1, R)$
   $d(s_0, \#) = (s_1, \#, L)$
   $d(s_1, 0) = (s_2, \#, H)$
   $d(s_1, 1) = (s_2, \#, H)$
   $d(s_2, \#) = (s_3, \#, L)$
   $d(s_3, 0) = (s_3, 0, L)$
   $d(s_3, 1) = (s_3, 1, L)$
   $d(s_3, \#) = (h, \#, R)$

7. $O(n^2)$ 9. $O(n)$

1. (a)

$\underline{a}abbbab/s_0$

$\underline{a}abbbab/q_0$ $\underline{a}abbbab/p_0$

| |
$a\underline{b}bbab/q_1$ $\underline{a}bbbab/p_0$

| |
$bb\underline{b}ab/q_0$ $b\underline{b}bab/p_0$

| |
$bb\underline{b}ab/q_0$ $bb\underline{b}ab/p_1$

| |
$b\underline{a}b/q_0$ $b\underline{a}b/p_0$

| |
$\underline{a}b/q_0$ $a\underline{b}/p_1$

| |
$\underline{b}/g_1$ $\underline{b}/p_1$

| |
$\underline{\#}/q_1$ $\underline{\#}/p_0$

|
$\underline{\#}/h$

(c)

$\underline{b}b/s_0$

$\underline{b}b/q_0$ $\underline{b}b/p_0$

| |
$b\underline{b}/q_0$ $b\underline{b}/p_1$

| |
$\underline{\#}/q_0$ $\underline{\#}/p_0$

| |
$\underline{\#}/h$ $\underline{\#}/h$

(b)

$\underline{a}aaaa/s_0$

$\underline{a}aaaa/q_0$ $\underline{a}aaaa/p_0$

| |
$a\underline{a}aaa/q_1$ $\underline{a}aaaa/p_0$

| |
$\underline{a}aaa/q_0$ $\underline{a}aaa/p_0$

| |
$a\underline{a}a/q_1$ $\underline{a}aa/p_0$

| |
$\underline{a}/q_0$ $\underline{a}/p_0$

| |
$\underline{\#}/q_1$ $\underline{\#}/p_0$

|
$\underline{\#}/h$

(d)

$\underline{a}babababab/s_0$

$\underline{a}babababab/q_0$ $\underline{a}babababab/p_0$

| |
$b\underline{a}babababab/q_1$ $b\underline{a}babababab/p_0$

| |
$\underline{a}babababab/q_1$ $\underline{a}babababab/p_1$

| |
$b\underline{a}babab/q_0$ $b\underline{a}babab/p_1$

| |
$\underline{a}babab/q_0$ $\underline{a}babab/p_0$

| |
$b\underline{a}bab/q_1$ $b\underline{a}bab/p_0$

| |
$\underline{a}bab/q_1$ $\underline{a}bab/p_1$

| |
$b\underline{a}b/q_0$ $b\underline{a}b/p_1$

| |
$\underline{a}b/q_0$ $\underline{a}b/p_0$

| |
$\underline{b}/q_1$ $\underline{b}/p_0$

| |
$\underline{\#}/q_1$ $\underline{\#}/p_1$

(e) $\underline{\#}/s_0$ (Does not accept empty string. What should be changed?)

3. The strings accepted are of the form $a*bb*baa*$.

5. States$(M) = \{s_0, p_0, q_0, q_1, q_2, h\}$
Symbols$(M) = \{\#, a, b\}$
Instructions(M)

$d(s_0, a) = \{(p_0, a, H), (q_0, a, H)\}$
$d(s_0, b) = \{(p_0, b, H), (q_0, b, H)\}$
$d(s_0, \#) = \{(p_0, \#, H), (q_0, \#, H)\}$
$d(p_0, a) = \{(h, a, H)\}$
$d(p_0, a) = \{(p_0, b, R)\}$
$d(q_0, a) = \{(q_1, a, R)\}$
$d(q_1, b) = \{(q_2, \#, R)\}$
$d(q_1, \#) = \{(h, \#, L)\}$
$d(q_2, \#) = \{(q_2, \#, R)\}$

9. Allow the use of a "fork" type instruction so as to let more than one step occur at the same time.

11. The process can be illustrated by the following tree for three colors.

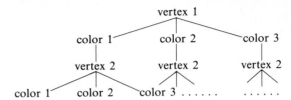

CHAPTER 9

Problem Set 1

1. ⟨Sentence⟩ → ⟨subject phrase⟩ ⟨predicate phrase⟩
       → ⟨subject phrase⟩ es ⟨adjective⟩
       → ⟨subject phrase⟩ es gris
       → ⟨subject⟩ ⟨possessive phrase⟩ es gris
       → ⟨article⟩ ⟨noun1⟩ ⟨possessive phrase⟩ es gris
       → El ⟨noun1⟩ ⟨possessive phrase⟩ es gris
       → El gato ⟨possessive phrase⟩ es gris
       → El gato de ⟨possessive⟩ ⟨noun2⟩ es gris
       → El gato de su ⟨noun2⟩ es gris
       → El gato de su madre es gris

3.

```
                          ⟨sentence⟩
              ⟨subject phrase⟩        ⟨predicate phrase⟩
                    |
               ⟨subject⟩
        ⟨article⟩    ⟨noun1⟩      es     ⟨adjective⟩
           |           |                     |
          El         perro                 blanco
```

5. Not in the language. ⟨predicate phrase⟩ can be broken into either 'es' ⟨adjective⟩, or ⟨verb⟩ ⟨adverb⟩, but 'rapido' cannot be derived from ⟨adjective⟩, and 'es' cannot be derived from ⟨verb⟩.

9. (a)

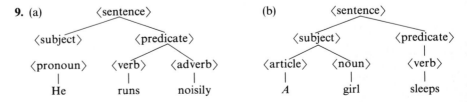

```
(a)            ⟨sentence⟩
        ⟨subject⟩       ⟨predicate⟩
       ⟨pronoun⟩    ⟨verb⟩   ⟨adverb⟩
          |           |         |
         He          runs     noisily
```

(b)
```
              ⟨sentence⟩
        ⟨subject⟩        ⟨predicate⟩
                             |
     ⟨article⟩   ⟨noun⟩     ⟨verb⟩
        |          |          |
        A         girl      sleeps
```

1. Derivations:

   (a) $A \to aB$ (b) $A \to aB$
          $\to aD$ $\to abB$
          $\to ab$ $\to abbB$
                      $\to abbbB$
                      $\to abbbbB$
                      $\to abbbbD$
                      $\to abbbbb$

3. $L(G_1) = \{s \mid s = ab^n a^m b, n \geq 0, m \geq 0\}$.

5. In $L(G_2)$ the number of a's in an accepted string is always two more than the number of b's. This fails for each of the three proposed strings.

7. ab^2: $S \to aAB$
                $\to abB$
                $\to abb$
  $a^2 bab^3$: $S \to aAB$
                $\to a^2 BAB$
                $\to a^2 bAB$
                $\to a^2 baBAB$
                $\to a^2 baBbB$
                $\to a^2 bab^2 B$
                $\to a^2 bab^3$

$ba^2 bab^3$ cannot belong to $L(G_2)$, because all strings in that language must end in an 'a'.

9. $S \to A$
   $A \to aB \mid B$
   $B \to bB \mid C$
   $C \to cC \mid c \mid \Lambda$

11. One example of a grammar which works is
   $S \to P$
   $P \to 0P0 \mid 1P1 \mid 0 \mid 1 \mid \Lambda$

Problem Set 3

1. Use of incorrect grammar.   3. Syntax: 1.5 is real, not integer.

5. Syntax: need $x[3]$.

7. Semantic: probably means 'tiny'; compiler may detect as a syntax error if tiny is not a variable.

9. Semantic: x will be -1 if $y = z$ and 0 if $y \lessgtr z$.

11. Semantic: will print misleading message.

13. Semantic: grammatically correct statement, but the effect is an infinite recursive call.

15. Line 10 has a semantic error: need to input C.
Line 20 has a syntax error: LET D is needed, not LED.
Line 30 has syntax errors and a semantic error: need R1, parentheses, and 2∗A.
Line 50 has two syntax errors: PRONT and a:.
The program as a whole does not test for a negative discriminant as it should.

17. Line 1 contains a semantic error: Output file should be 'Output'.
Line 2 contains a semantic error: A, B, C should be type REAL.
Line 3 contains a syntax error: identifier should be R1, not R!.
Line 5 contains a syntax error: need a ' following the message to be printed.
Line 6 contains a syntax error: the < needs to be replaced with a ,.
Line 8 contains a semantic error: should have (sqrt(D) − B).
Line 9 contains a semantic error: should have (b − sqrt(D)).
Line 10 contains a syntax error: identifier cannot be R!, also includes a superfluous ; which will cause no problems.

Problem Set 4

1. Type 0   **3.** Type 0   **5.** Type 2   **7.** Type 2   **9.** Type 3

11. $S \rightarrow \Lambda \,|\, Z \,|\, E$
$Z \rightarrow b \,|\, bZ$
$E \rightarrow bE \,|\, aO \,|\, b$
$O \rightarrow a \,|\, aE \,|\, bO$

13. $S \rightarrow \Lambda \,|\, A$
$A \rightarrow a \,|\, aA \,|\, B$
$B \rightarrow b \,|\, bB \,|\, C$
$C \rightarrow c \,|\, cC$

15. $S \rightarrow B1 \,|\, 1$
$B \rightarrow B0 \,|\, B1 \,|\, 1 \,|\, 0$

17. $\langle \text{Start} \rangle ::= a \langle A \rangle$
$\langle A \rangle ::= a \langle A \rangle \, \langle B \rangle \,|\, a$
$\langle B \rangle ::= b$

19. (1) $\langle \text{Start} \rangle ::= x \langle X \rangle$
$\langle X \rangle ::= y \langle X \rangle \, \langle Y \rangle \,|\, x$
$\langle Y \rangle ::= y$

(4) $\langle \text{Start} \rangle ::= x \langle Y \rangle$
$\langle Y \rangle ::= y \langle X \rangle \,|\, x \,|\, y$
$\langle X \rangle ::= x \langle Y \rangle \,|\, x$

(5) $\langle \text{Start} \rangle ::= x \langle X \rangle \, \langle Y \rangle$
$\langle X \rangle ::= \langle Y \rangle yx$
$\langle Y \rangle ::= y \langle Y \rangle \,|\, y$

(6) $\langle \text{Start} \rangle ::= x \langle Y \rangle$
$\langle X \rangle ::= x \langle Y \rangle$
$\langle Y \rangle ::= y \langle X \rangle \,|\, y$

(7) $\langle \text{Start} \rangle ::= x \langle \text{start} \rangle \, \langle X \rangle \,|\, y \langle \text{start} \rangle \,|\, x$
$\langle X \rangle ::= \langle X \rangle y \,|\, x \langle \text{start} \rangle \langle X \rangle \,|\, y$

Problem Set 5

5. $\langle S \rangle \rightarrow a \,|\, b \,|\, a \langle Q_2 \rangle \,|\, b \langle Q_2 \rangle$
$\langle Q_1 \rangle \rightarrow a \,|\, a \langle Q_2 \rangle \,|\, b \langle Q_3 \rangle$
$\langle Q_2 \rangle \rightarrow a \,|\, a \langle Q_2 \rangle \,|\, b \,|\, b \langle Q_1 \rangle$
$\langle Q_3 \rangle \rightarrow a \langle Q_3 \rangle \,|\, b \langle S \rangle$

7. $\langle S_0 \rangle \rightarrow \Lambda \,|\, a \,|\, \langle S_1 \rangle \,|\, b \,|\, b \langle S_0 \rangle$
$\langle S_1 \rangle \rightarrow a \,|\, a \langle S_2 \rangle \,|\, b \,|\, b \langle S_0 \rangle$
$\langle S_2 \rangle \rightarrow a \,|\, a \langle S_1 \rangle$

3. Algorithm size

Begin
{k is the number of states in the defining automaton}
For i := 0 to k − 1 **do**
    Construct all strings of length in Σ^*;
    Test each string for membership in L(G) **od**;
If all string fail membership test **then** Size := 'empty'
    else for i := k to 2∗k − 1 **do**
        Construct all strings of length i in Σ^*;
        Test each string for membership in L(G) **od**;
      If all fail **then** Size := 'finite'
           else Size := 'infinite' **fi**

fi
end.

13. $d(s_0, a) = (s_0, a, R)$
$d(s_0, b) = (s_1, b, R)$
$d(s_1, b) = (s_1, a, R)$
$d(s_1, c) = (s_2, c, R)$
$d(s_2, c) = (s_2, c, R)$
$d(s_2, \#) = (s_3, \#, L)$
$d(s_3, a) = (s_3, a, L)$
$d(s_3, b) = (s_3, b, L)$
$d(s_3, c) = (s_3, c, L)$
$d(s_3, \#) = (s_4, \#, R)$
$d(s_4, a) = (s_5\ \$, L)$
$d(s_4, \$) = (s_4, \$, R)$
$d(s_5, a) = (s_5, a, L)$
$d(s_5, \$) = (s_5, \$, L)$
$d(s_5, b) = (s_6, \$, L)$
$d(s_6, b) = (s_6, b, L)$
$d(s_6, \$) = (s_6, \$, L)$
$d(s_6, c) = (s_7, c, R)$
$d(s_7, \$) = (s_7, \$, R)$
$d(s_7, b) = (s_7, b, R)$
$d(s_7, a) = (s_8, a, R)$
$d(s_7, \#) = (s_9, \#, L)$
$d(s_8, a) = (s_8, a, R)$
$d(s_8, \$) = (s_4, \$, L)$
$d(s_9, \$) = (s_9, \$, L)$
$d(s_9, \#) = (h, \$, H)$

(Here $ is used as a 'dirty' blank.)

Index

Index

443